CORTICAL DEVELOPMENT: GENES AND GENETIC ABNORMALITIES

Novartis Foundation Symposium 288

CORTICAL DEVELOPMENT: GENES AND GENETIC ABNORMALITIES

BICENTENNIAL
1807
WILEY
2007
BICENTENNIAL

John Wiley & Sons, Ltd

Other Wiley Editorial Offices

John Wiley & Sons Inc., 111 River Street, Hoboken, NJ 07030, USA

Jossey-Bass, 989 Market Street, San Francisco, CA 94103-1741, USA

Wiley-VCH Verlag GmbH, Boschstr. 12, D-69469 Weinheim, Germany

John Wiley & Sons Australia Ltd, 33 Park Road, Milton, Queensland 4064, Australia

John Wiley & Sons (Asia) Pte Ltd, 2 Clementi Loop #02-01, Jin Xing Distripark, Singapore
129809

John Wiley & Sons Canada Ltd, 6045 Freemont Blvd, Mississauga, Ontario, Canada L5R 4J3

Wiley also publishes its books in a variety of electronic formats. Some content that appears
in print may not be available in electronic books.

Novartis Foundation Symposium 288

x + 282 pages, 45 figures, 3 tables

Anniversary Logo Design: Richard J Pacifico

British Library Cataloguing in Publication Data

A catalogue record for this book is available from the British Library

ISBN 978-0 470-06092-6

Typeset in 10½ on 12½ pt Garamond by SNP Best-set Typesetter Ltd., Hong Kong
Printed and bound in Great Britain by T. J. International Ltd, Padstow, Cornwall.
This book is printed on acid-free paper responsibly manufactured from sustainable forestry,
in which at least two trees are planted for each one used for paper production.

Contents

Participants

Colin Blakemore Medical Research Council, 20 Park Crescent, London W1B 1AL, UK

Jamel Chelly Institut Cochin (IC), Département de Génétique et Pathologie Moléculaire, Equipe de Génétique et Physiopathologie des Retards Mentaux, 24, rue du Faubourg St Jacques, F-75014 Paris, France

Peter B. Crino Hospital of the University of Pennsylvania, Department of Neurology, 3 West Gates Building, 3400 Spruce Street, Philadelphia, PA 19104, USA

Kay Davies MRC Functional Genetics Unit, Department of Physiology, Anatomy and Genetics, University of Oxford, South Parks Road, Oxford OX1 3QX, UK

Gordon Fishell NYU School of Medicine, Smilow Neuroscience Program and the Department of Cell Biology, 5th Smilow Bldg, 522 First Avenue, New York, NY 10016, USA

Gaëlle Friocourt Masse Laboratoire de Genetique Moléculaire et d'Histocompatibilité, INSERM U613, 46 rue Félix Le Dantec, 29200 Brest, France

André M. Goffinet University of Louvain Medical School, Developmental Neurobiology Unit, 73 avenue E. Mounier, Box DENE 7382, B-1200 Brussels, Belgium

François Guillemot Division of Molecular Neurobiology, NIMR, Mill Hill, London NW7 1AA, UK

Paul J. Harrison Neurosciences Building, Department of Psychiatry, Warneford Hospital, University of Oxford, Oxford OX3 7JX, UK

Robert Hevner Department of Pathology, University of Washington, Harborview, Box 359791, Seattle, WA 98104-9791, USA

David Keays St Anne's College, University of Oxford, Oxford OX2 6HS, UK

Henry Kennedy Stem Cell and Brain Research Institute, INSERM U846, 18 avenue Doyen Lépine, F-69675 Bron Cedex, France

Arnold R. Kriegstein Institute for Regeneration Medicine, UCSF School of Medicine, 513 Parnassus Avenue, HSW 1201, Campus Box 0525, San Francisco, CA 94143-0525, USA

Jeffrey D. Macklis Harvard Medical School, Massachusetts General Hospital, MGH-HMS Center for Nervous System Repair, 50 Blossom Street, EDR 410, Boston, MA 02114, USA

Antonello Mallamaci Laboratory of Cerebral Cortex Development, SISSA— Neurobiology Sector, Research Area of Basovizza, Box Q1, Floor 1, Basovizza S. S. 14, Km 163.5, I-34012 Trieste, Italy

Zoltán Molnár Department of Physiology, Anatomy and Genetics, Le Gros Clark Building, University of Oxford, South Parks Road, Oxford OX1 3QX, UK

Fujio Murakami Laboratory of Neuroscience, Graduate School of Frontier Biosciences, Osaka University, Yamadaoka 1-3, Suita, Osaka 565-0871, Japan

Dennis D. M. O'Leary Molecular Neurobiology Laboratory, The Salk Institute for Biological Studies, 10010 North Torrey Pines Road, La Jolla, CA 92037, USA

John G. Parnavelas *(Chair)* Department of Anatomy and Developmental Biology, Anatomy Building, University College London, Gower Street, London WC1E 6BT, UK

David Price Biomedical Sciences, Hugh Robson Building, University of Edinburgh, 1 George Square, Edinburgh EH8 9XD, UK

Pasko Rakic Department of Neurobiology, Yale University School of Medicine, 333 Cedar Street, SHM C303, PO Box 208001, New Haven, CT 06520-8001, USA

Linda J. Richards The University of Queensland, School of Biomedical Sciences and The Queensland Brain Institute, Department of Anatomy and Developmental Biology, Otto Hirschfeld Building (81), Brisbane, QLD 4072, Australia

John L. R. Rubenstein University of California at San Francisco, Department of Psychiatry, 1550 4th Street, 2nd Floor South, Room RH 284C, Box 2611, San Francisco, CA 94143-2611, USA

Anastassia Stoykova Max Planck Institute for Biophysical Chemistry, Am Fassberg 11, 37077 Göttingen, Germany

Seong-Seng Tan Howard Florey Institute, Cnr. Royal Parade and Grattan Street, University of Melbourne, Parkville VIC 3010, Australia

Christopher A. Walsh Division of Genetics, Children's Hospital Boston, Howard Hughes Medical Institute, BIDMC, Harvard Medical School, 77 Avenue Louis Pasteur, Boston, MA 02115, USA

Adam Wilkins BioEssays, 10/11 Tredgold Lane, Napier Street, Cambridge CB1 1HN, UK

Michael Wilson University of New Mexico, Department of Neurosciences, Basic Medical Sciences Building, Albuquerque, NM 87131-5223, USA

Nobuhiko Yamamoto Neuroscience Laboratories, Graduate School of Frontier Biosciences, Osaka University, Yamadaoka 1-3, Suita, Osaka 565-0871, Japan

Chair's introduction

John G. Parnavelas

Department of Anatomy and Developmental Biology, University College London, London WC1E 6BT, UK

This is the third symposium on cortical development to be held at the Novartis Foundation in recent years. The first, entitled *Development of the cerebral cortex*, was held in the winter of 1994. The second meeting, which focused on *Evolutionary development of the cerebral cortex*, took place in April 1999. The present symposium, proposed by Dr Zoltán Molnár, will concentrate on genes and genetic abnormalities of the developing cortex.

The first meeting in 1994 took place only a couple of years after the publication of the seminal papers by Antonio Simeone, Edoardo Boncinelli and colleagues in Naples on the nested expression domains of four homeobox genes in the developing forebrain (Simeone et al 1992a, b). At that meeting, there was a solitary paper by Edoardo Boncinelli on the expression of these *Emx* and *Otx* genes in the developing forebrain that included some faint suggestions about their roles in the developing cortex (Boncinelli et al 1995). In the second meeting, five years later, we heard a bit more about genes with papers by Edoardo Boncinelli and John Rubenstein focusing on the genetic control of regional identity in the developing cortex (Boncinelli et al 2000, Rubenstein et al 2000).

Interest in genetic and molecular mechanisms involved in the different phases of cortical formation and in genes associated with human cortical malformations has never been greater. The availability of state-of-the-art molecular techniques, transgenic mouse models and much improved techniques for imaging the human brain have been behind the explosion in the amount and complexity of new information to emerge in the past 10 years, and especially since the turn of the century. The purpose of this symposium is to bring together people with different backgrounds, both from the basic and clinical sciences, to review and discuss the present state of knowledge about the development of the normal cerebral cortex and of cortical developmental abnormalities associated with human disorders such as mental retardation, epilepsy and schizophrenia. As the title of the symposium suggests, the emphasis will be on genes and genetic abnormalities. This volume shows how much progress has been made in the field since the last Novartis Foundation symposium on cortical development in 1999.

References

Boncinelli E, Gulisano M, Spada F, Broccoli V 1995 *Emx* and *Otx* gene expression in the developing mouse brain. In: Development of the cerebral cortex (Ciba Found Symp 193). Wiley, Chichester, p 100–126

Boncinelli E, Mallamaci A, Muzio L 2000 Genetic control of regional identity in the developing vertebrate forebrain. In: Evolutionary developmental biology of the cerebral cortex (Novartis Found Symp 228). Wiley, Chichester, p 53–66

Rubenstein JLR 2000 Intrinsic and extrinsic control of cortical development. In: Evolutionary developmental biology of the cerebral cortex (Novartis Found Symp 228). Wiley, Chichester, p 67–82

Simeone A, Acampora D, Gulisano M, Stornaiuolo A, Boncinelli E 1992a Nested expression domains of four homeobox genes in developing rostral brain. Nature 358:687–690

Simeone A, Gulisano M, Acampora D, Stornaiuolo A, Rambaldi M, Boncinelli E 1992b Two vertebrate homeobox genes related to *Drosophila empty spiracles* gene are expressed in the embryonic cerebral cortex. EMBO J 11:2541–2550

Molecular development of corticospinal motor neuron circuitry[1]

Bradley J. Molyneaux, Paola Arlotta[2] and Jeffrey D. Macklis[3]

MGH-HMS Center for Nervous System Repair, Departments of Neurosurgery and Neurology, and Program in Neuroscience, Harvard Medical School; Nayef Al-Rodhan Laboratories, Massachusetts General Hospital, and Department of Stem Cell and Regenerative Biology and Harvard Stem Cell Institute, Harvard University, Boston, MA 02114, USA

Abstract. The organization, function and evolution of the brain depends centrally on the precise development of a wide diversity of distinct neuronal subtypes. Furthermore, given the heterogeneity of neuronal subtypes within the CNS and the complexity of their connections, attempts to functionally repair circuitry will require a detailed understanding of the molecular controls over differentiation, connectivity and survival of specific lineages. Toward these goals, we recently identified developmentally regulated transcriptional programmes for specific lineages of long-distance neocortical projection neurons as they develop *in vivo* (in particular, for corticospinal motor neurons; CSMN). We purified CSMN, a clinically important population of neocortical projection neurons, at distinct stages of development *in vivo*, and compared their gene expression to that of two other pure populations of neocortical projection neurons. We identified novel and largely uncharacterized genes that are instructive for CSMN development and implicated in key developmental processes. These include *Fezf2* (also known as *Fezl*), a regulator of subcerebral projection neuron identity, and *Ctip2* (also known as *Bcl11b*), a regulator of the fasciculation, outgrowth and pathfinding of CSMN axonal projections to the spinal cord. Loss-of-function and gain-of-function analysis for multiple identified genes reveal programmes of combinatorial molecular controls over the precise development of key neocortical and other forebrain projection neuron populations that elucidate organization and function of the forebrain, and that might be manipulated toward functional cellular repair of complex brain circuitry.

2007 Cortical development: genes and genetic abnormalities. Wiley, Chichester (Novartis Foundation Symposium 288) p 3–20

[1] The text of this review is modified and expanded from one section of a broader review published in *Nature Reviews Neuroscience* (Molyneaux et al 2007).

[2] Current address: Center for Regenerative Medicine, Department of Neurosurgery, Massachusetts General Hospital, Harvard Medical School, Boston, MA 02114 USA.

[3] This paper was presented at the symposium by Jeffrey Macklis, to whom correspondence should be addressed.

3

The neuronal diversity and precise connectivity of the neocortex underlies high-level cognition, integration and motor control. In addition, previous data from our laboratory demonstrate that new neurons can be added to adult neocortical circuitry from transplanted neural precursors or via manipulation of endogenous precursors *in situ* (Sheen & Macklis 1995, Magavi et al 2000, Shin et al 2000, Fricker-Gates et al 2002, Chen et al 2004). These data indicate that cellular repair of damaged neocortical and neocortical output circuitry (e.g. corticospinal circuitry) might be possible, if controls over specific lineage differentiation are understood. Given the heterogeneity of neuronal subtypes in the CNS and the complexity of their connections, a deeper understanding of CNS organization, function and evolution, as well as attempts to functionally repair circuitry will require detailed knowledge of the molecular controls over the differentiation, connectivity and survival of specific lineages.

Within the neocortex, some of the basic mechanisms that control general neuronal specification, migration and connectivity during development have been identified (Bertrand et al 2002, Marin & Rubenstein 2003, Guillemot 2005). More recently, the discovery of genes that have layer and neuronal subtype specificity within the neocortex has made it possible to investigate the mechanisms underlying the specification of individual projection neuron subtypes (Molyneaux et al 2007).

Corticospinal motor neurons (CSMN), also known as upper motor neurons, form the basis of voluntary motor control in humans, and are therefore a subtype of particular interest for investigation. Located in the cerebral cortex, CSMN extend extremely long axons that synapse on lower motor neurons and interneurons in the spinal cord. Clinically, they are an important population, as they degenerate in motor neuron degenerative diseases such as amyotrophic lateral sclerosis (ALS), and their injury contributes to the loss of motor function following spinal cord injury. The anatomical and morphological development of CSMN has been extensively characterized (Jones et al 1982, Terashima 1995, Bareyre et al 2005), but strategies to repair or prevent degeneration of CSMN are limited by a lack of understanding of the molecular controls over CSMN development, including neuron type-specific differentiation, survival, and connectivity (Arlotta et al 2005).

CSMN are one subtype within the broader class of subcerebral projection neurons (O'Leary & Koester 1993, Arlotta et al 2005), which are defined by possessing axons that project below the cerebrum to targets in the spinal cord or brainstem, including the tectum, red nucleus and pons (Wise & Jones 1977, Killackey et al 1989, Legg et al 1989, O'Leary & Koester 1993, Arlotta et al 2005, Molnar & Cheung 2006). Among the different types of cortical projection neurons, subcerebral projection neurons are an ideal model population for studying the mechanisms of subtype specification within the neocortex. They are a discrete, readily identifiable, prototypical projection neuron population, located within layer Vb of neocortex (Fig. 1). After being born in the germinal zone, all subcerebral projection neurons migrate to layer Vb and extend a primary axon through the internal capsule, cerebral peduncle, and pyramidal tract toward the spinal cord. Secondary collaterals sprout from the primary axon only after it has passed other targets such as the superior colliculus and pons (O'Leary & Terashima 1988). Inappropriate connections are later eliminated, leaving different populations of subcerebral projection neurons with specific patterns of connectivity. For example, CSMN in sensorimotor cortex project axons to the pons and spinal cord, while corticotectal neurons in the visual cortex project axons to the rostral pons and superior colliculus (O'Leary & Terashima 1988, Schreyer & Jones 1988, O'Leary & Koester 1993).

Given this common pattern of initial development, many of the genes controlling early specification and differentiation are likely to be shared among the

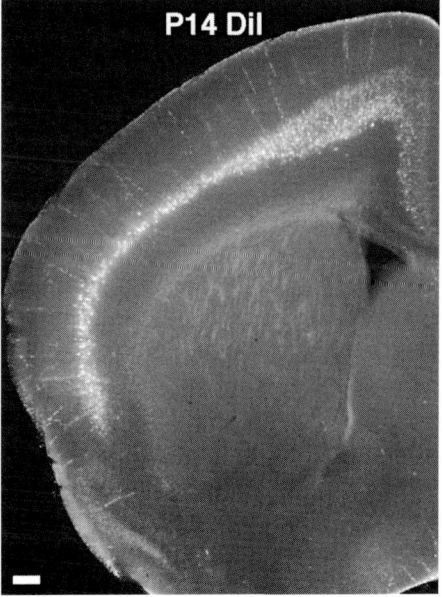

FIG. 1. Subcerebral projection neurons are located in layer V of neocortex. CSMN are one subpopulation of projection neurons within the broader class of neurons termed subcerebral projection neurons. All subcerebral projection neurons can be retrogradely labelled via an injection of DiI into the cerebral peduncle. This coronal section of a P14 mouse brain shows the distribution of DiI-labelled subcerebral projection neurons within the neocortex. Scale bar: 250 µm. Reproduced from Arlotta et al (2005).

different types of subcerebral projection neurons (Arlotta et al 2005). Over the last several years, substantial progress has been made towards understanding the molecular controls over the specification and development of CSMN and other subcerebral projection neurons, including the identification of a large number of genes expressed with varying degrees of specificity in CSMN (Arlotta et al 2005), the identification of *Fezf2* as a critical regulator of subcerebral projection neuron identity (Chen et al 2005a, b, Molyneux et al 2005), and the identification of *Ctip2* as a regulator of the fasciculation, outgrowth, and pathfinding of CSMN axonal projections to the spinal cord (Arlotta et al 2005).

Progressive specification of projection neurons

The recent identification of a large number of subcerebral and CSMN specific genes has enabled an expanding effort to decipher the programmes controlling CSMN development (Arlotta et al 2005). This was achieved through the comparison of purified neuronal subtypes by microarray analysis (Arlotta et al 2005). The expression patterns of the identified genes indicate that the fate specification and differentiation of subcerebral projection neurons in general, and CSMN in particular, is likely directed by a combinatorial code of transcription factors and other molecules. These molecules are expressed in a pattern that uniquely identifies CSMN. For example, a small number of CSMN genes appear restricted to sensorimotor cortex (e.g. *Diap3*, *Igfbp4* and *Crim1*), suggesting that they distinguish CSMN from other subcerebral projection neurons of layer V (Arlotta et al 2005). Other genes are expressed across the full extent of layer V (e.g. *Ctip2*, *Encephalopsin*, *Fezf2*, *Clim1*, *Pcp4* and *S100a10*) suggestive of restriction to most subcerebral projection neurons (Arlotta et al 2005). Thus far, the functions of only a few of these genes have been reported, but these studies are already revealing key roles for these genes in subcerebral specification and differentiation (Weimann et al 1999, Arlotta et al 2005, Chen et al 2005a, b, Molyneux et al 2005).

The expression pattern for only a small number of genes has been examined in detail via a combination of retrograde labelling and immunocytochemistry. These include: *Ctip2*, which is expressed at high levels in subcerebral neurons of layer V and at much lower levels in corticothalamic neurons of layer VI (Arlotta et al 2005); *Scip*, which is primarily expressed in subcerebral projection neurons of layer V, in addition to lower levels of expression in neurons of layers II–III (Frantz et al 1994b); *Otx1*, which is expressed in 40–50% of subcerebral neurons as well as a number of cells in layer VI (Weimann et al 1999); *Er81*, which is expressed in cortico–cortical as well as subcerebral projection neurons of layer V (Hevner et al 2003); and *Nfh*, which is expressed in subcerebral projection neurons of layer V (Voelker et al 2004). It will be important to perform similar careful investigation of neuronal subtype expression by immunocytochemistry for each of the other genes that have been identified as preferentially expressed in CSMN and other subcerebral projection neurons by microarray or *in situ* hybridization.

While significant progress has been made in identifying markers of post-mitotic subcerebral projection neurons once they have reached the cortical plate, it is unclear whether the same markers can be used to identify progenitors that will give rise to subcerebral projection neurons, or whether such lineage committed progenitors even exist. A number of neuronal subtype specific genes are expressed in what appears to be subpopulations of cells within the VZ and SVZ, where they might label progenitors or early post-mitotic neurons of that same neuronal subtype. For example, both *Fezf2* and *Otx1* are expressed in the VZ prior to and during the generation of subcerebral projection neurons, and are later expressed in post-mitotic subcerebral projection neurons.

However, it is important to be extremely cautious about inferring that a gene plays a role in the specification of subcerebral projection neurons at the progenitor level based on restricted expression that is later observed in a particular neuronal subtype; it is entirely possible that the gene has two independent functions during development (Alvarez-Bolado et al 1995). This is best illustrated by considering *Lhx2*, which is expressed in the VZ and SVZ prior to and during the generation of upper layers and is also expressed in post-mitotic neurons of the upper layers. The finding that the loss of *Lhx2* results in the absence of neurons of all layers (Bulchand et al 2001, Monuki et al 2001) suggests that *Lhx2* likely has two functions during development: in the ventricular zone, it is required to establish the neocortical identity of progenitors of all layers, while later in development it might control more specific aspects of upper layer differentiation. Therefore, further study is required for each of the subcerebral projection neuron specific genes to define the relationship between progenitors and post-mitotic neurons expressing the same genes.

CSMN and subcerebral projection neuron specification

Investigations into the function of some of the subcerebral projection neuron-specific layer and subtype-restricted genes are starting to provide insight into how CSMN and subcerebral projection neurons are specified in the neocortex. *Brn1* and *Brn2* are two genes that appear to play a role early in subcerebral projection neuron specification and migration. They are expressed primarily in neurons of layers II–V and are involved in directing the differentiation and migration of neurons within these layers (McEvilly et al 2002, Sugitani et al 2002, Hevner et al 2003). The *Brn1/Brn2* double knockouts possess decreased numbers of neurons of layers II–V, and those that are born exhibit abnormalities in migration, arresting in the VZ/SVZ (McEvilly et al 2002, Sugitani et al 2002). Additionally, some markers of upper layer neurons are expressed in these mutants, while others (e.g.

mSorLa) are absent, suggesting abnormalities in differentiation. In contrast, *Tle4* and *Tbr1* expressing neurons of layer VI appear to form and migrate normally into the cortical plate in the absence of *Brn1* and *Brn2* (McEvilly et al 2002, Sugitani et al 2002). Further analysis of *Brn1/Brn2* mutants with recently identified markers is needed to illuminate precisely which subtypes of neurons are affected in the absence of *Brn1* and *Brn2*.

Fezf2 (Fezl/Zfp312), a putative transcription factor that is expressed in all sub-cerebral projection neurons from the early stages of development through to adulthood (Inoue et al 2004, Arlotta et al 2005), was recently found to be required for the specification of all subcerebral projection neurons (Chen et al 2005a, b, Molyneaux et al 2005). In the absence of *Fezf2* function in null mutant mice, the entire population of subcerebral projection neurons is absent, and there are no projections from the cerebral cortex to either the spinal cord or the brainstem (Fig. 2; Chen et al 2005a, Molyneaux et al 2005). Layer VI neurons and subplate

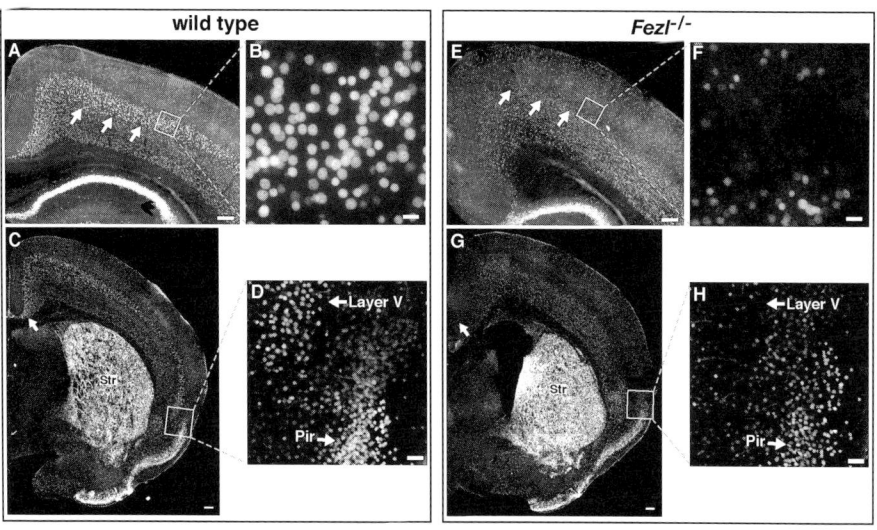

FIG. 2. *Fezl⁻ᐟ⁻* cortex lacks all CSMN and subcerebral projection neurons, indicated by CTIP2 expression. (A) Wild-type neocortex in coronal section, showing subcerebral projection neurons in layer V (arrows) intensely expressing CTIP2. (B) Boxed area in (A). (E) Coronal section matching that in (A) from a *Fezl⁻ᐟ⁻* brain showing absence of all CTIP2-positive neurons from layer V (arrows in F). (F) Boxed area in E. (C) Coronal section of wild-type brain showing normal CTIP2 staining across the entire mediolateral extent of layer V, in piriform cortex (Pir), and in striatum (Str). (D) Boxed area in C. (G) Section matching that in (C) from a *Fezl⁻ᐟ⁻* brain showing lack of subcerebral projection neurons from layer V, while CTIP2-expressing neurons in piriform cortex and striatum are not affected. (H) Boxed area in (G). Scale bars: (A, C, E, G) 200µm; (B, F) 20µm; (D, H) 50µm. Reproduced from Molyneaux et al (2005).

neurons, which express *Fezf2* at lower levels than subcerebral projection neurons, exhibit disorganization and abnormalities in gene expression, but are less affected (Hirata et al 2004, Chen et al 2005a, Molyneaux et al 2005). In contrast, upper layer pyramidal neurons are born correctly and appear normal (Chen et al 2005a, Molyneaux et al 2005). Importantly, without *Fezf2*, neocortical progenitors still produce similar numbers of layer V neurons (Chen et al 2005a, Molyneaux et al 2005), but morphologically they appear to be an expansion of layer VI, instead of exhibiting the distinctive appearance of layer V subcerebral projection neurons. Additionally, alkaline phosphatase expression from the *Fezf2* locus in null mutants labels an enlarged anterior commissure, further suggesting that a different type of projection neuron is generated in place of subcerebral projection neurons (Chen et al 2005a). Thus, *Fezf2* does not affect the ability of progenitors to generate glutamatergic neurons that position themselves in layer V; it likely acts to direct the next step in the programme of specification, defining the characteristics of a subcerebral projection neuron. As additional support for *Fezf2* in directing subcerebral projection neuron specification, the overexpression of *Fezf2* is sufficient to induce the birth of entirely new deep layer projection neurons that express *Ctip2* and *Tbr1* and extend axons through the internal capsule (Chen et al 2005b, Molyneaux et al 2005).

A second set of genes has been identified that control later aspects of subcerebral projection neuron development, likely acting downstream of factors such as *Fezf2*. These include the transcription factors *Ctip2* and *Otx1*. *Ctip2* is a transcription factor that is expressed at high levels in all subcerebral projection neurons while it is not expressed within callosal projection neurons of layer V (Arlotta et al 2005). However, prior loss of function experiments *in vivo* highlight an important role for CTIP2 in cell-type specification in the immune system (Wakabayashi et al 2003) and suggested that it might play a similar role in the development of CSMN and other subcerebral projection neurons. In the absence of *Ctip2*, subcerebral projection neuron axons exhibit defects in fasciculation, outgrowth, and pathfinding, with decreased numbers of axons reaching the brainstem (Fig. 3; Arlotta et al 2005). In addition, reduced *Ctip2* expression in *Ctip2* heterozygous mice results in abnormal developmental pruning of corticospinal axons (Arlotta et al 2005). These experiments identified *Ctip2* as a critical regulator of subcerebral axon extension and the refinement of collaterals as these neurons mature.

Another key transcription factor known to play a role in the target choice of subcerebral projection neurons is *Otx1*. This protein is expressed in putative deep layer progenitors in the VZ, exhibiting decreasing levels of expression in the VZ during the generation of upper layer neurons (Weimann et al 1999, Inoue et al 2004). As deep layer projection neurons mature, localization of OTX1 shifts from the cytoplasm to the nucleus, indicating a fine regulation of the activity of this protein (Frantz et al 1994a, Weimann et al 1999). Postnatally, within layer V, *Otx1*

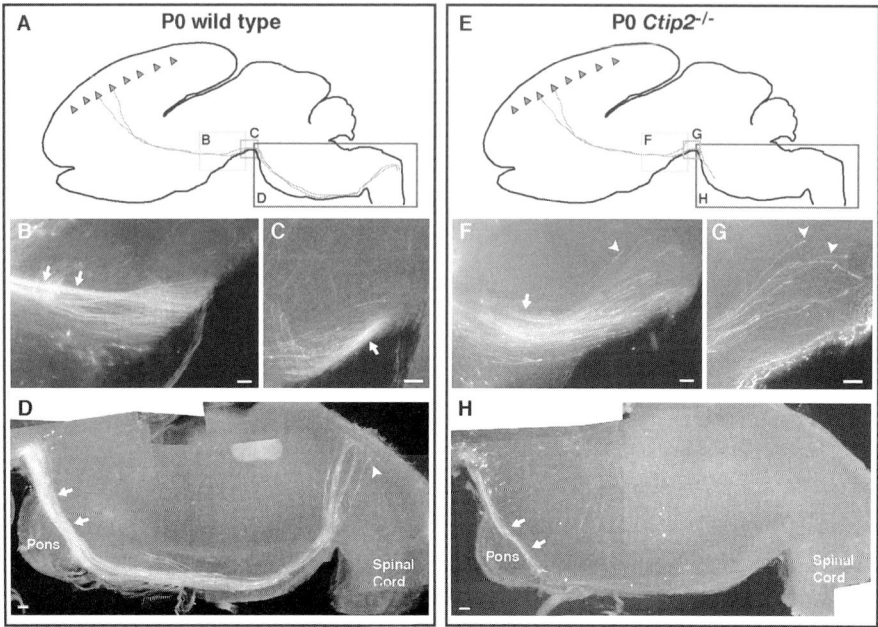

FIG. 3. CSMN in $Ctip2^{-/-}$ mice display pathfinding defects and fail to extend to the spinal cord. (A, E) Schematic representations of sagittal views of the brain and proximal spinal cord in wild type and $Ctip2^{-/-}$ mice, respectively, showing the location of CSMN somas in the cortex (triangles) and their axonal projections toward the spinal cord (lines). (B–D and F–H) Photomicrographs of boxed areas in (A) and (E) respectively. (B, F) Axonal projections by subcerebral projection neurons showing that (B) P0 wild-type axons are organized in typical axon fascicles (arrows), but (F) matched P0 $Ctip2^{-/-}$ null mutant axons are very disorganized, nonfasciculated (arrow), and display axonal projections that deviate from the normal pathway and extend to ectopic targets (arrowhead). (C, G) The same axonal fibres as (B) and (F), at a more caudal location. (C) Wild type axons are highly organized in tight bundles of fibres progressing unidirectionally toward the pons (arrow), while (G) $Ctip2^{-/-}$ axons are strikingly reduced in number with many individual fibres extending to ectopic sites (arrowheads). (D, H) Photomicrographic montages demonstrating (D) that P0 wild-type axons are abundant through the pons (arrows), and have already reached the pyramidal decussation entering the spinal cord (arrowhead). (H) A much smaller number of axons in $Ctip2^{-/-}$ mice enter the pons (arrows) and no axons extend into the medulla or reach the pyramidal decussation. Scale bars, 100 μm. Reproduced from Arlotta et al (2005).

is expressed in 40–50% of subcerebral neurons, primarily those within the visual cortex, while it is not expressed in callosal neurons (Weimann et al 1999). Mice lacking the gene for *Otx1* have defects in the development of corticotectal projection neurons. Without *Otx1*, corticotectal projection neurons maintain an axon to the spinal cord and caudal pontine nuclei, collaterals that are only appropriate for CSMN and are normally eliminated by corticotectal projection neurons

(Weimann et al 1999), indicating that *Otx1* might play a later role in subcerebral projection neuron development than *Fezf2* and *Ctip2*, controlling the refinement and pruning of axonal collaterals. Additional axon outgrowth and guidance molecules, such as IGF1 and RYK, have been described that play a role in the extension and guidance of subcerebral projection neuron axons to targets in the brainstem and spinal cord (Liu et al 2005, Harel & Strittmatter 2006, Ozdinler & Macklis 2006).

While a comprehensive understanding of the role played by additional subcerebral projection neuron-specific genes still awaits substantial experimental work *in vivo*, based on the data available thus far, a possible model for the generation of subcerebral projection neurons can be put forward that requires sequential steps of progressive differentiation. We propose that the concerted function of *FoxG1*, *Lhx2*, *Pax6* and *Emx2* first gives progenitors neocortical potential, setting the stage for the generation of multiple classes of glutamatergic projection neurons. It is conceivable that radial glia progenitors might then express a sequential series of transcription factors that are maintained in intermediate progenitors and post-mitotic neurons, imparting subtype identity. Thus, during the generation of subcerebral projection neurons, genes such as *Brn1* and *Brn2* might act on partially specified progenitors to determine aspects of laminar identity as individual subtypes of pyramidal neurons are generated. *Fezf2* then specifies the subcerebral projection neuron lineage within a layer (i.e. layer V), enabling the development of the molecular, morphological, and anatomical projection properties of subcerebral projection neurons. Finally, once this cascade is initiated, the expression of genes such as *Ctip2* and *Otx1*, which govern subcerebral axonal outgrowth and target selection, would act to establish the precise connectivity and later morphological features of subcerebral projection neurons. The direct relationships between these transcription factors and the many, yet functionally uncharacterized, genes that act in the cascade of subcerebral projection neuron development remain to be determined. Together, these molecules comprise the first elements of the molecular programme that drives the anatomical model of subcerebral projection neuron development described more than a decade ago (O'Leary & Koester 1993).

CSMN repopulation and repair of corticospinal circuitry

Knowledge of the molecular controls over CSMN development and survival might provide new approaches for the treatment of traumatic spinal cord injury and CSMN degenerative diseases such as ALS, primary lateral sclerosis (PLS) and hereditary spastic paraplegia (HSP). These signals could be manipulated to enhance the survival of degenerating CSMN or induce the re-growth of axons after injury. Alternatively, manipulation of the genes that control the progressive differentiation of progenitors along the CSMN lineage could potentially be used to induce the formation of CSMN from neural precursors *in vitro* and ultimately *in vivo*.

Previous experiments from our laboratory have shown that distinct subtypes of neurons can be induced to undergo neurogenesis from immature precursors, even in the normally inhibitory environment of the adult mammalian neocortex (Magavi et al 2000, Scharff et al 2000). More recent data indicate that endogenous neural precursors can also be induced to differentiate into CSMN *in vivo* (Chen et al 2004). However, the number of newborn CSMN is quite low, and only a small percentage of the newborn neurons survive long enough to establish permanent connections to distal targets in the spinal cord (Chen et al 2004). While these experiments demonstrate that new CSMN can be added to the normally inhibitory environment of the adult cortex and extend long-distance projections to the spinal cord, it is likely that the number of functional newborn CSMN could be substantially increased, and functional recovery might be effected by improved understanding of controls over CSMN specification, survival, and connectivity at the molecular level. Such information might enhance survival of developing CSMN and improve the ability of CSMN to connect to proper targets, which in turn might enhance functional connectivity and circuit repair.

One gene that might be a candidate for manipulation to generate CSMN from progenitors is *Fezf2*. *Fezf2* is expressed in the ventricular zone during the generation of deep layer neurons, and its expression is maintained in postmitotic neurons of layers V and VI. As development progresses, the expression of *Fezf2* in progenitors decreases and disappears by the time upper layer neurons are generated (Hirata et al 2004, Inoue et al 2004, Molyneaux et al 2005). As described above, it is required for the specification of CSMN and all other subcerebral projection neurons. Overexpression of *Fezf2* in progenitors soon after the generation of layer V and VI is completed (i.e. in progenitors that give rise to layer IV neurons) is sufficient to at least partially override this restriction and induce later-stage progenitors to produce neurons with some molecular and anatomical features of earlier-born neurons (Molyneaux et al 2005). Further analysis of *Fezf2* transfected neurons with additional positive and negative markers of subcerebral projection neurons is needed to determine the extent of *Fezf2*'s effect on neuronal phenotype. Interestingly, *Fezf2* appears to, at least in part, affect progenitor plasticity late in development, as suggested by the fact that forced expression of *Fezf2* in E17 progenitors results in the generation of upper layer neurons that inappropriately express *Tbr1* at a higher frequency than is normally observed in upper layer neurons and extend axonal projections to the pons (a feature of deep layer neurons) (Chen et al 2005b). However, *Fezf2*-overexpressing neurons migrate to the layer appropriate for this late birth date instead of layer V, suggesting some limitations to the ability of *Fezf2* alone to alter the fate of late-stage progenitors. While it remains to be elucidated to what extent these late born neurons change their identity in response to *Fezf2* overexpression, together, these experiments indicate that cortical progenitors might be more plastic than previously suspected, even late in neurogenesis, if manipulated by the appropriate control molecules. In the future, such neocortical progenitor plasticity might be manipulated much later in development, or even during adulthood, in order to repair damaged cortical circuitry.

Acknowledgements

This work was partially supported by grants from the NIH (NS45523, NS49553, NS41590), the Harvard Stem Cell Institute, the Spastic Paraplegia Foundation, and the ALS Association to JDM. PA was partially supported by a Claflin Distinguished Scholar Award and a grant from the ALS Association. BJM was supported by the Harvard M.S.T.P. and the United Sydney Association.

References

Alvarez-Bolado G, Rosenfeld MG, Swanson LW 1995 Model of forebrain regionalization based on spatiotemporal patterns of POU-III homeobox gene expression, birthdates, and morphological features. J Comp Neurol 355:237–295

Arlotta P, Molyneaux BJ, Chen J, Inoue J, Kominami R, Macklis JD 2005 Neuronal subtype-specific genes that control corticospinal motor neuron development in vivo. Neuron 45:207–221

Bareyre FM, Kerschensteiner M, Misgeld T, Sanes JR 2005 Transgenic labeling of the corticospinal tract for monitoring axonal responses to spinal cord injury. Nat Med 11:1355–1360

Bertrand N, Castro DS, Guillemot F 2002 Proneural genes and the specification of neural cell types. Nat Rev Neurosci 3:517–530

Bulchand S, Grove EA, Porter FD, Tole S 2001 LIM-homeodomain gene Lhx2 regulates the formation of the cortical hem. Mech Dev 100:165–175

Chen J, Magavi SS, Macklis JD 2004 Neurogenesis of corticospinal motor neurons extending spinal projections in adult mice. Proc Natl Acad Sci USA 101:16357–16362

Chen B, Schaevitz LR, McConnell SK 2005a Fezl regulates the differentiation and axon targeting of layer 5 subcortical projection neurons in cerebral cortex. Proc Natl Acad Sci USA 102:17184–17189

Chen JG, Rasin MR, Kwan KY, Sestan N 2005b Zfp312 is required for subcortical axonal projections and dendritic morphology of deep-layer pyramidal neurons of the cerebral cortex. Proc Natl Acad Sci USA 102:17792–17797

Frantz GD, Weimann JM, Levin ME, McConnell SK 1994a Otx1 and Otx2 define layers and regions in developing cerebral cortex and cerebellum. J Neurosci 14:5725–5740

Frantz GD, Bohner AP, Akers RM, McConnell SK 1994b Regulation of the POU domain gene SCIP during cerebral cortical development. J Neurosci 14:472–485

Fricker-Gates RA, Shin JJ, Tai CC, Catapano LA, Macklis JD 2002 Late-stage immature neocortical neurons reconstruct interhemispheric connections and form synaptic contacts with increased efficiency in adult mouse cortex undergoing targeted neurodegeneration. J Neurosci 22:4045–4056

Guillemot F 2005 Cellular and molecular control of neurogenesis in the mammalian telencephalon. Curr Opin Cell Biol 17:639–647

Harel NY, Strittmatter SM 2006 Can regenerating axons recapitulate developmental guidance during recovery from spinal cord injury? Nat Rev Neurosci 7:603–616

Hevner RF, Daza RA, Rubenstein JL, Stunnenberg H, Olavarria JF, Englund C 2003 Beyond laminar fate: toward a molecular classification of cortical projection/pyramidal neurons. Dev Neurosci 25:139–151

Hirata T, Suda Y, Nakao K, Narimatsu M, Hirano T, Hibi M 2004 Zinc finger gene fez-like functions in the formation of subplate neurons and thalamocortical axons. Dev Dyn 230:546–556

Inoue K, Terashima T, Nishikawa T, Takumi T 2004 Fez1 is layer-specifically expressed in the adult mouse neocortex. Eur J Neurosci 20:2909–2916

Jones EG, Schreyer DJ, Wise SP 1982 Growth and maturation of the rat corticospinal tract. Prog Brain Res 57:361–379

Killackey HP, Koralek KA, Chiaia NL, Rhodes RW 1989 Laminar and areal differences in the origin of the subcortical projection neurons of the rat somatosensory cortex. J Comp Neurol 282:428–445

Legg CR, Mercier B, Glickstein M 1989 Corticopontine projection in the rat: the distribution of labelled cortical cells after large injections of horseradish peroxidase in the pontine nuclei. J Comp Neurol 286:427–441

Liu Y, Shi J, Lu CC et al 2005 Ryk-mediated Wnt repulsion regulates posterior-directed growth of corticospinal tract. Nat Neurosci 8:1151–1159

Magavi SS, Leavitt BR, Macklis JD 2000 Induction of neurogenesis in the neocortex of adult mice. Nature 405:951–955

Marin O, Rubenstein JL 2003 Cell migration in the forebrain. Annu Rev Neurosci 26:441–483

McEvilly RJ, de Diaz MO, Schonemann MD, Hooshmand F, Rosenfeld MG 2002 Transcriptional regulation of cortical neuron migration by POU domain factors. Science 295:1528–1532

Molnar Z, Cheung AF 2006 Towards the classification of subpopulations of layer V pyramidal projection neurons. Neurosci Res 55:105–115

Molyneaux BJ, Arlotta P, Hirata T, Hibi M, Macklis JD 2005 Fezl is required for the birth and specification of corticospinal motor neurons. Neuron 47:817–831

Molyneaux BJ, Arlotta P, Menezes JR, Macklis JD 2007 Neuronal subtype specification in the cerebral cortex. Nat Rev Neurosci 8:427–437

Monuki ES, Porter FD, Walsh CA 2001 Patterning of the dorsal telencephalon and cerebral cortex by a roof plate-lhx2 pathway. Neuron 32:591–604

O'Leary DD, Terashima T 1988 Cortical axons branch to multiple subcortical targets by interstitial axon budding: implications for target recognition and 'waiting periods'. Neuron 1:901–910

O'Leary DD, Koester SE 1993 Development of projection neuron types, axon pathways, and patterned connections of the mammalian cortex. Neuron 10:991–1006

Ozdinler PH, Macklis JD 2006 IGF-I specifically enhances axon outgrowth of corticospinal motor neurons. Nat Neurosci 9:1371–1381

Scharff C, Kirn JR, Grossman M, Macklis JD, Nottebohm F 2000 Targeted neuronal death affects neuronal replacement and vocal behavior in adult songbirds. Neuron 25:481–492

Schreyer DJ, Jones EG 1988 Axon elimination in the developing corticospinal tract of the rat. Brain Res 466:103–119

Sheen VL, Macklis JD 1995 Targeted neocortical cell death in adult mice guides migration and differentiation of transplanted embryonic neurons. J Neurosci 15:8378–8392

Shin JJ, Fricker-Gates RA, Perez FA, Leavitt BR, Zurakowski D, Macklis JD 2000 Transplanted neuroblasts differentiate appropriately into projection neurons with correct neurotransmitter and receptor phenotype in neocortex undergoing targeted projection neuron degeneration. J Neurosci 20:7404–7416

Sugitani Y, Nakai S, Minowa O et al 2002 Brn-1 and Brn-2 share crucial roles in the production and positioning of mouse neocortical neurons. Genes Dev 16:1760–1765

Terashima T 1995 Anatomy, development and lesion-induced plasticity of rodent corticospinal tract. Neurosci Res 22:139–161

Voelker CC, Garin N, Taylor JS, Gahwiler BH, Hornung JP, Molnar Z 2004 Selective neurofilament (SMI-32, FNP-7 and N200) expression in subpopulations of layer V pyramidal neurons in vivo and in vitro. Cereb Cortex 14:1276–1286

Wakabayashi Y, Watanabe H, Inoue J et al 2003 Bcl11b is required for differentiation and survival of alphabeta T lymphocytes. Nat Immunol 4:533–539

Weimann JM, Zhang YA, Levin ME, Devine WP, Brulet P, McConnell SK 1999 Cortical neurons require Otx1 for the refinement of exuberant axonal projections to subcortical targets. Neuron 24:819–831

Wise SP, Jones EG 1977 Cells of origin and terminal distribution of descending projections of the rat somatic sensory cortex. J Comp Neurol 175:129–157

DISCUSSION

O'Leary: What is the source of IGF1 that is influencing the axon outgrowth *in vivo*?

Macklis: We don't know this for sure by direct experiments. From other work we know that it is highly expressed in the forebrain itself during these periods. We have done experiments placing the beads at different distances and angles from the growing axons. It doesn't seem to be an axon growth guidance molecule. It is active at the cell body as an activator, saying 'grow out'. IGF1 is present, but we think it takes more than simply the presence of IGF for axon outgrowth. There are a number of molecules that are specifically expressed on CSMN that are IGF binders and amassing proteins. It is combinatorial. The growth factor can be present, but if a neuron population pulls it in or excludes it or doesn't listen to it, this can lead to differential effects.

Tan: It is wonderful that your techniques are discriminative enough to give you such specific genes in the layer V neurons. Is this corroborated by the Paul Allen brain atlas? Are there any other parts of the brain that express these genes?

Macklis: The cerebral cortex is as complex as the rest of the body. *Ctip2* can be expressed in T cell progenitors; this doesn't worry me. It is different in the forebrain than in T cells. Similarly, *Ctip2* is expressed in cerebral cortex only in subcerebral neurons. It is expressed in striatum, but in different combinatorial interactions with areal specific genes. If you look at the Allen brain atlas, genes that in the cortex are specific can be expressed in different populations. You might say that this means that they are non-specific. I disagree: within that area they are highly specific. This is just the complexity of the forebrain. The same transcriptional cassette is being used for different populations. In work we hope to be publishing soon we have shown that the same *Ctip2* that is by our reckoning relatively downstream at the cortical plate projection outgrowth control for sub-cerebral neurons, is high upstream for medium spiny neurons. Without its func-tion, medium spiny neurons don't develop correctly, patch matrix organization is eliminated, and repulsive cues that allow the striatum to sit at that complex place and exclude migratory populations, and allow piercing corticospinal neurons is totally eliminated. We have found repeatedly that the same gene can be in a couple of places, but in combination with another it is specific. Also, some of these genes are on at X dose in one population and half X dose in another population. This seems to be quite important as well.

Molnár: At the beginning of your talk you mentioned that you were working backwards. You identified a specific projection neuron population by exploiting the fact that you can access them through their specific target sites. Then you mentioned that you do this at P8, P6, P3 and E18.5. So how far did you go back? Surely your ultimate goal must be to understand what cocktail of transcription factors has to be present when these cells are born.

Macklis: The earliest that we could get at their projections was E17 injection, E18. Then we take those 'identity' genes that are on from E18 and not in any other populations, and then work back by *in situ* or immunoflourescence to find, for example, that *Fezf2* was on at E11 in ventricular zone progenitors. Then we can march forwards. We use the strategy of marching back to get a foothold.

Molnár: Are these steps of target selection, somatodendritic morphology or physiological signatures specified at the very early stages when the cells leave the ventricular zone, or are they imposed slightly later?

Macklis: There will be lots of inputs and tweaks. We think that certain elements are there right from the progenitors. We will find this from our fate mapping and some of our other work. People talk about the neural stem cells (what we prefer to term 'precursors') as though they are all one population, but I don't think this is true. There are some which are already partially fate restricted. We think it goes way back, but within the population there may be precursors that really don't want to cross certain lines. There are precursors that want to be a subcortical neuron. They really want to be a corticospinal neuron, but we can modify them more easily to be a corticothalamic neuron. Our working model is that there are dendritic trees of close evolutionary association.

Mallamaci: I have a question about *Ctip2*. In *Ctip2* knockouts, are only corti-cospinal projections impaired, or does the defect also apply to corticotectal projections?

Macklis: We came up with this term 'subcerebral' projection neurons. These were genes that were on in at least the two populations we looked at as our starting point, in both the corticotectal and corticospinal neurons. Then when we looked more closely they were on in all subcerebral neurons: all the neurons to brainstem and spinal cord. *Ctip2* is a subcerebral axon elongation and fasciculation control gene. Similarly, *Fez* is a subcerebral control gene. Now we have identified genes that arealize. A subset of the ones we originally published were areal specific only to those 5000–6000 corticospinal neurons. Neither *Fez* or *Ctip2* define those populations alone; it is in combination with one or two other genes.

Mallamaci: I have a question about *Fezf2*. Do we have any evidence that any discrepancy takes place between transcription of this gene and nuclear availability of the corresponding protein? In other words, is *Fezf2* post-transcriptionally regulated? Is the protein regulated in its cytoplasmic or nuclear localization?

Macklis: While Paola Arlotta was in my lab, we initiated fate mapping and looked at some downstream target genes. This is a direction that we are pursuing collaboratively, and that Paola is pursuing in her own lab.

Hevner: You mentioned the idea of guiding the cell fate decisions. This depends so much on transcription factors. You can change transcription factor expression in embryos, but how can you do this in the adult nervous system?

Macklis: It will not be a simple process of taking generic 'neural stem cells' out of a cup; there are many incredible complexities and limitations to thinking about nervous system repair. I wouldn't think about a disease with diffuse biology, like Alzheimer's or stroke. But I would think about a disease with only one or two neuron types involved, e.g. ALS with one in the forebrain and one in the spinal cord. For the last 20 years, Tom Jessell, Sam Pfaff and others have studied the spinal cord in depth, where there is a focal concentration of a few cell types. I think the relatively focal location of corticospinal neurons means that there is the possibility of identifying fate-restricted progenitors, which might allow local intervention or transplantation. Can we turn on a small set of endogenous quiescent progenitors that are already partially fate restricted and guide them down this differentiation direction? It is not just instruction: we have found that there are also active suppressive factors. Motor output disorders also make sense because we could do really 'lousy' job with them, get them to the wrong segment, with bad precision, in tiny numbers, and this could still lead to important restoration of function.

Rakic: At the first Novartis Foundation symposium on the cerebral cortex in 1994 I was arguing for specificity of neurons and neuronal connections, from much more primitive methods than those that are available today. I wish someone could have given a talk like you did now, back then. At the next meeting, perhaps someone will give a talk about the specificity of interneurons, because I think they are also very specialized, we just have to be patient to determine their unique properties. There is another problem that could arise: not just that these neurons are specific, but also that some of them have to migrate to cortex after layer V neurons have settled and before layer III neurons have been generated. This is a problem that will be difficult to solve.

Macklis: I agree that it will be hard, and I don't think it is going to happen with perfect precision. There is a reasonable likelihood that it is now possible, but I don't think it will be easy.

Fishell: When do they know what they know, and when do they start breeding true? The neural plate is initially comprised of a pseudostratified epithelium that it self replicating, concomitant with the initiation of neurogenesis, neuroepithelial cells become radial glia. In addition, this is coincident with SVZ and VZ progenitors separating out. Do you have any insights from the genes you are looking at about their influence on these early segregation events?

Macklis: I think this will be an evolving set of answers over the next few years, but it is a lot earlier than some people might have guessed. We have identified a gene that is involved with the pallial/subpallial progenitor stage, and at the Mash/Neurogenin control stage. There are others that are at the rostrocaudal sensori-motor cortex and barrel arrealization field stage. It will be a combinatorial system setting up fates. I am sure there will be other decisions as the cells reach the target and consolidate their identities. From fate switching experiments, we know that it is not just that a neuron shares a little bit of confused outgrowth like a corticospinal motor neuron. In some cases, they decide to become one type and leave their other fate behind. In other cases, neurons with mixed or 'confused' phenotype can be formed by genetic manipulation.

Fishell: I guess that the genes you are talking about are acting initially upon the pseudostratified epithelium, prior to the emergence of radial glia?

Macklis: Yes, it appears so, but we will know more in the coming year or so.

Richards: How specific is that decision to become subcortically projecting, as opposed to callosally projecting? Are you saying that the fate decision is made at the time when the cells are migrating and differentiating; that there is no decision made as the axon arrives at the subplate to project medially or laterally.

Macklis: I think it is clear from our work and that of others that those neurons know what they will become before they leave the ventricular zone.

Richards: So the knockouts don't have any callosal phenotypes that you have seen.

Macklis: Not a loss-of-function phenotype, but one can have a fate shift.

Rakic: I agree. For example, as soon as callosal neurons leave the ventricular zones in larger brains, such as in the rhesus monkey, they have already sent their axons to the cerebral hemisphere on the opposite side (Schwartz et al 1991). These neurons haven't come near their final destination in the cortex, but they already have information that they are going to be callosal neurons.

Macklis: I think all this will be firmed up over the next few years.

Yamamoto: I want to ask about the possibility that several transcription factors work together. As you know well, *Otx1* is also specifically expressed in layer V neurons. Is there any possibility of this working in combination with the genes that you found?

Macklis: If for even a second I suggested to anyone that our favourite genes already known in the field are not playing a central role, I take that back. They are all working in combination, I'm sure.

Price: What might happen to the expression of genes like *Fezf2* if you do a heterochronic transplant? Transplanting cortical precursors from one animal into the cortex of an animal of a different age can turn precursors of the deep layers into superficial layer cells (McConnell & Kaznowski 1991).

Macklis: Although we have done those experiments for other reasons, we haven't done a physical heterochronic transplant. But we have done a genetic heterochronic CSMN fate recruitment set of experiments. We know from these that *Fez* is upstream of at least many of those decisions.

Price: Would you predict that, if you did a heterochronic transplant, there would be *Fezf2*-positive cells sitting in the superficial layers but they wouldn't develop the morphology and connections of superficial layer cells?

Macklis: We already know that one can genetically heterotopically place neurons that otherwise we would all think are in a superficial position and are CSMN. They have fate shifted and sent an axon down; they don't send a callosal axon. We think this is fairly high up in the combinatorial transcription factor programme.

Molnár: No one has mentioned the other factors involved in layer V differentiation, such as ER81 and OTX1 (Frantz et al 1994, Weimann et al 1999, Hevner et al 2003, Yoneshima et al 2006). There are also some structural proteins, neurofilament markers that are specifically expressed in the subcortically projecting layer V populations (Voelker et al 2004). Susan McConnell, Sasha Nelson, Mriganka Sur and Etienne Audinat had a major contribution in addition to your group to this particular field (Chen et al 2005, Nelson et al 2006, Christophe et al 2005). Due to these works we are beginning to have a considerable group of biomarkers for subsets of projection neurons. When can we start connecting the dots and start clustering these genes into various differentiation pathways? What happens to a specific subtype of projection neurons when a particular transcription factor is present and another is missing? The understanding of these issues would have a huge influence not only on stem cell research but also on how we classify cortical neurons. We are entering a new era: so far we have been looking at Golgi-stained sections, but now we can really combine specific gene expression patterns and molecular markers together with somatodendritic morphology, targets and firing patterns.

Macklis: Yes, after our *Fezf2* paper was published, both Susan McConnell's lab and Nenad Sestain's lab published almost identical biology and results in two papers later in the year. These are reproducible results, therefore. The emerging biology from all our labs is getting co-ordinated and the dots are being connected.

I think the result will be a new molecular anatomy of cortical neuron subtypes. It won't be solely by layer.

Guillemot: When you discussed the role of *Fez* factors in specifying CSMN, you put forward two ideas that seem to me a bit conflicting. One is that they work combinatorially, and the other is that a gene such as *Fezf2* works as a master regulator. Can you elaborate on this? When you ectopically express *Fezf2* in the cortex and force upper layer neurons to acquire a CSMN fate, what are the neurons that are competent to adopt this fate?

Macklis: This is if we express it at the right place and time in cortex. It is not if we express it in the substantia nigra at E16. We know it works in that area of motosensory cortex and within the time of cortical development. It can fate-shift superficial layer neurons: we have done this experiment at multiple times throughout cortical development. It can fate-shift cortical neuron progenitors, but they are all within the domains of some of these other cortical neuron subtype control genes.

References

Chen B, Schaevitz LR, McConnell SK 2005 Fez1 regulates the differentiation and axon targeting of layer 5 subcortical projection neurons in cerebral cortex. Proc Natl Acad Sci USA 102:17184–17189

Christophe E, Doerflinger N, Lavery D, Molnár Z, Charpak S, Audinat E 2005 Two populations of layer V pyramidal cells of the mouse neocortex: Development and sensitivity to anesthetics. J Neurophysiol 94:3357–3367

Frantz GD, Bohner AP, Akers RM, McConnell SK 1994 Regulation of the POU domain gene SCIP during cerebral cortical development. J Neurosci 14:472–485

Hevner RF, Daza RA, Rubenstein JL, Stunnenberg H, Olavarria JF, Englund C 2003 Beyond laminar fate: toward a molecular classification of cortical projection/pyramidal neurons. Dev Neurosci 25:139–151

McConnell SK, Kaznowski CE 1991 Cell cycle dependence of laminar determination in developing neocortex. Science 254:282–285

Nelson SB, Sugino K, Hempel CM 2006 The problem of neuronal cell types: a physiological genomics approach. Trends Neurosci 29:339–345

Schwartz ML, Rakic P, Goldman-Rakic PS 1991 Early phenotype expression of cortical neurons: Evidence that a subclass of migrating neurons have callosal axons. Proc Natl Acad Sci USA 88:1354–1358

Voelker CC, Garin N, Taylor JS, Gahwiler BH, Hornung JP, Molnár Z 2004 Selective neurofilament (SMI-32, FNP-7 and N200) expression in subpopulations of layer V pyramidal neurons in vivo and in vitro. Cereb Cortex 14:1276–1286

Weimann JM, Zhang YA, Levin ME, Devine WP, Brulet P, McConnell SK 1999 Cortical neurons require Otx1 for the refinement of exuberant axonal projections to subcortical targets. Neuron. 24:819–831

Yoneshima H, Yamasaki S, Tsuji T et al 2006 ER81 expressed in a subpopulation of layer 5 projection neurons in rodent and primate cerebral cortex. Neuroscience 137:401–412

Perspectives on the developmental origins of cortical interneuron diversity

Gordon Fishell

Smilow Neuroscience Program and the Department of Cell Biology, 5th Floor Smilow Research Center, New York University School of Medicine, 522 First Avenue, New York, NY 10016, USA

Abstract. Cortical GABAergic interneurons in mice are largely derived from the subpallium. Work from our laboratory and others over the past five years has demonstrated that a developmental logic in space and time underlies the emergence of specific cortical interneuronal subtypes. Following on from the seminal work of the Rubenstein laboratory, we set out to fate map the output of the subpallial ganglionic eminences. Our initial approach utilized ultrasound backscatter microscopy to perform homotopic and heterotopic transplants of genetically marked progenitors from the lateral, medial and caudal ganglionic eminences (LGE, MGE and CGE, respectively) to unmarked host brains. The LGE, at least in the context of our transplant studies, did not appear to generate cortical interneurons. By contrast, we found that that approximately eighty percent of cortical interneurons arise from the MGE, while the remaining twenty percent were generated by the CGE. Hence, the majority of interneuron subtypes, including all fast spiking parvalbumin-positive basket cells and somatostatin-positive Martinotti cells appear to arise from the MGE. A more restricted set of cortical interneurons seems to be generated in the CGE, the majority of which are bipolar calretinin/VIP-positive interneurons. Complementing these results, we have recently demonstrated using inducible genetic fate mapping that the MGE produces specific cortical interneuron subtypes at discrete timepoints during development. These studies demonstrate that cortical interneurons arise from a precise developmental programme that acts in both space and time. Beyond this however, it seems likely that postmitotic events influence the specific function of subclasses of cortical interneurons. A primary challenge in the future will be to investigate what aspects of interneuron diversity are determined by intrinsic genetic programmes within each lineage versus those properties imposed by the local environment in the cortex.

2007 Cortical development: genes and genetic abnormalities. Wiley, Chichester (Novartis Foundation Symposium 288) p 21–44

Since the original descriptions by Ramon y Cajal a hundred years ago, cortical interneurons have been recognized to be a remarkably diverse population (Molnar et al 2006). Although they make up only a fifth of the neurons within the cerebral cortex, their function is thought to be central to learning and their dysfunction is correlated with disease. A starting point for understanding their role in both normal and abnormal contexts is determining how cortical interneuron diversity is generated. Work from my laboratory and others over the past five years has revealed that the place and time of origin of cortical interneurons predicts their intrinsic physiological properties. Here I will review findings that suggest that the broad classes of cortical interneurons are specified at the place and time of their generation. Moreover, I will argue that this is achieved through the action of transcriptional codes within the ventral progenitor zones that give rise to them. Understanding the logic of this transcriptional code holds the promise of both unravelling the molecular means by which different cortical interneurons subtypes are generated and providing genetic tools for their prospective identification and manipulation.

Cortical interneurons are generated within the subpallium

The realization that cortical interneurons originate within the subpallium provided the first clue that understanding how cortical interneurons are generated would require looking beyond the cortex (Anderson et al 1997a, b). In a series of landmark papers, the Rubenstein laboratory revealed that a massive ventral to dorsal migration of cortical interneurons occurred during late embryonic development. Near coincident work from the laboratory of Arturo Alvarez-Buylla suggested that the medial ganglionic eminence (MGE) might be the source of these interneurons (Wichterle et al 1999). This idea came from studies showing that MGE explants are remarkably migratory and perinatal transplants of embryonic day (E) 13.5 MGE to postnatal day (P) 0 recipients, resulted in the wide dispersion of cells within the telencephalon.

In an attempt to explore this issue further, I was approached by Arturo Alvarez-Buylla with the suggestion of using *in utero* ultrasound backscatter microscopy (UBM)-guided transplantation to fate map the MGE and LGE (lateral ganglionic eminence). My laboratory had used this method as a means of doing gain of function viral experiments (Gaiano et al 1999), after Daniel Turnbull had pioneered this method several years earlier (Olsson et al 1997). Lacking however was a precise method to introduce the nanolitre quantities of cells required for these transplants, a difficulty overcome through the innovative adaptation of a X-directional micromanipulator as a microinjector by Arturo Alvarez-Buylla. Our collaborative effort provided the first definitive proof that large numbers of cortical interneurons arose from the MGE, while the striatum and olfactory bulb neurons were derived from the LGE (Wichterle et al 2001, Fig. 1). Importantly, this work also showed that the site of origin of the donor tissue and not the site of transplantation predicted the fate of the transplanted tissue. Specifically LGE donor tissue transplanted into the MGE was indistinguishable from homotopic transplants of LGE progenitors. This provided the first strong evidence that the fates of progenitors

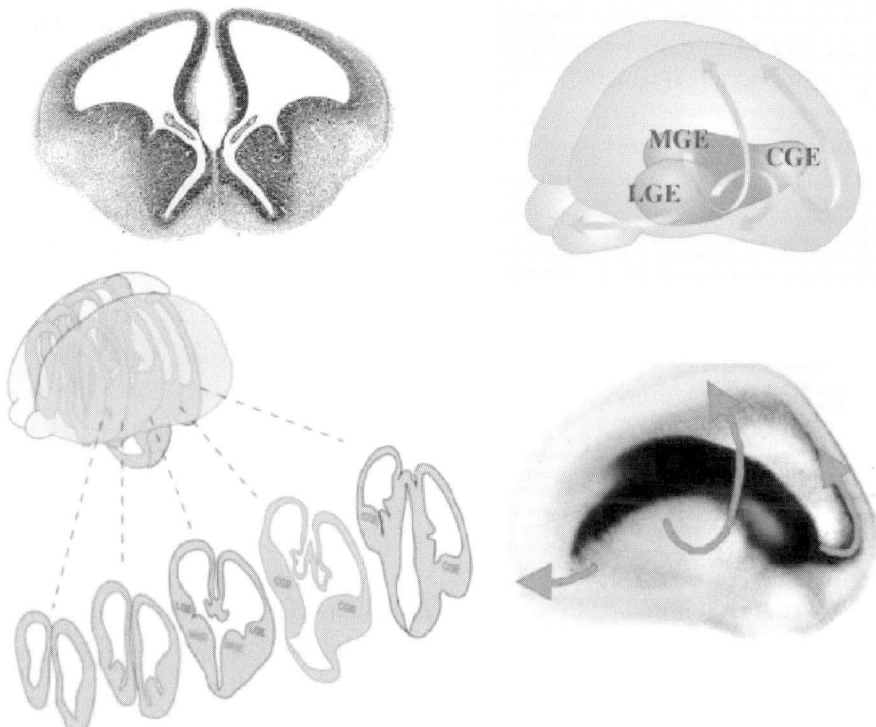

FIG. 1.   Cortical interneurons arise from the three progenitor zones within the subpallium, the lateral, the medial and the caudal ganglionic eminences (LGE, MGE and CGE, respectively). The brain section on the top left, shows a coronal section through the telencephalon at the level of the MGE and LGE. This is equivalent to the centre section in the series shown on the bottom left (taken from Danglot et al 2006 with permission). On the top right is a schematic showing the three ganglionic eminences and the general paths of migration from each of them, with tangentially migrating neurons from the LGE going to the olfactory bulb, while the CGE and the MGE send their progeny to the cortex. The saggital section below this shows a $Dlx2^{tauLacZ}$ brain histochemically stained for beta galactosidase. The labelled cells can be seen tangentially dispersing along the paths shown in the figure above it.

within the ventral eminences were cell autonomously specified in their respective proliferative zones.

The success of this approach encouraged Susanna Nery and Joshua Corbin, a student and postdoctoral fellow in my laboratory, to examine the caudal ganglionic eminence (CGE). While the MGE and LGE are distinct progenitor zones within the subpallium, it had been observed by us and others (Anderson et al 2001) that the posterior aspect of these eminences is fused. First referred to as the CGE by Anderson and colleagues (2001), it was unclear whether this was a posterior extension of the MGE and/or the LGE, or an entirely distinct structure in and of itself. The purpose of our study was to determine which structures within the mature telencephalon the CGE contributed to and indeed whether in fact it should be considered a separate progenitor zone. While the MGE can be distinguished from the CGE and LGE by its expression of *Nkx2.1*, there are currently no molecular markers to distinguish between CGE and LGE. Indeed, other than their differences in fate, the best evidence to date that the LGE and CGE are genetically distinct comes from the observation that in *Gsh2* null mutants, pallial genes such as *Ngn2* encroach on the LGE but not the CGE (Nery et al 2002).

MGE and CGE progenitors give rise to distinct populations of cortical interneurons

Our original goal in fate mapping the CGE, MGE and LGE was to determine the contribution of these eminences to structures in the telencephalon. While ultimately this focused on the cortical interneuron population, it is important to remember that a significant proportion of neurons derived from these eminences contribute to subpallial structures, including the striatum, the nucleus accumbens and the amygdaloid complex (Nery et al 2002). Importantly, as with our previous study, the fate of transplanted MGE and CGE progenitors appeared to be cell autonomous restricted: the pattern of integration and differentiation of MGE-derived cells was the same regardless of whether the graft was targeted to the MGE or the CGE. The resulting fate mapping study (which also re-examined the fate of LGE-progenitors for comparison) revealed that the three ganglionic eminences each contributed neurons to distinct and largely non-overlapping structures within the telencephalon. For instance, while the CGE-derived neurons populated the shell of the nucleus accumbens, LGE-derived neurons contributed to the core of this structure. Alternatively, while the CGE gave rise to neurons in the medial amygdala nuclei, the MGE contributed primarily to the lateral amygdala nuclei. In the cortex however, the areal distribution of MGE and CGE neurons largely overlapped. Moreover, in both cases all neurons within the cortex derived from these structures were GABAergic and hence inhibitory interneurons. However, in both their laminar distribution and their morphology, the inter-neurons derived from the MGE and CGE were markedly different. It was these observations that first focused my laboratory on the question of cortical interneuron diversity.

At the time we did our MGE/CGE fate mapping experiments, the literature describing the diversity of cortical interneurons was divided into the groupers (such as Kawaguchi 1993) and the dividers (such as Markram et al 2004). The groupers divided interneurons into five basic categories, while the dividers posited the existence of at least a hundred distinct cortical interneuronal subtypes. These subdivisions were based largely on three criteria: morphology, marker expression and intrinsic physiological properties. Notably absent from the debate was a clear notion as to how this diversity was generated. Our results suggested that at least some aspects of cortical interneuron diversity are established by their place and time of origin. While the work presented in Nery et al (2002) demonstrated that the distribution, morphology and immunomarkers from interneurons originating in the MGE and CGE differed, the extent to which specific subtypes are generated in space and time was less certain. Studies from the Anderson laboratory published shortly afterward (Xu et al 2004), clarified these divisions by showing how specific interneuron populations, identified on the basis of their marker expression, arose from the CGE and MGE. By studying the intrinsic properties of these cells, such as firing rate and adaptation, as well as their morphological characteristics (Fig. 2), we were able to relate a cortical interneuron's site of origin to their physiological subclass.

In 2001, shortly before we published our CGE fate map study, the first of a series of cortical development meetings was held in Delphi. In discussing our data with Arnold Kriegstein at this meeting, I asked whether they were interested in helping us explore the physiological properties of interneurons originating from the MGE and the CGE. This began an extremely fruitful collaboration between our laboratories initiated by Susanna Nery from my laboratory and Stephen Noctor from the Kriegstein laboratory. The proof of principle for this work came from UBM-guided transplants of GFP-labelled donor MGE or CGE cells done at NYU Medical Center, with the postnatal patch clamp analysis being done at Columbia Medical Center. However, transferring animals between laboratories proved cumbersome and Susanna Nery shortly departed my laboratory for a postdoctoral fellowship in England. The work was continued in my laboratory by a postdoctoral physiologist Simon Butt and a graduate student Marc Fuccillo. Working together they refined the protocols for both the transplants and the physiological analysis of cortical interneurons in P14–P24 slice preparations (Butt et al 2005). Broadly speaking, our classification schemes were based on those outlined by Kawaguchi and Kubota (1997) in their analysis of frontal cortex, as well as Cauli et al (1997) in their analysis of sensory and motor cortex. This was greatly aided by our interactions with Bernardo Rudy, who guided us in developing meaningful protocols for examining the intrinsic physiological properties of cortical interneurons. Moreover, Bernardo Rudy's focus on the biophysical functions of specific ion channels (Rudy & McBain 2001, Goldberg et al 2005) led us to consider how developmentally expressed transcription factors could control interneuron function through modifying their membrane properties.

Our studies revealed that at E13.5 while twenty percent of cortical interneurons are bipolar Calretinin and VIP-expressing interneurons and come from the CGE, the remaining eighty percent of cortical interneurons arose from the MGE

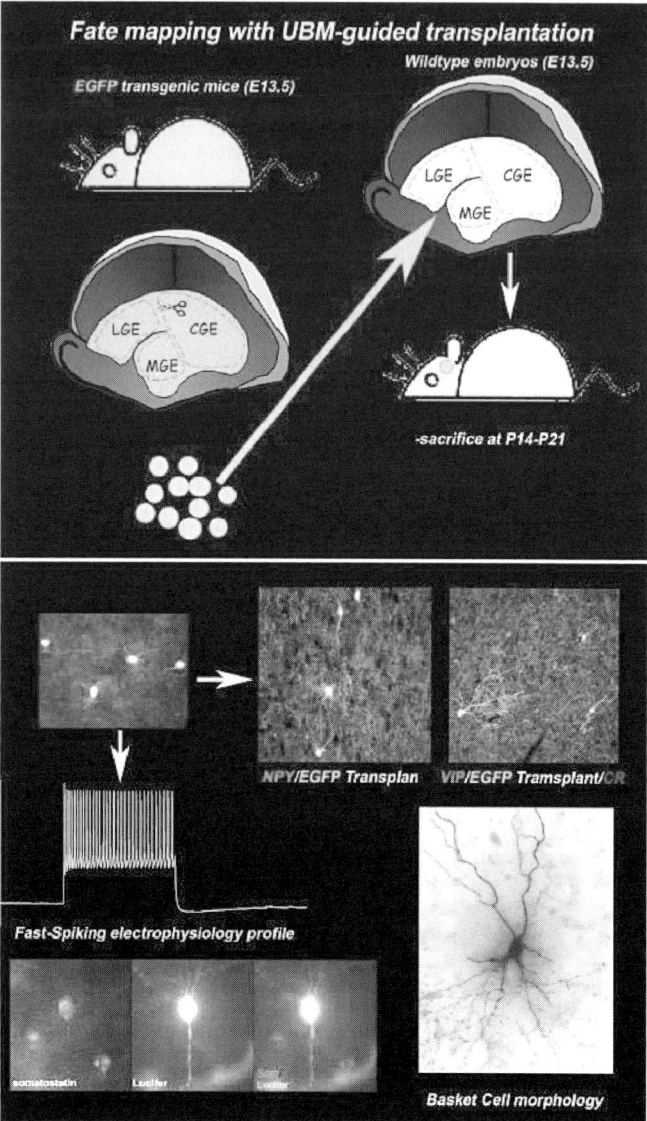

FIG. 2.   Schematic showing the procedure used to do ultrasound backscatter microscopy (UBM) guided injection *in utero* and the subsequent analysis of the genetically marked donor cells within the cerebral cortex. In the upper panel, donor tissue is dissected from one of the three ganglionic eminences from an EGFP-transgenic mouse. The EGFP-positive donor cells are introduced into an embryonic recipient host brain using UBM-guided transplantation. The animals receiving grafts are allowed to give birth and brain sections from these animals are analysed between P14 and P24 days postnatally. Donor cells that have entered the cortex and become mature cortical interneurons are visualized and examined for their expression of neural markers either immediately (top right panels in lower part of this figure, which unfortunately is indiscernible in the present B/W figure) or following electrophysiological analysis (three panels at the bottom left). Shown in the bottom panel is a trace of the firing pattern of a typical FS basket cell and its appearance after Lucifer yellow fill and immuno-visualization.

(Fig. 3). Within this latter population were included both the parvalbumin (PV)-positive, fast-spiking (FS), basket cell population, as well as the somatostatin (Sst)-positive, burst-spiking (BS) Martinotti cells. Therefore, the populations arising from these structures are distinct, demonstrating that the CGE and the MGE gave rise to entirely non-overlapping populations of cortical interneurons. Furthermore, through repeating the heterotopic transplantation of MGE progenitors to the CGE, we were able to demonstrate that the morphology and intrinsic physiological properties of cortical interneurons is also cell autonomously specified within the progenitor domains.

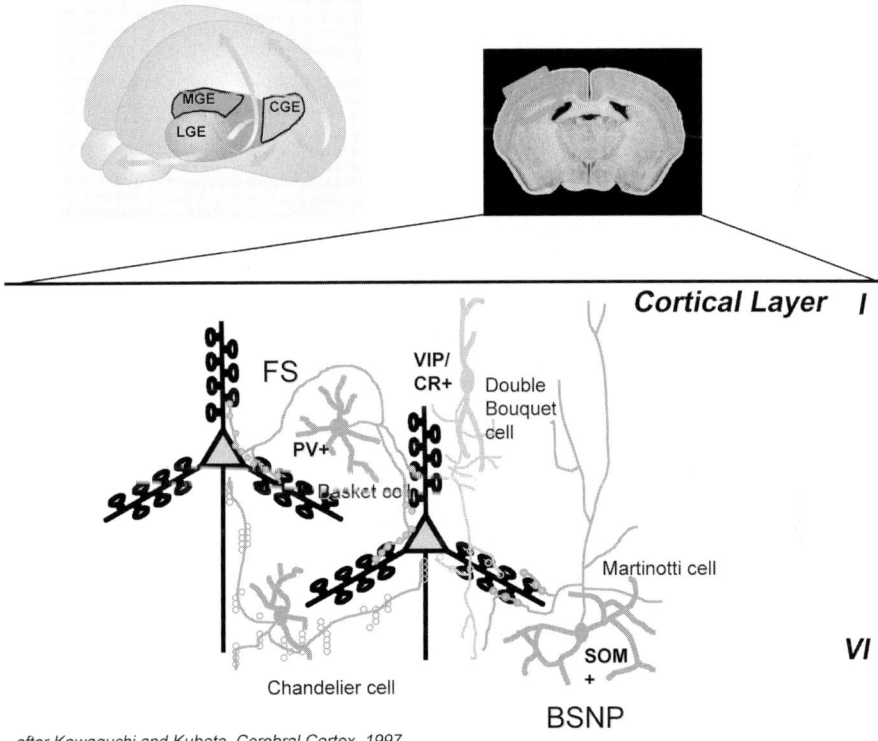

after Kawaguchi and Kubota, Cerebral Cortex, 1997

FIG. 3.   The MGE and CGE gives rise to distinct subsets of cortical interneurons. This figure shows a summary of the transplantation study by Butt et al (2005). In the schematic shown in the top left is shown the MGE as it appears at E13.5. The boxed area of cortex shown at the top right is schematically shown in the figure below. In this schematized view of the cortex (adapted from Kubota & Kawaguchi 1997) is shown a basket, a chandelier cell and a Martinotti cell, all of which are derived from the MGE at this timepoint. The indicated region in the schematic at the top left, shows the CGE. In the cortical schematic shown below is double bouquet cell, representative of the CGE-derived bipolar cortical interneuron population.

Different MGE-derived interneuron subtypes are generated at distinct developmental timepoints

The recognition that distinct subsets of cortical interneurons arise from the MGE and CGE only partially addresses the question of how the diversity in this population is generated. This is particularly pertinent with regards to the MGE from which 80% of all cortical interneuronal subtypes originate. The E13.5 transplant studies described above indicated that at this timepoint approximately half of all cortical interneurons derived from the MGE are FS basket cells or chandelier cells. An additional twenty percent are of the BS Martinotti interneuron subtype. While together these account for a majority of the interneuronal subtypes that arise from the MGE at E13.5, the remaining MGE-derived populations are quite diverse. In the longitudinal analysis of MGE-derived interneurons described below, we found that between E9 and E15.5 an additional eight subclasses of cortical interneurons are generated. Taken together, this suggests that the MGE generates ten distinct interneuronal subtypes. In principle, two broad developmental strategies could underlie the generation of different cortical interneuron subtypes from the MGE. Different interneuron classes might be generated within the MGE from discrete spatial subdomains. Alternatively, they might be generated through a shifting developmental fate of the MGE to sequentially give rise to discrete interneuron subtypes over time.

To date there is little evidence for the existence of subdomains within the MGE akin to those observed in the spinal cord. Recently Goichi Miyoshi and Simon Butt in my laboratory have explored the alternative possibility that the MGE produces different interneuron subtypes at discrete developmental timepoints (Miyoshi et al 2007). They did this using a genetic fate mapping method developed by our colleague Alexander Joyner (Zervas et al 2004, Joyner & Zervas 2006). This method relies on two components, a driver allele, coupled with a reporter allele (reviewed in Miyoshi & Fishell 2006). In our case we combined the use of an $Olig2^{CreER}$ driver (Takebayashi et al 2002) with the Z/EG reporter (Novak et al 2000). Fortuitously, we discovered that when $Olig2^{CreER}$; Z/EG mice are induced using 4 mg of tamoxifen it results in the labelling of nascent neurons as they exit the MGE. Hence, in effect this provides a genetic method for fate mapping MGE neurons based on their birthdate. From this analysis we found that each of the 10 classes of interneurons are born with a distinct temporal signature (Fig. 4). Some classes such as the Martinotti neurons are born only at early times during development (E9.5–E13.5), while others such as the delayed FS interneurons are only born at late times during neurogenesis (~E15.5) (for details of this analysis see Miyoshi et al 2007). While this work demonstrates that within the MGE the production of different cortical interneuronal subtypes are linked to their time of generation, it says nothing of the molecular means by which this diversity is created. Indeed,

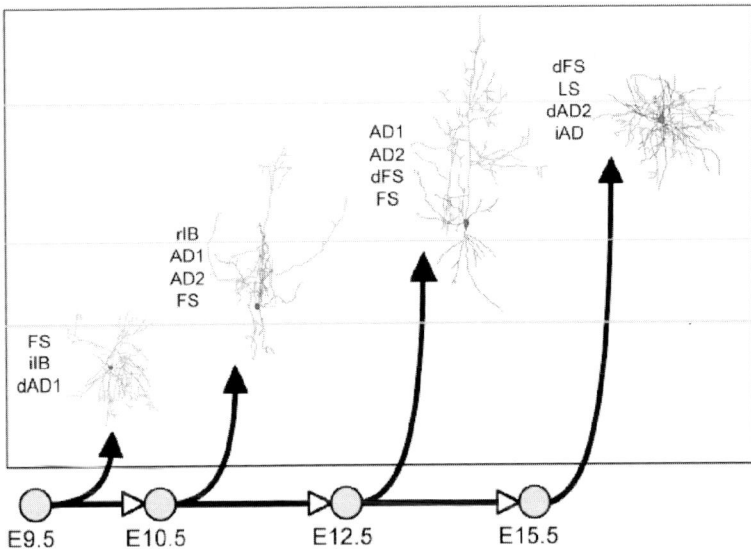

FIG. 4. Schematized results of the findings of Miyoshi et al (2007), where the cortical interneurons derived from the MGE were genetically determined by fate mapping the *Olig2*-expression population (see Miyoshi et al 2007 for details). In this study we found that the 10 different MGE-derived populations are born at precise developmental timepoints between E9.5 and E15.5. Their position in this schematic of cortex reflects their laminar position within the cortex. As interneurons, like pyramidal neurons, are born and populate the cortex in an inside out manner, the earliest born neurons are found in the deep layers, while the latest born neurons are found in the superficial layers.

our preliminary analysis of conditional *Olig2* loss of function mice indicate that this gene is not required for the generation of different MGE-derived interneuronal subclasses within the MGE.

Developmental genes involved in the development of cortical interneurons

What then is the developmental genetic basis by which cortical interneuron diversity is generated? Work over the past fifteen years has identified a number of transcription factors that are required for the generation of cortical interneurons, most notably the *Dlx* family of genes, *Mash1*, *Nkx2.1*, *Lhx6* and *Lhx7* (reviewed in Corbin et al 2001, Cobos et al 2006, Wonders & Anderson 2006). While analysis of null alleles of each of these genes supports that they play a role in the development of cortical interneurons, it is less clear how they are causal in the diversification of this population. A chief impediment to interpreting the phenotypes of these knockouts is that loss of many of these genes leads to lethality. Hence although the numbers of cortical interneurons seen at birth in *Dlx1/2* compound mutants is reduced by 60% (Anderson et al 1997b), and by 40% in either *Nkx2.1* (Sussel et al 1999) or *Mash1* mutants (Casarosa et al 1999), the fact that these animals do not survive postnatally makes assessing their contribution to subtype diversity difficult to gauge. The increasing availability of conditional alleles, as well as driver lines to genetically fate map these populations provides an attractive approach to begin addressing the role of these genes in generating cortical inter-neuron diversity. In addition, it is clear that many of the genes involved in this process have yet to be identified. To this end, Renata Brito in my laboratory has begun a microarray screen for genes expressed in cortical interneurons as a whole. We have achieved this by using a Dlx5/6^{EGFP} transgenic allele generated by Kenneth Campbell (Stenman et al 2003). While in accordance with the normal expression of the *Dlx5* and *Dlx6* genes, there is widespread expression of EGFP within the subpallium, it appears to selectively label the entire cortical interneuron population. By isolating embryonic cortex at E13.5 and E15.5 and FACS sorting EGFP-positive cells, we have succeeded in isolating large numbers of cortical interneuron precursors. Our preliminary microarray analysis of these cells fortifies our confidence that we have been successful in this endeavour, as *GAD1*, *Dlx2*, *Lhx6*, *NPY* and *Sst* are all highly enriched in this population. We are currently examining the expression of candidate genes in the hope they will provide clues as to the identity of specific determinants that control the fate of interneuron subpopulations.

What aspects of cortical interneuron identity are specified in the ganglionic eminences?

As noted above, the central issue surrounding cortical interneuron diversity is the question of how many subclasses of this cell type exist. At its essence, an inter-neuron's identity is defined by it connectivity coupled with its input/output func-tion. Even within the confines of cortical interneurons that share morphology, immunomarkers and intrinsic physiological properties, one can imagine that depending on their precise afferent input and efferent targets, they function quite differently. Similarly, relatively subtle changes in their firing-threshold, firing-rate or adaptation could allow very similar cortical interneuron populations to have distinctly different functions within the cortical network (e.g. Goldberg et al 2005). Which of these properties are controlled by the developmental genetic history of the progenitors that give rise to them? For the foreseeable future, this question will at some level remain one of 'nature versus nurture'.

Our analyses to date demonstrate that at least with regard to broad subclass, specification of cortical interneurons is established in the progenitor zones. However the extent to which an interneuron's subsequent interactions with its environment further shape its identity remains an open question. In the spinal cord, studies of motor neurons have to date provide us with our best understanding of the means by which specific neuronal subclasses in vertebrates are generated (Dasen et al 2003). Motor neurons while derived from a relative small progenitor region soon diverge into specific 'pools' as a result of their combinatorial gene expression (e.g. Dasen et al 2003, 2005). This lends itself to the idea that neuronal diversity in motor neurons is established very early in development, prior to the functional innervation of targets. However, the data as yet have failed to reveal a similar precision of gene expression in the MGE and CGE. While it is appealing to think that the logic used to create this phylogenically older spinal cord population would be maintained in the brain, the selective pressures refining these populations are likely quite distinct. Notably, the primary function of motor neurons is execution, while the central role of cortical interneurons is the filtering and processing of 'complex sensory' information. The necessity of precisely innervating a well-defined set of peripheral muscles may demand a fidelity that is neither required nor desirable in the structuring of cortical circuits. Also consistent with the idea that strategies vary between motor neurons is the observation that their final position of specific neuronal subtypes in the spinal cord is stereotyped, while that of cortical interneurons are not. Indeed, although the laminar fate of cortical interneurons is predicted by their birthdate, we as yet see little evidence that cortical interneurons derived at a given position within the subpallium are destined to occupy a precise cortical area. If one then imagines that a particular FS basket cell might stochastically 'choose' to reside in layer four of the visual cortex or layer five of motor cortex (Fig. 5), one can see significant pressure for the precise wiring of a given cortical interneuron to be shaped by its environment. At present, I suspect that cortical interneurons while relegated to a broad class, such as Martinotti cell or basket cell, retain significant plasticity to adapt to cues within the region of cortex to which they ultimately contribute. I envision that their specification beyond the adoption of a particular shape and firing pattern, amounts to a 'look-up table' of hardwired responses that dictate how they will react to the particular cues they might encounter during their migration and integration into the developing cortex. My view is that gene expression within progenitors undergoes an 'epigenetic lockin' upon exiting the cell cycle. The stability in cortical precursor identity that we observe after heterotopic transplants is, I believe, a result of an imprinted genetic programme. My guess is that this programme is maintained through chromatin modification, which is established concurrent with cortical interneurons becoming postmitotic.

At present our understanding of the mechanisms that control the fate of cortical interneurons is too rudimentary to understand this process at a molecular level. Nonetheless, it would seem that sorting out the level of specification of cortical interneurons in the progenitor zones is a prerequisite to understanding the subsequent influences of

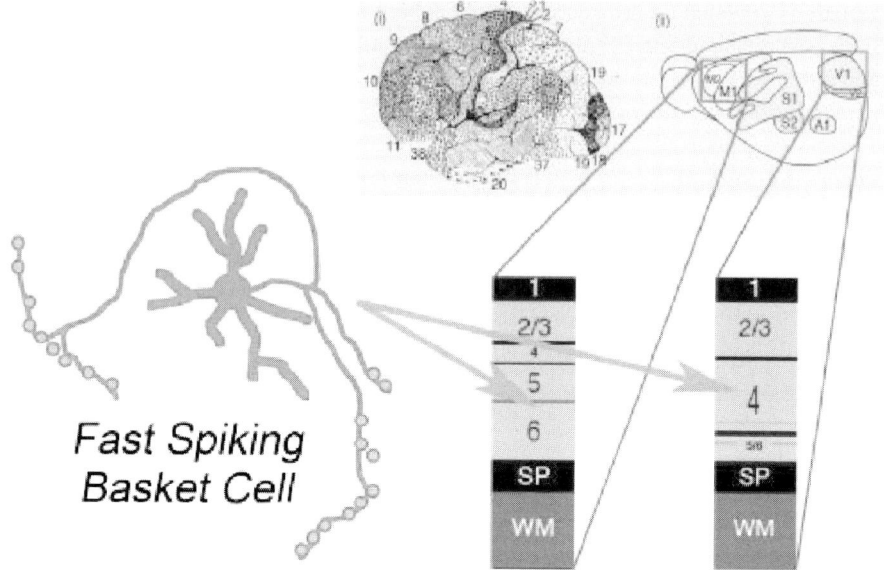

FIG. 5. Do specific cortical interneuron subtypes vary their function integration depending on the area of cortex into which they integrate? (FS schematic adapted from Kawaguchi & Kubota 1997). Our results suggest that the areal position of interneurons within the cortex is stochaistically determined. This raises the intriguing question of whether the function of particular subclass of cortical interneuron (such as the basket cell shown in this figure) varies according to whether it ends up in layers V or VI of the motor cortex versus layer IV of the visual cortex.

their postmitotic environment. Furthermore, as parallel understanding of the means by which interneuron diversity is generated in the spinal cord, it will be interesting to compare how local inhibitory neurons are generated in these two systems.

The role of cortical interneurons in the refinement of the cortical cell assembly

The cell assembly hypothesis proposed by Donald Hebb (1949) suggested that the CNS was comprised of a well-ordered circuitry with intrinsic function independent of sensory input. He suggested that learning was accomplished through the activity dependent reorganization of this cell assembly. The realization that inhibitory cells originate within the ventral telencephalon and enter the cortex *en masse* perinatally raises the question of how their integration into the developing cortex is regulated. Understanding this process is particularly pressing, as shortly after their arrival, activity-dependent refinement results in the reorganization of (third order) sensory afferents. Moreover it has been recently shown that this so-called 'critical period' of remodelling is dependent on cortical interneurons and can be delayed through genetic or pharmacological interference with their normal function (Fagiolini et al 2004, Hensch & Stryker 2004, Hensch 2005). Studies where the cortex has been genetic deprived of all afferents demonstrates that the areal organization of the cortex is intrinsically encoded (Miyashita-Lin et al 1999, Nakagawa et al 1999). Therefore, cortical interneurons while not obviously predestined to occupy a particular functional region of cortex are required soon after their arrival for refinement. Taken together, I hypothesize that a 'handshaking' occurs during development where pyramidal neurons bestow positional information onto cortical interneurons. Hence, insights into the initial interactions between these two populations during development will be central to understanding how the cortical cell assembly is established.

Acknowledgements

I would like to thank Robert Machold, Jens Hjerling Leffler, Renata Brito, Goichi Miyoshi, Natalia DeMarco, Vitor Sousa, Jeremy Dasen and Bernardo Rudy for reading this manuscript and offering many helpful suggestions. I would also like to thank Drs Kawaguchi and Danglot for allowing me to use work from their papers as a basis for some of my figures. Finally, I would like to thank Simon Butt for letting me include reconstructions from his work in Miyoshi et al 2007 in Fig. 4 of this manuscript. The work in this manuscript was supported by NIH grants R01MH071679 and R01NS032993.

References

Anderson SA, Eisenstat DD, Shi L, Rubenstein JL 1997a Interneuron migration from basal forebrain to neocortex: dependence on Dlx genes. Science 278:474–476

Anderson SA, Qiu M, Bulfone A et al 1997b Mutations of the homeobox genes Dlx-1 and Dlx-2 disrupt the striatal subventricular zone and differentiation of late born striatal neurons. Neuron 19:27 37

Anderson SA, Marin O, Horn C, Jennings K, Rubenstein JL 2001 Distinct cortical migrations from the medial and lateral ganglionic eminences. Development 128:353–363

Butt SJ, Fuccillo M, Nery S et al 2005 The temporal and spatial origins of cortical interneurons predict their physiological subtype. Neuron 48:591–604

Casarosa S, Fode C, Guillemot F 1999 Mash1 regulates neurogenesis in the ventral telencephalon. Development 126:525–534

Cauli B, Audinat E, Lambolez B et al 1997 Molecular and physiological diversity of cortical nonpyramidal cells. J Neurosci 17:3894–3906

Cobos I, Long JE, Thwin MT, Rubenstein JL 2006 Cellular patterns of transcription factor expression in developing cortical interneurons. Cereb Cortex 16 Suppl 1:i82–i88

Corbin JG, Nery S, Fishell G 2001 Telencephalic cells take a tangent: non-radial migration in the mammalian forebrain. Nat Neurosci 4 Suppl:1177–1182

Danglot L, Triller A, Marty S 2006 The development of hippocampal interneurons in rodents. Hippocampus 16:1032–1060

Dasen JS, Liu JP, Jessell TM 2003 Motor neuron columnar fate imposed by sequential phases of Hox-c activity. Nature 425:926–933

Dasen JS, Tice BC, Brenner-Morton S, Jessell TM 2005 A Hox regulatory network establishes motor neuron pool identity and target-muscle connectivity. Cell 123:477–491

Fagiolini M, Fritschy JM, Low K, Mohler H, Rudolph U, Hensch TK 2004 Specific GABAA circuits for visual cortical plasticity. Science 303:1681–1683

Gaiano N, Kohtz JD, Turnbull DH, Fishell G 1999 A method for rapid gain-of-function studies in the mouse embryonic nervous system. Nat Neurosci 2:812–819

Goldberg EM, Watanabe S, Chang SY et al 2005 Specific functions of synaptically localized potassium channels in synaptic transmission at the neocortical GABAergic fast-spiking cell synapse. J Neurosci 25:5230–5235

Hebb DO 1949 The organization of behavior: a neuropsychological theory. Wiley, New York

Hensch TK 2005 Critical period plasticity in local cortical circuits. Nat Rev Neurosci 6:877–888

Hensch TK, Stryker MP 2004 Columnar architecture sculpted by GABA circuits in developing cat visual cortex. Science 303:1678–1681

Joyner AL, Zervas M 2006 Genetic inducible fate mapping in mouse: establishing genetic lineages and defining genetic neuroanatomy in the nervous system. Dev Dyn 235:2376–2385

Kawaguchi Y 1993 Groupings of nonpyramidal and pyramidal cells with specific physiological and morphological characteristics in rat frontal cortex. J Neurophysiol 69:416–431

Kawaguchi Y, Kubota Y 1997 GABAergic cell subtypes and their synaptic connections in rat frontal cortex. Cereb Cortex 7:476–486

Markram H, Toledo-Rodriguez M, Wang Y, Gupta A, Silberberg G, Wu C 2004 Interneurons of the neocortical inhibitory system. Nat Rev Neurosci 5:793–807

Miyashita-Lin EM, Hevner R, Wassarman KM, Martinez S, Rubenstein JL 1999 Early neocortical regionalization in the absence of thalamic innervation. Science 285:906–909

Miyoshi G, Fishell G 2006 Directing neuron-specific transgene expression in the mouse CNS. Curr Opin Neurobiol 16:577–584

Miyoshi G, Butt SJB, Takebayashi H, Fishell G 2007 Physiologically distinct temporal cohorts of cortical interneurons arise from telencephalic olig2 expressing precursors. J Neurosci 27:7786–7798

Molnar Z, Metin C, Stoykova A et al 2006 Comparative aspects of cerebral cortical development. Eur J Neurosci 23:921–934

Nakagawa Y, Johnson JE, O'Leary DDM 1999 Graded and areal expression patterns of regulatory genes and cadherins in embryonic neocortex independent of thalamocortical input. J Neurosci 19:10877–10885

Nery S, Fishell G, Corbin JG 2002 The caudal ganglionic eminence is a source of distinct cortical and subcortical cell populations. Nat Neurosci 5:1279–1287

Novak A, Guo C, Yang W, Nagy A, Lobe CG 2000 Z/EG, a double reporter mouse line that expresses enhanced green fluorescent protein upon Cre-mediated excision. Genesis 28:147–155

Olsson M, Campbell K, Turnbull DH 1997 Specification of mouse telencephalic and midhindbrain progenitors following heterotopic ultrasound-guided embryonic transplantation. Neuron 19:761–772

Rudy B, McBain CJ 2001 Kv3 channels: voltage-gated K+ channels designed for high-frequency repetitive firing. Trends Neurosci 24:517–526

Stenman J, Toresson H, Campbell K 2003 Identification of two distinct progenitor populations in the lateral ganglionic eminence: implications for striatal and olfactory bulb neurogenesis. J Neurosci 23:167–174

Sussel L, Marin O, Kimura S, Rubenstein JL 1999 Loss of Nkx2.1 homeobox gene function results in a ventral to dorsal molecular respecification within the basal telencephalon: evidence for a transformation of the pallidum into the striatum. Development 126:3359–3370

Takebayashi H, Nabeshima Y, Yoshida S, Chisaka O, Ikenaka K, Nabeshima Y 2002 The basic helix-loop-helix factor olig2 is essential for the development of motoneuron and oligodendrocyte lineages. Curr Biol 12:1157–1163

Wichterle H, Garcia-Verdugo JM, Herrera DG, Alvarez-Buylla A 1999 Young neurons from medial ganglionic eminence disperse in adult and embryonic brain. Nat Neurosci 2:461–466

Wichterle H, Turnbull DH, Nery S, Fishell G, Alvarez-Buylla A 2001 In utero fate mapping reveals distinct migratory pathways and fates of neurons born in the mammalian basal forebrain. Development 128:3759–3771

Wonders CP, Anderson SA 2006 The origin and specification of cortical interneurons. Nat Rev Neurosci 7:687–696

Xu Q, Cobos I, De La Cruz E, Rubenstein JL, Anderson SA 2004 Origins of cortical interneuron subtypes. J Neurosci 24:2612–2622

Zervas M, Millet S, Ahn S, Joyner AL 2004 Cell behaviors and genetic lineages of the mesencephalon and rhombomere 1. Neuron 43:345–357

DISCUSSION

Parnavelas: In the 1980s, Marion Cavanagh and I published three papers about the origin of peptide-containing subpopulations of cortical interneurons (Cavanagh & Parnavelas 1989a, b, 1990). We used autoradiography and cell-specific markers to look at somatostatin/VIP/NPY-containing interneurons. The data are solid and show that different subtypes are indeed generated at different time points during corticogenesis. The somatostatin cells, for example, are generated early with a peak around E16 in the rat. NPY and VIP are produced later; the former around E17 and the latter show a sharp peak at E19.

Fishell: I agree in that even if you look at a constellation of the neurons' various attributes—morphology, immunocytochemistry and physiology—it certainly isn't as simple as one particular cell type being generated at one particular time. However, using multiple criteria does provide for an improved ability to parce the heterogeneity and their corresponding temporal origins. For instance, parvalbumin, while biased towards fast spiking cells, also marks multipolar burst cells. Similarly, somatostatin is expressed in both Martinotti cells and regular spiking non-pyramidal cells. Hence, if you are just relying on immunocytochemistry alone, you don't see the wealth of diversity or the precision of the system. This is precisely why you need a multipronged approach to get at the question of interneuron diversity.

Parnavelas: But, there is a problem with your approach. You seem to use different criteria (markers) to identify different subpopulations of interneurons. For example, you are talking about parvalbumin cells and somatostatin cells. You are using Ca^{2+} binding proteins on the one hand, and peptides on the other, and it is known that some of the parvalbumin cells contain somatostatin. The majority of interneurons are parvalbumin positive. The calretinin subpopulation is the smallest; the calbindin group is in-between.

Fishell: I don't think that calbindin, at least in mouse, is all that useful for getting at the diversity question. I'd argue that somatostatin is better. We took considerable effort to do double and triple immunocytochemistry to see how these various populations intersect with one another. There are distinct populations that are parvalbumin only, though. Hence, I agree with you that this is simply not a precise way to define an interneuron. How Ca^{2+} binding proteins relate to function is vague in my mind. I'm not implying that physiology is the panacea, but it is closer to function than Ca^{2+} binding proteins. In the end it is just a more stringent assay, and as developmental biologists we'd do well to use adopt more physiological approaches classifying neurons.

Parnavelas: You proposed that there is handshaking between interneurons and pyramidal neurons. What is the evidence for this?

Fishell: My suggestion that this occurs is based on work from Michael Stryker and Tako Hensch (Hensch & Stryker 2004). They observed that if they knocked down GABAergic signalling (either pharmacologically or genetically) during the first postnatal week, this delays the activity-dependent refinement of thalamic afferents. Hence, the GABAergic cells must play a key role in activity-dependent remodelling of cortical afferents. This in turn suggests that there must be some communication to transfer that information.

Parnavelas: Pasko Rakic's lab has shown that a large population of interneurons come into the cortical plate through the marginal zone, without ever coming close to pyramidal neurons.

Fishell: Ultimately they do come into contact.

Parnavelas: Yes, but by then their fate is determined.

Fishell: Interneurons integrate into the cortical plate during late embryogenesis and continue to do so during the first postnatal week. In essence, it is a question of taking a look at who is there, when and who do they interact with. I don't know what we will find, but it is highly likely that there is a strong genetic component to the logic that dictates the circuitry between pyramidal cells and interneurons. This is what we are trying to get at.

Rakic: In the marginal zone, migrating interneurons can come in contact with apical dendrites of pyramidal cells (Ang et al 2003). As they grow, at one point they must touch the apical dendrites of the previously generated pyramidal neurons in the marginal zone, but as you say, we don't have good markers, so we don't know how they interact and whether this interaction is critical for their positioning. In non-human primates and humans there are some cytoarchitectonic regions where there are four times more interneurons than in the adjacent areas. The only way these cells can recognize the appropriate regions is by cues in the marginal zone that are situated on the processes of cortical pyramids that extend to the marginal zone.

Goffinet: There is a large proportion of interneurons that are reelin positive. Where do they come from?

Fishell: We haven't looked at this.

Goffinet: Has anyone looked at the origin of GABAergic interneurons in dien-cephalons. Could some of them come from the ganglionic eminence?

Fishell: Michele Studer has some evidence arguing for this: that there is a migra-tion back towards the diencephalons (Tripodi et al 2004). There are hints of that argument from the work from Pasko Rakic and Richard Sidman many years ago (Sidman & Rakic 1973).

Rakic: At that time we recognized that the ganglionic eminence generates inter-neurons but we did not have any specific markers. Nevertheless, we did observe that most of these cells migrate to the temporal lobe and parietal lobes from the ganglionic eminence (Rakic 1975). However, in addition, we have observed using classical Golgi and electron microscopic methods, a stream of neurons migrating from the ganglionic eminence to the posterior thalamus, particularly pulvinar (Rakic & Sidman 1969). About three decades later we comfirmed these results with modern methodologies including retroviral labelling (Letinic & Rakic 2001). This population of interneurons exists only in the human and could not be observed in non-human primates. In general, humans not only have much more but also different classes of interneurons that are originating from the subventricular zone, than rodents. Some interneurons in human perhaps even come from the ventricu-lar zone (Letinic et al 2002). In mouse there are no or very little interneurons of this origin. It could be that these types of interneurons are involved, for example, in human-specific psychiatric disorders such as schizophrenia and other diseases that don't occur spontaneously in mice.

Fishell: As far as I know the only major difference in the diversity of mouse versus human cortical interneuron diversity is in the chandelier cells. These are 15% in mice but 30% in humans. I don't think there are any absolute *de novo* classes in human. I think that there are elaborations on what there is in mouse. I'd also argue that the observation by Dr Rakic that interneurons within the cortex are generated locally in the subventricular zone may be a result of seeding of progeni-tors from subcortical areas, which in turn proliferate within the cortex. The idea that there is a fundamentally different mechanism for making interneurons in human and mice strikes me as surprising.

Rakic: It doesn't surprise me. There's a fundamental difference in what we can do compared with mouse. There must be some biological basis for this.

Parnavelas: There is evidence from retrovirus work that some interneurons are generated in the cortical ventricular zone.

Fishell: At least in mouse the genetic evidence for this is minimal.

Parnavelas: When we injected retrovirus into the lateral ventricles and looked one day or two days later, we found GABA-containing cells in the developing cortex that had arisen from infected progenitors in the ventricular zone.

Fishell: Without knowing the precise populations you are infecting, knowing the origin of the labelled populations is impossible.

Macklis: Is your interpretation that those retroviruses are infecting those cells that invade from ventrally and then divide?

Fishell: That would be the most straightforward explanation.

Parnavelas: With this in mind, we looked in the cortex at short times after the injection of the virus and found GABA-containing cells, suggesting that they arose from progenitors in the cortical ventricular zone.

Walsh: With the genetics, you see what you see, but you don't tend to see what you don't see. Presence of absence is not absence of presence. Just not seeing something doesn't mean that it is not there. You have a good label for particular classes of cells, and you are labelling them in a class-specific way, but you can't claim that you have been able to determine the origin of every cell.

Fishell: I see your point. That said, I would argue that most cortical interneurons, at least in mouse, originate within the ventral telencephalon.

Parnavelas: There is no doubt that the vast majority of cortical interneurons do come from the ventral forebrain, but the evidence that some, perhaps 5%, come from the cortical ventricular zone is very good.

Rakic: You are correct. We have some unpublished data on the similarity between mouse and human in this respect. We have identified a population of interneurons in the mouse that can be double-labelled with CCK and cannabinoid receptor (Morozov et al 2007). These cells are generated in either the ventricular or subventricular zone, and then travel to the hippocampus. There are few of them, but we were able to follow them. So, there are some populations of interneurons in the mouse that are very similar to human. However, there are some phylogentically new types that are developed and amplified in humans.

Fishell: Some years ago we knocked the *Tau-LacZ* reporter into the *Dlx2* locus, a gene that appears to be expressed developmentally in the majority of cortical interneurons. In a small number of cases we saw these cells in the cortical ventricular zone. This would support the arguments that have been proposed in this discussion.

Molnár: You made an evolutionary argument about primates and rodents. It is also interesting to look in the other direction and study the generation and migration of GABAergic interneurons in sauropsids: John Rubenstein and Luis Puelles (Cobos et al 2001a, b) and John Parnavelas (Tourto et al 2003) have looked at the avian (chick) brain, and Christine Métin and I have studied the reptilian (turtle) brain. These studies revealed that GABAergic neurons are generated in the subpallium using identical approaches as in mice (Bellion & Métin 2005). It would be nice to investigate the expression of several genes in these species. In addition to generating GABAergic neurons in subpallium, new cortical mitotic compartments could have emerged in the primate cortical subventricular zone. Colette Dehay and Henry Kennedy described additional subdivisions of the subventricular zone in the primate; the inner and outer subventricular zones (Smart et al 2002). Perhaps these additional compartments generate a new opportunity to line up these transcription factors in unique combinations, and then they produce interneurons locally in the primate cortex.

O'Leary: Gord, you mentioned that about 20% of all interneurons are generated in the CGE. Do they have a differential distribution in the adult from one area to another, or one region of the cerebral cortex to another? This question is based not only on function, but also because of our results from experiments involving electroporation to block the migration pathway of interneurons from the MGE. We find that the remaining cells are largely distributed in the occipital lobe and the hippocampal formation (Li et al, unpublished results). Is such a CGE-specific distribution maintained into the adult?

Fishell: I would have loved to have seen areal differences along the AP axis in terms of their receipt of CGE and MGE-derived cortical interneurons, but at least at present, this has not been very evident. Interpretation of this is, however, complicated because it seems that some of the somatostatin cells sweep caudally back thorough the CGE while on route to the cortex. In addition, these experiments were all done with transplantation and I am not sure they would have been able to pick up the areal differences. Certainly, work from Colette Dehay (Lukaszewicz et al 2006) has argued for strong differences in organization of interneurons across areas of cortex. I am sure they exist, but I am not sure how they arise. Migration would be the obvious mechanism.

Macklis: How much information is there now about the instructional ability of GABAergic interneurons, which as they migrate up could be instructing the differentiation of their projection neuron partners, and vice versa?

Fishell: The point Jeff raises is the salient one: what is the evidence for any sort of handshaking at all? The answer at present is not much, but we need to look for it.

Tan: You talk about how the layer fate of interneurons is dependent on their birthdate. Our transplant experiments show that interneurons have a layer address when they are transplanted at different times. But when we look at the reelin mutants, we find that early interneurons invert as their projection neurons but the late-born interneurons are scattered everywhere and don't seem to invert. We also find this difference between early and late born interneurons in rescue experiments, i.e. late born interneurons can be rescued when we place them in embryos where there is proper reelin signalling. This suggests to me that the intrinsic mechanism for layering perhaps relies on extrinsic signals such as the reelin pathway, at least for the late born interneurons. Does this sit well with your hypothesis?

Fishell: I am pretty agnostic on this point. The argument that reelin mutants affect the distribution of cortical interneurons seems compelling. Whether this is reflective of the pyramidal cells being in the wrong position and the interneurons hence not having the right information, or whether reelin signalling is required in the interneurons, is an open question.

Tan: We have tested this question precisely in p35 mutants (Hammond et al 2006). In these mutants, reelin signalling is normal. We found that despite inversion of projection neuron layers, the late-born interneurons in the p35 mutant are normally distributed in the upper layers. This would suggest that although projection neuron positions are inverted (being close to the ventricle), the layer II/III interneurons are correctly positioned close to the pial surface, suggesting that projection neuron positions do not necessarily influence interneuron positioning.

Goffinet: I have a technical question regarding your temporal fate mapping experiments. If I understood correctly, you concluded that CreER expression from the *Olig2* locus is expressed in transient cells with limited proliferation potential. It is surprising that *Olig2* is expressed solidly in the ventricular zone. What is your interpretation?

Fishell: The point you make is exactly why we were surprised by these results as well. *Olig2* is expressed widely within subpallial progenitors, not just in the medial ganglionic eminence (MGE) but also at lower levels within caudal and lateral ganglionic eminences as well. It is not just the driver line that matters but also the reporter. The reporter we are using is the Z/EG reporter, which has a number of idiosyncrasies. I think that because of the position of the Z/EG allele, it is not accessible for recombination until the progenitors are just about to become post-mitotic. What we observe when we cross the Z/EG reporter to *Olig2^{CreER}* mice and give tamoxifen is that at the level we are using we only get cells in the MGE that are about to become post-mitotic, despite the fact that *Olig2* is directed to the ventricular zone.

Goffinet: Have you tried other reporter mice?

Fishell: Yes, and we don't see this. If we do *Rosa^{stopYFP}* we end up getting progenitors, and cells in all layers as you might predict on the basis of the *Olig2* expression. It is a balancing game that in this case works in our favour.

Harrison: I want to return to the chandelier cells, with reference to the human brain. What is known about their origin in terms of timing and location, compared with the other parvalbumin-positive cells?

Fishell: The only way we know they are chandelier cells versus basket cells is their distinctive morphology. Fast-spiking cells parvalbumin-positive cortical interneurons are born throughout neurogenesis. The number of cells we have to study is very small, because for every 100 fast spikers, only 5 at most are chandelier cells.

Rubenstein: In the neocortex, there is a gradient in the maturation of projection neurons —rostral cells mature earlier than the caudal cells. In the tangential migrations you have shown, cells migrate from the CGE over the relatively immature caudal cortex. At the stage that you showed there is very little staining in the frontal cortex. There seems to be a temporal mismatch in the birth dating and migration such that the frontal cortex, which is maturing earlier, would have a paucity of neurons, yet there seems to be a tight correlation in the birth date of the interneurons and the cortical layer where they integrate.

Fishell: I had never thought of the problem in this way. It is a great way of thinking about it. The notion that the interneurons have different roles in different places is intriguing. The posterior cortex probably sees interneurons migrating up before anterior cortex, which may mean something.

Parnavelas: If the chandelier cells, double bouquet cells and bipolar cells are generated in the caudal ganglionic eminence, which sends interneurons predominantly to the caudal cortex (Yozu et al 2005), how is it that we have a fairly homogeneous distribution of double bouquet cells throughout the cortex?

Fishell: We don't see them preferentially in occipital cortex. We see those cells going throughout.

Parnavelas: The data published from the Nakajima lab (Yozu et al 2005) are pretty compelling. They performed electroporation in the whole brain kept in culture.

Molnár: Perhaps we don't have the resolution we need at the moment to answer some of these questions. If we want to test the idea that a subset of pyramidal cells will determine the fate and proportions of different subtypes of interneurons, then we should look into models like those Jeff Macklis was presenting (Arlotta et al 2005, Molyneaux et al 2005). In some of his models (*Ctip2* knockout, *Fezf2* knockout mice) a specific group of layer V projection neurons are missing. We should study these models in much more detail to understand how the lack of a subset of pyramidal cells will influence a subset of interneurons. We don't know which ones to study in what combinations at the moment.

Rakic: Along those lines, there might be a much larger difference between the various classes of cells that what we currently think. If we look in the occipital lobe with old-fashioned Golgi staining, as Jenny Lund and many other investigators of the primate cortex did, there is a layer IV A, B, Cα and Cβ, in the primary visual cortex and each one has specific classes of interneurons (Lund 1988). Each class has a quite different distribution of axons in various infra- and supra-granular layers, even though they are associated with the same pyramidal cells in each layer. We don't have methods of high enough resolution to see how this diversity is generated.

Fishell: You raised a good point which we haven't discussed: this population of interneurons definitely comes from the cortex. We don't know much about their origin.

Parnavelas: How do we know that they definitely come from the cortex?

Fishell: I'd argue this because everything I know about them says that they are glutamatergic, and I haven't seen any evidence to say that subcortical cells give rise to glutamatergic cells in the cortex. From this I would infer they came from the cortex.

Parnavelas: Has anyone shown that spiny stellates are glutamatergic?

Fishell: There is work from Larry Katz which looked at the development of the spiny stellate cells. He showed that initially they send processes out much like pyramidal cells, and then they withdraw them to become the spiny stellate cells. At first principles this would argue that they are cortically derived, and would not be incompatible with them being glutamatergic.

Parnavelas: Dennis O'Leary, what do you think?

O'Leary: I agree. All cortical plate neurons generated in the ventricular or sub-ventricular zone and presumably glutamatergic initially have a radial pyramidal type morphology, and then take on their unique adult morphology (Koester & O'Leary 1992).

Tan: The idea that there are subpopulations of interneurons involved in the healthy situation is probably true. But seeing that we know that there are co-signatures of Ca^{2+} binding proteins in the same interneuron, is it possible that in a particular physiological state, perhaps in disease, we could have a switch from one phenotype to another?

Fishell: Yes, there is some wonderful work from Nick Spitzer (Borodinsky et al 2004) arguing that activity can change the neurotransmitter type. The question of how activity and timing change the phenotype is one we should not lose sight of.

Tan: I'm not just thinking about a chemical level. Perhaps also at an electrophysiological level.

Fishell: The size of the study can't rule out this sort of plasticity.

Hevner: Earlier on someone asked about whether we know more about the excitatory or inhibitory neurons. It seems that we know a lot more about the transcription factors in the projection neurons than we do in the interneurons. The interneuron signature seems to focus on Ca^{2+}-binding proteins and morphology. Has there been any transcription factor identified that can change a Martinotti cell to a double bouquet cell, for example? I know John Rubenstein's lab has studied transcription factors and interneurons. What is the state of that field?

Rubenstein: There isn't much progress in that area. My opinion is that there are two types of cells: ones that are *Dlx1* positive and those that are *Lhx6* positive. At least in the adult, the *Lhx6*-positive cells tend to be associated with parvalbumin, whereas the *Dlx1*-positive cells are not. Beyond that, we haven't made much progress.

Price: Gord, you brought up *FoxG1*. Are you suggesting this transcription factor is important in the specification and migration of interneurons?

Fishell: A reason, among others, for doing the conditional *FoxG1* knockout is that the null completely loses the ventral telencephalon. It is wiped out. It is important down there, but what it is doing is a bit of a puzzle.

References

Ang ESBC Jr, Haydar TF, Gluncic V, Rakic P 2003 Four-dimensional migratory coordinates of GABAergic interneurons in the developing mouse cortex. J Neurosci 23:5805–5815

Arlotta P, Molyneaux BJ, Chen J, Inoue J, Kominami R, Macklis JD 2005 Neuronal subtype-specific genes that control corticospinal motor neuron development in vivo. Neuron 45:207–221

Bellion A, Métin C 2005 Early regionalisation of the neocortex and the medial ganglionic eminence. Brain Res Bull 66:402–409

Borodinsky LN, Root CM, Cronin JA, Sann SB, Gu X, Spitzer NC 2004 Activity-dependent homeostatic specification of transmitter expression in embryonic neurons. Nature 429:523–530

Cavanagh ME, Parnavelas JG 1989a Development of somatostatin immunoreactive neurons in the rat occipital cortex: a combined immunocytochemical-autoradiographic study. J Comp Neurol 268:1–12

Cavanagh ME, Parnavelas JG 1989b Development of vasoactive-intestinal-polypeptide-immunoreactive neurons in the rat occipital cortex: a combined immunohistochemical-auto-radiographic study. J Comp Neurol 284:637–645

Cavanagh ME, Parnavelas JG 1990 Development of neuropeptide Y (NPY) immunoreactive neurons in the rat occipital cortex: a combined immunohistochemical-autoradiographic study. J Comp Neurol 297:553–563

Cobos I, Puelles L, Martinez S 2001a The avian telencephalic subpallium originates inhibitory neurons that invade tangentially the pallium (dorsal ventricular ridge and cortical areas). Dev Biol 239:30–45

Cobos I, Shimamura K, Rubenstein JL, Martinez S, Puelles L 2001b Fate map of the avian anterior forebrain at the four-somite stage, based on the analysis of quail-chick chimeras. Dev Biol 239:46–67

Hammond V, So E, Gunnersen J, Valcanis H, Kalloniatis M, Tan S-S 2006 Layer positioning of late-born cortical interneurons is dependent upon reelin but not p35 signaling. J Neurosci 26:1646–1655

Hensch TK, Stryker MP 2004 Columnar architecture sculpted by GABA circuits in developing rat visual cortex. Science 303.1678–1681

Koester SE, O'Leary DD 1992 Functional classes of cortical projection neurons develop dendritic distinctions by class-specific sculpting of an early common pattern. J Neurosci 12:1382–1393

Letinic K, Rakic P 2001 Telencephalic origin of human thalamic GABAergic neurons. Nat Neurosci 4:931–936

Letinic K, Zoncu R, Rakic P 2002 Origin of GABAergic neurons in the human neocortex. Nature 417:645–649

Lukaszewicz A, Cortay V, Giroud P et al 2006 The concerted modulation of proliferation and migration contributes to the specification of the cytoarchitecture and dimensions of cortical areas. Cereb Cortex 16 Suppl 1:i26–i34

Lund JS 1988 Anatomical organization of macaque monkey striate visual cortex. Annu Rev Neurosci 11:253–288

Molyneaux BJ, Arlotta P, Hirata T, Hibi M, Macklis JD 2005 Fezl is required for the birth and specification of corticospinal motor neurons. Neuron 47:817–831

Morozov MY, Torii M, Rakic P 2007 Long complex journey of cannabinoid type 1 receptor containing interneurons to the developing hippocampus. in press

Rakic P 1975 Timing of major ontogenetic events in the visual cortex of the rhesus monkey. In: Buchwald NA, Brazier M (eds) Brain mechanisms in mental retardation. Academic Press, New York, p 3–40

Rakic P, Sidman RL 1969 Telencephalic origin of pulvinar neurons in the fetal human brain. Z Anat Entwicklungsgesch 129:53–82

Sidman RL, Rakic P 1973 Neuronal migration, with special reference to developing human brain: a review. Brain Res 62:1–35

Smart IH, Dehay C, Giroud P, Berland M, Kennedy H 2002 Unique morphological features of the proliferative zones and postmitotic compartments of the neural epithelium giving rise to striate and extrastriate cortex in the monkey. Cereb Cortex 12:37–53

Tripodi M, Filosa A, Armentano M, Studer M 2004 The COUP-TF nuclear receptors regulate cell migration in the mammalian basal forebrain. Development 131:6119–6129

Tuorto F, Alifragis P, Failla V, Parnavelas JG, Gulisano M 2003 Tangential migration of cells from the basal to the dorsal telencephalic regions in the chick. Eur J Neurosci 18:3388–3393

Yozu M, Tabata H, Nakajima K 2005 The caudal migratory stream: a novel migratory stream of interneurons derived from the caudal ganglionic eminence of the developing mouse forebrain. J Neurosci 25:7268–7277

Genetic determinants of neuronal migration in the cerebral cortex

Pasko Rakic, Kazue Hashimoto-Torii and Matthew R. Sarkisian

Department of Neurobiology and Kavli Institute of Neuroscience, Yale University School of Medicine, New Haven, CT 06520, USA

Abstract. The techniques and concepts of modern molecular biology and experimental neurobiology give new insights into molecular mechanisms involved in neuronal migration within the cerebral cortex. The findings collectively indicate that a diverse family of genes and transcription factors co-operate in orchestrating the multistage, interrelated phenomena that include control of mode of cell proliferation, fate determination, establishment of polarity, detachment from the local substrate and migration to the proper laminar and areal position in the cortex. Herein, we will review some new data from our laboratory on the initiation of migration, nuclear translocation and attainment of final positions. We can now propose working models of the sequence of gene expression, cascade of multiple molecular pathways and cell–cell interactions that are involved in neuronal migration. Disruption or even slowing down of any step in neuronal migration during embryonic development can result in either gross or subtle abnormalities in neuronal positioning that may only later, during postnatal life, affect the formation of synaptic circuits and eventually behaviour.

2007 Cortical development: genes and genetic abnormalities. Wiley, Chichester (Novartis Foundation Symposium 288) p 45–58

It is not necessary to remind participants of this meeting that cortical neurons are not generated locally, and that they acquire their appropriate areal and laminar position by active migration from the sites of origin. It is generally agreed that neuronal migration depends critically on sequential activation of genes, and co-ordination of multiple molecular events and complex cell–cell interactions (Ayala et al 2007, Fishell & Kriegstein 2003, Hatten & Mason 1990, Rakic 1990, 2006). Recent methodological advances in molecular, cellular and developmental neurobiology have given us an opportunity to study mechanisms involved in neuronal migration at a level that was not believed possible even a few years ago. The progress can be appreciated by examining content of chapters at the Novartis Foundation symposium on cortical development and evolution just seven years ago (Novartis Foundation 2000). Although the basic concepts have not significantly changed, we now know much more about genetic and molecular mechanisms. For example, contemporary research revealed that the process of neuronal migration can be divided into several distinct but closely interrelated components, such as the exit from the cell cycle, detachment from the ventricular surface, establishment of neuronal identity, establishment of cell polarity (including extension of the leading process and selection of its migratory pathway by association with local substrates), followed by nuclear and somatic translocation, and ending with the cessation of migration at the correct final position. Each of these steps requires specific reciprocal signalling and complex intercellular interactions. The illustration of this type of data require multicolour images of the molecules that fluoresce at different wavelengths that is not effective in black and white figures, and thus the readers should consult the original research papers for the detailed documentation. Furthermore, we are likely aware of only a fraction of what may be involved in these complex cellular events. Thus, I will somewhat arbitrarily focus on our recent findings on some of these steps, with the hope that the participants will provide their input and helpful suggestions for future directions.

Factors involved in neural proliferation, establishment of polarity and detachment from the VZ surface

The process of corticoneurogenesis starts when the dividing neural stem cells in the ventricular zone (VZ) produce more committed precursor types that may divide again in the subventricular zone (SVZ) and then never divide again. During and shortly after the exit from the asymmetric cell division, postmitotic daughter neurons become polarized with the leading process directed toward the cortical plate aligned along the adjacent radial glial shafts (Fishell & Kriegstein 2003, Rakic 1972, Sidman & Rakic 1973). The endfeet of the apical processes of radial glial cells (RGC) are interconnected via cadherin-based adherens junctions keeping interkinetic (up and down) nuclear movement restricted to the VZ (Gotz & Huttner 2005, Rašin et al 2007). Migration begins only after the last cell division has been completed and the cell extends a leading process and establishes its polarity. We found that Numb (classically characterized as an inhibitor of Notch) (Spana & Doe 1996) segregates at the basolateral side of dividing cells and becomes enriched during interphase along the apical-most end at the adherens junctions associated with cadherins (Rašin et al 2007). Notch signalling is restricted to the communication between the adjacent cells since both ligand and receptor are anchored on the adjacent cell surface. Previous publications have reported an apical Numb crescent in dividing VZ cells (Zhong et al 1996). However after closer analysis using combination of confocal microscopy and immunoDAB electron microscopy, we have shown that this previous observation was an artifact of Numb (or Numblike) expression in neighbouring apical endfeet. This is an instructive example of how use of the fluorescence alone, even when combined with confocal microscopy, can lead to a false cellular and subcellular localization without verification by electron microscopy. Using a combination of transgenic technology and electroporation, we recently found that interfering with the expression and/or function of Numb disturbs radial organization of the VZ leading to the formation of rosettes and heterotopically localized neurons (Rašin et al 2007). This function is completely independent of the putative role for Numb in antagonizing Notch during cell fate determination. Thus, Numb may contribute to the determination of symmetric/asymmetric divisions with Notch, but it also has a distinct role in the detachment of cells from the ventricular lining and the establishment of apico-basal polarity of RGCs (Rašin et al 2007). Only after that has been achieved, the nucleus begins translocation within the extended leading process.

Factors involved in the initiation of neuronal migration

Neuronal migration, which involves relocating the position of the cell body, can not be achieved without translocation of the nucleus and surrounding cytoplasm. Nuclear translocation is perhaps the least studied and most misunderstood component of neuronal migration. In the initial reports of neuronal migration in the cerebral and cerebellar cortices, the term 'nuclear and/or soma translocation' was used to differentiate between extension of the leading process and subsequent movement of the nucleus and surrounding cytoplasm (Rakic 1971, 1972). Substantial progress deciphering the molecular mechanisms of this process was not made until the resurgence of interest in the role of cytoskeleton and contractile proteins in regulation of nuclear translocation (Rakic et al 1996, Rivas & Hatten 1995). Analysis of the polarity of microtubule assemblies (tubulin polymerization) within the leading and trailing processes revealed that the positive ends of the newly assembled microtubules situated in the leading process face the growing tip, while their disintegrating negative ends face the nucleus; while in the trailing process the polarity of microtubule arrays are mixed (Rakic et al 1996). These findings indicated that the extension of the leading process, followed by translocation of the nucleus within it, requires synchronized polymerization and disintegration of the microtubules that form cytoskeletal scaffolding that is orchestrated by calcium fluxes through the activity of various receptor/channel complexes (e.g. Behar et al 1999, Komuro & Rakic 1993, 1998, Owens & Kriegstein 2002, Schaar & McConnell 2005). Discoveries have since led to key molecules, such as Doublecortin and Filamin-A (FLNa) (Fox et al 1998, Gleeson et al 1998), which are involved in microtubule stabilization and cytoskeletal rearrangement, and are frequently associated with human cortical malformation. Recently, we have explored the regulation of FLNa by the mitogen-activated protein kinase (MAPK) pathway for its role in neuronal migration (Sarkisian et al 2006).

FLNa is an actin-binding protein expressed in the developing CNS (Fox et al 1998, Sheen et al 2002). FLNa reorganizes the actin cytoskeleton and generates the forces for cell motility (Stossel et al 2001). Mutations resulting in *FLNa* loss-of-function in human leads to periventricular nodular heterotopia (PH): a condition marked by heterotopic neuronal clusters along the VZ surface (Fox et al 1998). Whether it is the simple loss of FLNa in migrating neurons that causes their arrest or other defects at the VZ surface remains unclear. Surprisingly, two recent reports of *Flna* null mice revealed no PH, though one mouse showed a thinner cortical plate but no neuronal accumulations within the VZ (Feng et al 2006, Hart et al 2006). These data suggest the loss of FLNa alone may not be sufficient to inhibit migration or cause PH. FLNa has over 40 interacting molecules, and putatively serves as a scaffolding protein for the MAPK signalling pathway since FLNa binds the MAP2K, MKK4 (SEK1) (Feng & Walsh 2004, Marti et al 1997, Robertson 2005). Recent work from our laboratory has extended these findings to the MAP3K, MEKK4, an upstream regulator of SEK1, whose loss of function not only results in PH but also alters the expression and phosphorylation of FLNa and affects neuronal migration initiation (Sarkisian et al 2006).

Our understanding of FLNa regulation and PH pathogenesis has been limited by the lack of reported genetic models. Remarkably, $Mekk4^{-/-}$ mice have an increased incidence of PH. PH contained mostly postmitotic neurons and fewer radial glia; consistent with cellular compositions found within human fetal PH (Santi & Golden 2001). BrdU pulse-chase labelling studies revealed that postmitotic neurons within PH were unable to migrate to the cortical plate, and failed migration was confirmed by knocking down MEKK4 in postmitotic neurons using RNA interference (RNAi) (Sarkisian et al 2006). Why are $Mekk4^{-/-}$ mice prone to PH? Radial glial endfeet form the lining of both the pial and ventricular surface (Levitt & Rakic 1980), and defects in the adhesion or survival of radial glia could predispose $Mekk4^{-/-}$ mice to PH, similar to the loss of Numb (Rašin et al 2007). We found that migration defects were commonly associated with focal disruptions in the radial glial endfeet lining the VZ surface (Sarkisian et al 2006). Therefore, MEKK4 appears to be essential for both initiation of migration and the integrity of the VZ surface.

Considering that PH is most commonly associated with *FLNa* mutations, we hypothesized that PH in $Mekk4^{-/-}$ forebrain may arise due to disrupted FLNa. Surprisingly, we observed abnormally increased levels of FLNa, especially where neuronal migration was impaired. While increased FLNa at first seemed paradoxical to a loss of FLNa function, we also showed that over-expression of wild-type human FLNa alone can delay neurons migrating into the cortical plate (Sarkisian et al 2006). Furthermore, MEKK4 signalling can influence FLNa phosphorylation at Ser^{2152} and this phosphorylation could be blocked by dominant-negative SEK1.

This suggests SEK1 (the downstream target of MEKK4) mediates FLNa phosphorylation (Sarkisian et al 2006).

The factors that control the spatiotemporal expression of FLNa and direct it to the appropriate intracellular location are poorly understood in CNS. The graded expression of FLNa (i.e. low in the VZ compared to high in CP) suggests an enrichment in postmitotic neurons, but FLNa is also present along the VZ surface suggesting likely expression within radial glia (Feng et al 2006, Lu et al 2006, Sarkisian et al 2006, Sato & Nagano 2005). FILIP (Filamin-A interacting protein), a potent degrader of FLNa, is thought to maintain this gradient in neuronal precursors (Sato & Nagano 2005). FILIP siRNA (which increases FLNa) caused morphological changes in neurons and promoted more bipolar morphology as cells transition from the SVZ into the IZ (Nagano et al 2004). Interestingly, similar to FILIP siRNA, more radially-oriented, thickened processes were observed in $Mekk4^{-/-}$ forebrain (Sarkisian et al 2006). Considering data from $Mekk4^{-/-}$ forebrain and FILIP studies (Sarkisian et al 2006, Sato & Nagano 2005), we propose that variability amongst neurons in intracellular FLNa could influence the VZ surface and migrating neurons. Moderate increases may influence the leading process morphology that would be consistent with FLNa overexpression and FILIP siRNA studies (Nagano et al 2004, Sarkisian et al 2006). Likewise, prematurely dramatic increases in FLNa may cause migration arrest at the VZ surface. Because FLNa cross-links and gelatinizes actin filaments (Stossel et al 2001), excessive FLNa may cause migrating neurons to become rigid or fixed and consequently unable to translocate the nucleus and soma. Indeed, *in vitro* studies showed that excess FLNa inhibits cell migration (Cunningham et al 1992). In addition, if FLNa is expressed at radial glial endfeet, altered rigidity may impair the integrity of the VZ lining. Collectively, these new data suggest that genetic mutations affecting the level or function of FLNa may impact both postmitotic neurons initiating their migration as well as the VZ surface integrity.

Factors that co-ordinate migrating neurons to their proper destination

It is generally agreed that multiple classes of putative recognition and adhesion molecules are involved in the process of neuronal migration across the widening cerebral wall and there are numerous reviews on this subject (e.g. Ayala et al 2007, Hatten 2002, Marin & Rubenstein 2003, Rakic et al 1994). The implicated molecules are usually selectively and transiently expressed in the cell soma and/or membranes of its leading process and the adjacent shafts of RGCs and are reported to be involved in a cascade of interactions all of which are necessary for achieving a cell's final position. Perhaps best known amongst these molecules is reelin, which is considered essential for proper inside-out settling of neurons in the laminated neocortex. Another frequently studied gene is *Notch*, known to be integral in progenitor cell fate determination in the VZ as well as a regulator of dendritic branching in postmitotic neurons (Sestan et al 1999). Although we have been working for many years on both of these molecular pathways, only recently have we discovered that they closely co-operate during neuronal migration and final positioning (Hashimoto-Torii et al 2007).

Reelin was the first gene reported to be involved in positioning of neurons in the cerebral cortex (Falconer 1951). Initially it was thought that the sequence of neurons in the cortex was simply 'inverted' in *Reelin*-deficient (*Reeler*) mice compared to wild-type, but more thorough examination showed that the pattern of neuronal mispositions is more complex (e.g. Caviness & Rakic 1978). Furthermore, we now know that mutation of the human *Reelin* (*RLN*) gene is linked to a type of lissencephaly (Hong et al 2000). VLDLR and ApoER2 serve as canonical receptors for the Reelin signal, and Dab1 binds to these receptors, which mediates Reelin signalling. Although Dab1 interacts with multiple molecules (Chen et al 2004, Suetsugu et al 2004), what mediates its downstream signals remains unclear. Since a physical interaction between Disabled (a *Drosophila* homologue of mammalian Dab1) and Notch has been reported (Giniger 1998), Notch became a logical candidate molecule for interaction with mammalian Dab1. Recently, we have found a genetic and physical interaction of Dab1 and Notch is critical for the migration of cortical neurons (Hashimoto-Torii et al 2007). To examine whether Notch has a functional role in neuronal migration, we deleted *Notch* genes specifically within postmitotic migrating neurons in the VZ using Tα1 promoter. Surprisingly, *Notch* deficiency produced similar defects to *Reelin* deficiency, including multifarious leading processes of migrating neurons, and premature arrest of upper layer-bound neurons in deeper layers. Remarkably, replenishing the active form of Notch in migrating neurons mitigates migration defects in *Reeler*. Furthermore, we found that Notch activity is significantly reduced in *Reeler*, and activated Dab1 can stabilize the active form of Notch by blocking a specific ubiquitination complex for Notch. Collectively, these results indicate that Reelin signalling controls neuronal migration, in part through stabilization of the active form of Notch, but molecular mechanisms require further investigation.

Despite over half a century of research since the first report of *Reeler* (Falconer 1951), the underlying pathogenetic mechanisms have remained unclear. We now implicate Reelin signalling in the regulation of Notch by affecting the degradation of the active form of Notch. These new findings point to unappreciated interaction between Reelin and Notch that influences the neuronal migration process and ultimate laminar positioning of neurons in the neocortex. Disruption of this signalling interaction could be involved in the pathogenesis of human neurological disorders such as lissencephaly and autism.

Conclusion

The examples of recent progress made in understating initiation of neuronal migration and attaining their final position in the cortex illustrates the remarkable power of contemporary molecular and developmental neurobiology. I presented several examples from my laboratory, but I anticipate that every participant in this meeting has something new and exciting to report. Although our knowledge is still fragmentary, we know enough to propose realistic models of the sequence of gene activation, cascade of multiple molecular pathways and cell–cell interactions that are involved in normal and abnormal neuronal migration. We can also use similar approaches to examine in more detail the effect of various environmental factors on neuronal migration and their positioning in the cerebral cortex. For example, we have begun to explore the effect of various physical agents and drugs that affect calcium fluctuations and thus indirectly influence rate of neuronal migration and their proper positioning in the cortex (Ang et al 2006). Our working hypothesis is that diverse environmental factors can affect neuronal migration by interfering with intercellular communication mediated by pathways such as intercellular ATP signalling and gap junctions (Weissman et al 2004). Research in this area may uncover pathogeneses of disorders of neuronal position that are not detectable by routine pathological examinations, but are involved in neuropsychiatric disorders whose causes and substrates are still largely unknown.

References

Ang ES Jr, Gluncic V, Duque A, Schafer ME, Rakic P 2006 Prenatal exposure to ultrasound waves impacts neuronal migration in mice. Proc Natl Acad Sci USA 103:12903–12910

Ayala R, Shu T, Tsai LH 2007 Trekking across the brain: the journey of neuronal migration. Cell 128:29–43

Behar TN, Scott CA, Greene CL et al 1999 Glutamate acting at NMDA receptors stimulates embryonic cortical neuronal migration. J Neurosci 19:4449–4461

Caviness VS Jr, Rakic P 1978 Mechanisms of cortical development: a view from mutations in mice. Annu Rev Neurosci 1:297–326

Chen K, Ochalski PG, Tran TS et al 2004 Interaction between Dab1 and CrkII is promoted by Reelin signalling. J Cell Sci 117:4527–4536

Cunningham CC, Gorlin JB, Kwiatkowski DJ et al 1992 Actin-binding protein requirement for cortical stability and efficient locomotion. Science 255:325–327

Falconer DS 1951 Two new mutants, trembler and reeler, with neurological actions in the house mouse (mus-musculus l) J Genet 50:192–201

Feng Y, Walsh CA 2004 The many faces of filamin: a versatile molecular scaffold for cell motility and signalling. Nat Cell Biol 6:1034–1038

Feng Y, Chen MH, Moskowitz IP et al 2006 Filamin A (FLNA) is required for cell–cell contact in vascular development and cardiac morphogenesis. Proc Natl Acad Sci USA 103:19836–19841

Fishell G, Kriegstein AR 2003 Neurons from radial glia: the consequences of asymmetric inheritance. Curr Opin Neurobiol 13:34–41

Fox JW, Lamperti ED, Eksioglu YZ et al 1998 Mutations in filamin 1 prevent migration of cerebral cortical neurons in human periventricular heterotopia. Neuron 21:1315–1325

Giniger E 1998 A role for Abl in Notch signalling. Neuron 20:667–681

Gleeson JG, Allen KM, Fox JW et al 1998 Doublecortin, a brain-specific gene mutated in human X-linked lissencephaly and double cortex syndrome, encodes a putative signalling protein. Cell 92:63–72

Gotz M, Huttner WB 2005 The cell biology of neurogenesis. Nat Rev Mol Cell Biol 6:777–788

Hart AW, Morgan JE, Schneider J et al 2006 Cardiac malformations and midline skeletal defects in mice lacking filamin A. Hum Mol Genet 15:2457–2467

Hashimoto-Torii K, Torii M, Sarkisian MR et al 2007 Reelin regulates neuronal migration in the cerebral cortex by modulating Notch activity. Nat Neurosci, in review

Hatten ME 2002 New directions in neuronal migration. Science 297:1660–1663

Hatten ME, Mason CA 1990 Mechanisms of glial-guided neuronal migration in vitro and in vivo. Experientia 46:907–916

Hong SE, Shugart YY, Huang DT et al 2000 Autosomal recessive lissencephaly with cerebellar hypoplasia is associated with human RELN mutations. Nat Genet 26:93–96

Komuro H, Rakic P 1993 Modulation of neuronal migration by NMDA receptors. Science 260:95–97

Komuro H, Rakic P 1998 Orchestration of neuronal migration by activity of ion channels, neurotransmitter receptors, and intracellular Ca2+ fluctuations. J Neurobiol 37:110–130

Levitt P, Rakic P 1980 Immunoperoxidase localization of glial fibrillary acidic protein in radial glial cells and astrocytes of the developing rhesus monkey brain. J Comp Neurol 193:815–840

Lu J, Tiao G, Folkerth R, Hecht J, Walsh C, Sheen V 2006 Overlapping expression of ARFGEF2 and Filamin A in the neuroependymal lining of the lateral ventricles: insights into the cause of periventricular heterotopia. J Comp Neurol 494:476–484

Marin O, Rubenstein JL 2003 Cell migration in the forebrain. Annu Rev Neurosci 26: 441–483

Marti A, Luo Z, Cunningham C et al 1997 Actin-binding protein-280 binds the stress-activated protein kinase (SAPK) activator SEK-1 and is required for tumor necrosis factor-alpha activation of SAPK in melanoma cells. J Biol Chem 272:2620–2628

Nagano T, Morikubo S, Sato M 2004 Filamin A and FILIP (Filamin A-Interacting Protein) regulate cell polarity and motility in neocortical subventricular and intermediate zones during radial migration. J Neurosci 24:9648–9657

Novartis Foundation 2000 Evolutionary developmental biology of the cerebral cortex. Wiley, Chichester (Novartis Found Symp 228)

Owens DF, Kriegstein AR 2002 Is there more to GABA than synaptic inhibition? Nat Rev Neurosci 3:715–727

Rakic P 1971 Neuron-glia relationship during granule cell migration in developing cerebellar cortex. A Golgi and electron microscopic study in Macacus Rhesus. J Comp Neurol 141:283–312

Rakic P 1972 Mode of cell migration to the superficial layers of fetal monkey neocortex. J Comp Neurol 145:61–83

Rakic P 1990 Principles of neural cell migration. Experientia 46:882–891

Rakic P 2006 A century of progress in corticoneurogenesis: from silver impregnation to genetic engineering. Cereb Cortex 16 Suppl 1:13–17

Rakic P, Cameron RS, Komuro H 1994 Recognition, adhesion, transmembrane signalling and cell motility in guided neuronal migration. Curr Opin Neurobiol 4:63–69

Rakic P, Knyihar-Csillik E, Csillik B 1996 Polarity of microtubule assemblies during neuronal cell migration. Proc Natl Acad Sci USA 93:9218–9222

Rašin M, Gazula V, Bruenig JJ et al 2007 Numb and Numbl are required for maintenance of cadherin-mediated adhesion and polarity of neural progenitors Nat Neurosci 10:819–827

Rivas RJ, Hatten ME 1995 Motility and cytoskeletal organization of migrating cerebellar granule neurons. J Neurosci 15:981–989

Robertson SP 2005 Filamin A: phenotypic diversity. Curr Opin Genet Dev 15:301–307

Santi MR, Golden JA 2001 Periventricular heterotopia may result from radial glial fiber disruption. J Neuropathol Exp Neurol 60:856–862

Sarkisian MR, Bartley CM, Chi H et al 2006 MEKK4 signalling regulates filamin expression and neuronal migration. Neuron 52:789–801

Sato M, Nagano T 2005 Involvement of filamin A and filamin A-interacting protein (FILIP) in controlling the start and cell shape of radially migrating cortical neurons. Anat Sci Int 80:19–29

Schaar BT, McConnell SK 2005 Cytoskeletal coordination during neuronal migration. Proc Natl Acad Sci USA 102:13652–13657

Sestan N, Artavanis-Tsakonas S, Rakic P 1999 Contact-dependent inhibition of cortical neurite growth mediated by Notch signalling. Science 286:741–746

Sheen VL, Feng Y, Graham D, Takafuta T, Shapiro SS, Walsh CA 2002 Filamin A and Filamin B are co-expressed within neurons during periods of neuronal migration and can physically interact. Hum Mol Genet 11:2845–2854

Sidman RL, Rakic P 1973 Neuronal migration, with special reference to developing human brain: a review. Brain Res 62:1–35

Spana EP, Doe CQ 1996 Numb antagonizes Notch signalling to specify sibling neuron cell fates. Neuron 17:21–26

Stossel TP, Condeelis J, Cooley L et al 2001 Filamins as integrators of cell mechanics and signalling. Nat Rev Mol Cell Biol 2:138–145

Suetsugu S, Tezuka T, Morimura T et al 2004 Regulation of actin cytoskeleton by mDab1 through N-WASP and ubiquitination of mDab1. Biochem J 384:1–8

Weissman TA, Riquelme PA, Ivic L, Flint AC, Kriegstein AR 2004 Calcium waves propagate through radial glial cells and modulate proliferation in the developing neocortex. Neuron 43:647–661

Zhong W, Feder JN, Jiang MM, Jan LY, Jan YN 1996 Asymmetric localization of a mammalian numb homolog during mouse cortical neurogenesis. Neuron 17:43–53

DISCUSSION

Walsh: What is the mechanism of the action of connexin 26 in migration? I think of it as a hemi-channel.

Rakic: You are correct, connexin is a hemi-channel protein and interference with its function disturbs Ca^{2+} transients. Ca^{2+} transients then don't give a message to contractile proteins, so cells can't translocate the nucleus and move from the ventricular zone (Liu and Rakic, unpublished)

Kriegstein: We have been looking at connexins in migration, seeing which ones are expressed in the ventricular zone, with loss and gain of function. Connexin 26 is involved in migration, and connexin 43 even more so. We have looked with a rescue paradigm specifically at what aspect of the connexin molecule is responsible for migration. There are ways of putting a mutation in the channel-forming part of the connexin, or the C-terminal end which is the signalling part, or in the part of the molecule that is involved in adhesion. For the two hemi-channels to come together and form a pore they first have to adhere. We can rescue the migration defect if we restore adhesion. Restoring the channel function, for example, of a hemi-channel isn't sufficient to restore migration. We think the contact point between the radial glial cell and the migrating neuron is an adhesion where there are plaques of connexins. We have taken a fusion molecule that labels connexin 43 or 26 and electroporated it into these cells. Time lapse images show the cell migrating. The plaque moves, the nucleus moves, and then the plaque moves and the nucleus moves: this seems to be the way these cells migrate. Our current thinking, therefore, is that connexins may work as adhesion molecules.

Rakic: You mentioned connexin 43. We did perform X-irradiation of the monkey fetus and saw that in addition to eliminating cells that are dividing at this time, many neurons didn't migrate to the cortex and were stacked in the subcortical white matter (Algan & Rakic 1997). This phenomenon is also well known from the studies in human, but no one knows the molecular mechanism of this abnormality. Our working hypothesis is that X-rays affect connexin 43.

Crino: One of the purposes of this meeting is to establish some translational link between the mouse genetics we are discussing and human disease. A recent study by Lee and colleagues reported 89 patients who had intractable epilepsy that required surgery (Lee et al 2005). The patients had preoperative brain magnetic resonance imaging (MRI) and in most cases the MRIs were normal. The surgeons felt they were operating in unknown territory. When the investigators evaluated the histopathological post-surgical specimens, about 90% of the patients had subtle evidence of cortical laminar disorganization or frank cortical dysplasia. These results suggest that subtle abnormalities of cortical development may be much more pervasive disorders that relate to human epilepsy than the overt malformations we see on preoperative neuroimaging. In any of the malformations you describe, in addition to alterations in lamination, do you see alterations in cell soma size?

Rakic: If cells don't come to the right positions and don't make totally the right connections, they don't get proper trophic influences which affects the size of the dendrites and number of branches and so on.

Macklis: The flip side is that we have found over and over that premature differentiation is another cause for dysmigration disorders, at least in the mouse. Heterotopy is sometimes the result when a neuron says, 'I am a grown up corticospinal neuron, why should I migrate any more?' If you decide you are something too early, you might not get the right connections.

Rakic: I agree with you. We are now working on this very problem in non-human primates, going back to examine the type of the interneurons that you have mentioned. In some of these cases heterotopic neurons are GABAergic cells. It is possible that these cells didn't move. In the mouse we have shown four different ways of stopping movement of cells. Then we can study how they are connected and what they do in their aberrant positions. For example, excitatory, glutaminergic neurons in aberrant positions can cause epilepsy, whereas GABAergic interneurons may not. There are also several strains of transgenic mice with defects of neuronal migration that can provide an answer to these types of clinically relevant questions (e.g. Sarkisian et al 2006).

Molnár: There has been spectacular progress in understanding these molecular pathways in migration. You decided to talk about rodents today, but you are very much aware that there are enormous species differences in some of the genes you mentioned. Also, there are great differences between distinct cortical areas. Sometimes one area is affected but not another, e.g., in bitemporal polymicrogyria syndromes. So how do you explain that the same kind of general molecular machinery is implicated, yet there are such large species and areal differences?

Rakic: Evolution always proceeds by building the next step upon an existing foundation. If you want to look at and understand the significance of the difference, you can't diminish it by saying that there is only a 1% or 0.1% difference in genomic sequence. We can't quantify the functional significance of this difference. We obviously have different brains than the mouse. The only way we can address this is to study mouse and then look at the very specific functional difference compared to human. We have been looking at very early stages at what are the differences in cortical organization and gene expression between mouse, monkey and human (Bystron et al 2006). We were surprised that in the ventricular zone and telencephalon of the human and mouse before any neurons are present, there is a big difference in size. Neural stem cells have information that they are going to be mouse or human. The first step is to look at the number of neurons. The next is to examine the types of cells and we think the predecessor neurons have something to do with evolutionary advances. If I insist that we have to study primates or humans to understand the advent of the human cortex, it doesn't mean that I think work in mouse is not important.

Fishell: When I think of Ca^{2+} influx I usually think of vesicular release of neurotransmitters. Much has been said recently about the effect of GABA on migration. Do you have any evidence that those Ca^{2+} influxes correlate with vesicular release of GABA in those migrating neurons?

Rakic: We and others have evidence that neurotransmitters already play a role at these non-synaptic levels between cells (Haydar et al 2000). The mechanism is still unclear.

Kriegstein: It is useful to inject a bit of physiology here. The progenitor cells all express certain receptors, including the $GABA_A$ receptor. Because of the high Cl^- in these cells, when the $GABA_A$ receptor is activated and the chloride channel opens, they depolarize, which leads to voltage-gated Ca^{2+} entry. GABA causes significant increases in Ca^{2+} in these cells and young neurons that they are migrating along. Gord Fishell is right: GABA could cause an increase in intracellular Ca^{2+} through depolarization and entrance through voltage-gated Ca^{2+} channels. This could propel these cells across some of that salutatory movement. However, while some studies claim that $GABA_A$ receptor activation can promote migration (Behar et al 1996, 2000, Manent et al 2005), there are also reports that it can inhibit migration (Behar et al 2000, Bolteus & Bordey 2004, Heck et al 2007).

Rakic: When I give a talk about the role of glutamate or GABA in neuronal proliferation and migration, physiologists usually say 'how come this neurotransmitter can have an effect when it is known that it works only when synapses are present?' My answer is that GABA doesn't know that it is a neurotransmitter. It is an amino acid and plays a role in cell division in animals that don't even have a nervous system. GABA is not a neurotransmitter: it is an amino acid that happens to have been used by cells at a certain point. It plays a role in early stages where cells divide, in, say, the sea urchin.

Kriegstein: The interesting implication from what we have been discussing today is that the GABA-containing cells, the interneurons, generally migrate in from the ventral telencephalon. Given the fact that the cortical progenitors and radial glia have GABA receptors, one can imagine that the circuit involves interneurons coming from extracortical locations, reaching the cortical cells, and signalling to them through their GABA receptors in a paracrine manner prior to synapse formation. This process can regulate either processes of neurogenesis or migration. This makes an interesting circuit.

Yamamoto: You have shown the involvement of E-cadherin in cell migration. You have also shown the existence of adherens junctions. Can such a structure easily appear and disappear during migration? I thought adherens junctions or tight junctions were stable structures.

Rakic: Hopefully, they are stable, because they are essential to normal brain development. We have seen that if you disturb the function of E-cadherin by electroporation, cell polarity is lost. Then cells don't get properly polarized and fail to migrate (Rasin et al 2007). This is one other way to get heterotopy. This is why it is difficult to give a simple answer about mechanisms of neuronal migration, and select just one or two molecules. I thought it was simple when I entered the field, but now I know it isn't. We know more but understand less.

Murakami: In the developing brain GABA causes depolarization. This causes opening of Ca^{2+} channels. Membrane depolarization can also be caused by glutamate. So, glutamate could do the same thing.

Kriegstein: Exactly.

Rakic: Yes. We have worked on the role of NMDA receptor in glia-guided neuronal migration (Komuro & Rakic 1993).

Wilson: Zoltán Molnár and I have looked at mutants in mice that have eliminated regulated exocytosis (e.g. Snap25 gene knockouts). They no longer can release neurotransmitter on demand. In some cases there is still spontaneous transmission of neurotransmitter. And yet the layering forms correctly. If there is release of GABA, it may take place through constitutive fusion of vesicles, but then there are other mutants such as those generated by Tom Südhof and colleagues in which neurons would have no vesicular release of GABA at all. How GABA is being released becomes a question. What is going on in terms of the leading edge of the glial process? You have the leading edge for these radial glial cells. Are they delivering any materials besides membrane? Are certain proteins being secreted from the leading edge?

Rakic: We do not have data. Some people have suggested that the leading edge of the cell produces plasminogen activator, which will dissolve cell adhesion around the tip of the leading process (Soreq & Miskin 1981). However it is clear that somehow membranes have to be added at the top.

Wilson: Is the leading edge of the pseudopodium delivering anything else besides membrane? One assumes so.

Rakic: I don't know. More than three decades ago Dennis Bray discovered that new membrane is added at the tip of growth cones, and I then transferred this idea to my work on the leading tip of migrating neurons. He discovered this in a clever way by spraying glass or carmine particles over cultures of neurons, and the particles attached to the growing processes. He then examined their distribution over time and found that the most growth occurred at the tip of the growth cone as the relative positions of the particles did not change near the cell body while the tips were free of them (Bray 1970). His model was published in *Nature* (Bray 1973) but only now, for the first time, are we able using specific molecular markers to confirm the fact that growth of the leading process occurs at the tip of filopodia (Letinic and Rakic, unpublished results).

Goffinet: You referred to the junction between radial glia close to the pial surface. They are not full adherens junctions between radial end feet. They are much smaller and less solid than those located at the apical, ventricular pole.

Rakic: I agree.

O'Leary: Pasko, you mentioned that you can rescue the Reeler phenotype, and you showed a linkage between Notch signalling and reelin signalling. What do the reelin experts in this room think about this particular mechanism?

Goffinet: It is all new to me! Yesterday was the first time I heard about a link between Notch and reelin. I think it is interesting.

O'Leary: Are the spatial distributions of the various proteins required for this interaction appropriate for your proposal to be effective?

Rakic: It fits. Initially, Notch was thought just to be involved in cell division. However, Nenad Sestan, who was a graduate student in my lab at that time, showed that as brain cells become post-mitotic and attain their final position in the cortex activated Notch transfers to the nucleus and continues to play a role in regulation of dendritic arborization long after cell division (Sestan et al 1999). When we recently examined the distribution of Notch in the Reeler mouse, there was a dramatically decreased expression (Hashimoto-Torii et al 2007).

References

Algan O, Rakic P 1997 Radiation-induced area- and lamina-specific deletion of neurons in the primate visual cortex. J Comp Neurol 381:335–352

Behar TN, Li Y, Tran HT et al 1996 GABA stimulates chemotaxis and chemokinesis of embryonic cortical neurons via calcium-dependent mechanisms. J Neurosci 16:1808–1818

Behar TN, Schaffner AE, Scott CA, Greene CL, Barker JL 2000 GABA receptor antagonists modulate postmitotic cell migration in slice cultures of embryonic rat cortex. Cereb Cortex 10:899–909

Bolteus AJ, Bordey A 2004 GABA release and uptake regulate neuronal precursor migration in the postnatal subventricular zone. J Neurosci 24:7623–7631

Bray D 1970 Surface movements during growth of single explanted neurons. Proc Natl Acad Sci USA 65:905–910

Bray D 1973 Model for membrane movements in neural growth cone. Nature 244:93–96

Bystron I, Rakic P, Blakemore C 2006 Neuronal stem cells and predecessor neurons in the primordium of the human forebrain. Nat Neurosci 9:880–885

Hashimoto-Torii K, Torii M, Sarkisian MR et al 2007 Reelin regulates neuronal migration in the cerebral cortex by modulating Notch activity. Nat Neurosci, in review

Haydar TF, Wang F, Schwartz ML, Rakic P 2000 Differential modulation of proliferation in the neocortical ventricular and subventricular zones. J Neurosci 20:5764–5774

Heck N, Kilb W, Reiprich P et al 2007 GABA-A receptors regulate neocortical neuronal migration in vitro and in vivo. Cereb Cortex 17:138–148

Komuro H, Rakic P 1993 Modulation of neuronal migration by NMDA receptors. Science 260:95–97

Lee SK, Lee SY, Kim KK, Hong KS, Lee DS, Chung CK 2005 Surgical outcome and prognostic factors of cryptogenic neocortical epilepsy. Ann Neurol 58:525–532

Manent JB, Demarque M, Jorquera I et al 2005 A noncanonical release of GABA and glutamate modulates neuronal migration. J Neurosci 25:4755–4765

Rasin M-R, Valeswara-Rao, Gazula V-R et al 2007 Numb and Numblike Regulate the Basolateral Membrane Localization of E-cadherin and Neural Stem Cell Polarity in the Cerebral Proliferative Zones. Cell submitted

Sestan N, Artavanis-Tsakonas S, Rakic P 1999 Contact-dependent inhibition of cortical neurite growth by Notch signaling. Science 286:741–745

Sarkisian MR, Bartley CM, Chi H, Nakamura F, Flavell RA, Rakic P 2006 MEKK4 regulates neural migration and survival in the developing cerebral cortex. Neuron 52:789–801

Soreq H, Miskin R 1981 Plasminogen activator in the rodent brain. Brain Res 216:3612–374

Neural stem and progenitor cells in cortical development

Stephen C. Noctor, Veronica Martinez-Cerdeño and Arnold R. Kriegstein

Department of Neurology, UCSF School of Medicine, 513 Parnassus Avenue, San Francisco, CA 94143-0525, USA

Abstract. Recent work has begun to identify neural stem and progenitor cells in the embryonic and adult brain, and is unravelling the mechanisms whereby new nerve cells are created and delivered to their correct locations. Radial glial (RG) cells, which are present in the developing mammalian brain, have been proposed to be neural stem cells because they produce multiple cell types. Furthermore, time-lapse imaging demonstrates that RG cells undergo asymmetric self-renewing divisions to produce immature neurons that migrate along their parent radial fibre to reach the developing cerebral cortex. RG cells also produce intermediate progenitor (IP) cells that undergo symmetric division in the subventricular zone of the embryonic cortex to produce pairs of neurons. The symmetric IP divisions increase cell number within the same cortical layer. This two-step process of neurogenesis suggests new mechanisms for the generation of cell diversity and cell number in the developing cortex and supports a model similar to that proposed for the development of the fruit fly CNS. In this model, a temporal sequence of gene expression changes in asymmetrically dividing self-renewed RG cells could lead to the differential inheritance of cell identity genes in cortical cells generated at different cell cycles

2007 Cortical development: genes and genetic abnormalities. Wiley, Chichester (Novartis Foundation Symposium 288) p 59–78

The mature human cerebral cortex comprises approximately 20 billion neurons (Pakkenberg & Gundersen 1997), and 40 billion glial cells (Pakkenberg et al 2003). Most cortical neurons are generated before birth, while glial cell production extends into the postnatal period (Bhardwaj et al 2006). The proliferative cells that produce cortical neurons and glial cells are located in two zones that surround the ventricular lumen during development: the ventricular zone (VZ), which is adjacent to the ventricle, and the subventricular zone (SVZ), which is superficial to the VZ (Boulder Committee 1970). Radial glial (RG) cell bodies are located in the embryonic VZ, and intermediate progenitor (IP) cells reside in the embryonic SVZ. Recent work has demonstrated that both of these embryonic cell types generate neurons during cortical development. However, RG and IP cells exhibit important differences in mitotic behaviour and morphology. RG cells are bipolar neuroepi-thelial cells that have a long thin pial-contacting process, undergo interkinetic nuclear migration during the cell cycle, and divide at the margin of the lateral ventricle. IP cells lack the pial process, do not undergo interkinetic nuclear migra-tion, and most divide away from the margin of the ventricle in the SVZ (reviewed in Noctor et al 2007). In addition to the aforementioned differences, RG and IP cells appear to adopt distinct modes of division (Haubensak et al 2004, Noctor et al 2004).

Neural stem cells have been isolated *in vitro* from several regions of the embry-onic and adult CNS, including the embryonic neocortex. Neural stem cells iso-lated from the embryonic neocortex have the capacity to undergo self-renewing divisions and to generate multiple cell types (Davis & Temple 1994). Recent work has shown that embryonic neural stem cells share properties with RG cells (Conti et al 2005), suggesting that RG cells may be a form of neural stem cell in the embryonic neocortex. Here we review evidence supporting the hypothesis that RG cells are a neural stem cell type, and that IP cells are committed progenitor cells with restricted potential. We consider the features that RG and IP cells *in situ* share with neural stem cells that have been isolated from embryonic and adult brain, and whether unique features that identify RG and IP cells serve as useful indicators of neural stem, or neural progenitor cell status in the developing cerebral cortex. For the purposes of this discussion we define stem cells as proliferative cells that undergo self-renewing divisions and produce multiple daughter cell types, and progenitor cells as proliferative cells that have limited potential for self-renewal and that are dedicated to the production of single cell types.

Radial glial cells as neural stem cells

Recent research has conclusively demonstrated that RG cells generate cortical neurons during embryonic development. This has been shown in dissociated cell cultures (Malatesta et al 2000), through time-lapse imaging of fluorescently labelled precursor cells in organotypic slice cultures (Miyata et al 2001, Noctor et al 2001, 2004, Tamamaki et al 2001), and through the expression of fluorescent reporter genes in mutant mice lines (Anthony et al 2004, Haubensak et al 2004). The neuronal identity of RG daughter cells has been verified through immunolabelling with neuronal specific markers, and also through electrophysiological recordings obtained from RG daughter cells and progeny (Noctor et al 2001, 2004). The evidence that RG cells function as neural stem cells, rather than restricted potential neural progenitor cells, rests on two properties: the production of multiple cell types, and self-renewal of the RG cells.

Generation of multiple cell types

RG cells in the dorsal telencephalon are generally thought to produce only excitatory glutamatergic neurons, while progenitor cells in the ventral telencephalon produce inhibitory GABAergic neurons (Parnavelas 2000). Nonetheless, RG cells produce cortical neurons that are destined to populate each of the cortical layers. It is not yet known if there are multiple subtypes of RG cells that are committed to the production of specific cortical neuron subtypes, whether cortical neuron subtype is determined by environmental factors after birth, or whether cell fate is specified earlier during the cell cycle as reported for laminar fate specification (McConnell & Kaznowski 1991). However, it remains possible that individual RG cells generate multiple subtypes of excitatory cortical neurons. RG cells also generate several additional cell types during cortical development: IP cells (Noctor et al 2004), astroglial cells (Schmechel & Rakic 1979, Voigt 1989, Noctor et al 2004), additional RG cells (Noctor et al 2004), and at later stages of development, ependymal cells (Spassky et al 2005), and neurogenic astrocytes that reside in the adult SVZ (Merkle et al 2004). RG cells in the developing spinal cord can produce both neurons and oligodendrocytes (Fogarty et al 2005), but it has yet to be determined if neocortical RG cells can also produce oligodendrocytes. Thus RG cells have been shown to produce at least five distinct cell types, and may generate additional cell types. The potential of RG cells *in vivo* matches that described for neural stem cells that have been isolated from the embryonic cerebral cortex; lineage studies performed *in vivo* and *in vitro* have described clones that consist of the same multiple cell types (e.g. Noctor et al 2004, Shen et al 2006). These data strengthen the argument that RG cells function as neural stem cells, but we cannot rule out the possibility that subsets of RG cells are dedicated to the production of specific lineages.

Self-renewal of radial glial cells

A requisite for claiming stem cell identity is the demonstration that a particular cell type undergoes self-renewing divisions: candidate stem cells must produce daughter cells that are identical to, or share key characteristics with, their parental cell in order to be classified as a stem cell. Following this line of reasoning, RG cells must produce daughter cells that are identical or very similar to the mother cell to be classified as a type of neural stem cell. RG cells can be identified based on several criteria such as morphology, marker expression, mitotic behaviour, and function. These identifying characteristics can therefore be used to determine whether mother and daughter cells are each radial glial cells.

RG cells have several unique morphological characteristics, but the pial fibre is a classical feature that has identified these cells in the developing CNS for more than 100 years (Ramón y Cajal 1995). This long, thin process ascends from the cell body in the VZ through overlying structures to terminate in the pial basement membrane (reviewed in Bentivoglio & Mazzarello 1999). The pial fibre is relatively short during early stages of development when the neocortex consists of a single neuroepithelial layer, but grows progressively longer throughout development and spans distances that are greater than 5 mm in primates (Rakic 1972). RG cells can also be identified by the expression of markers such as vimentin (Dahl et al 1981), nestin (Hockfield & McKay 1985), RC2 (Misson et al 1988b), BLBP (Feng et al 1994), Pax6 (Gotz et al 1998), and GLAST (Malatesta et al 2000). The combination of morphological and marker expression criteria provide a useful standard for RG cell classification, but additional unique characteristics can be used to identify RG cells. For example, RG cell bodies undergo a to-and-fro movement during the cell cycle that is termed interkinetic nuclear migration (Misson et al 1988a, Noctor et al 2001, 2004). These nuclear movements are correlated with the cell cycle such that during S-phase RG cell bodies are situated away from the ventricle in the superficial portion of the VZ. During G2-phase the RG cells move toward the ventricle and during M-phase RG cells undergo division at the margin of the ventricle. Interkinetic nuclear migration distinguishes RG cells from other mitotic cells in the developing neocortex that do not display this behaviour, such as the IP cells (Takahashi et al 1995). Finally, RG cells guide the migration of newborn neurons and thus RG cell processes can be identified by their association with migrating neurons (Rakic 1972).

Identifying individual RG cells can be difficult because VZ cells are tightly packed in the embryonic cortex. In addition, other cell types, such as tangentially migrating interneurons, are found in the VZ (Nadarajah et al 2002), and other mitotic cell types may also be present in the VZ. Pioneers of CNS research in the 19th century, such as Wilhelm His, believed that the VZ included two cell classes, 'spongioblasts' (today known as radial glial cells), and non-process bearing 'germinal cells' that divided at the ventricular margin (see Ramón y Cajal 1995). Later work determined that spongioblasts and germinal cells were actually the same cell type at different stages of interkinetic nuclear migration (Sauer 1935). Nonetheless, the idea that the ventricular zone contains two distinct populations of progenitor cells, one dedicated to the production of neurons, and a second dedicated to the production of glial cells remained widely accepted, and was bolstered by the report that GFAP positive and negative cells coexist in the primate VZ (Levitt et al 1983). Studies of the developing neocortex that utilized scanning electron microscopy (EM) of transverse sections, or reconstruction of ultra-thin material concluded that all ventricular cells undergoing division at the ventricle, whether neuronal or glial in nature, lacked identifiable processes (e.g. Hinds & Ruffett 1971). Subsequent work showed that RG cells are mitotic and divide at the ventricular margin (Misson et al 1988a), where mitotic 'non-process bearing' cells were identified in

EM studies. This led to two assumptions: (1) that VZ cells are a mixed population of neuronal and glial precursor cells, and (2) since EM work did not detect pial fibres it was assumed that RG cells must lose or retract the pial process during division, and then presumably extend a new process to the pial surface at the start of the next cell cycle. The loss and re-growth of a pial fibre would be a formidable task given the energy expense and time that would be required to repeat this process at each successive cell cycle, especially in large cortices where pial fibres span long distances. In addition, the loss of the pial fibre during mitosis would raise important questions concerning migratory guidance. If young cortical neurons were repeatedly deprived of pial fibre guidance during the M-phase of each cell cycle, would they temporarily lose their 'sense' of direction and halt migration, or re-associate with the pial fibre of a neighbouring RG cell that was not dividing? These problems were potentially resolved by more recent studies that utilized fluorescent dyes or fluorescent reporter genes to label mitotic cells in the developing neocortex. Fluorescent markers produce an extremely bright signal that reveals fine cellular processes in great detail (Chamberlin et al 1998), making it possible to study the morphology, physiology, marker expression and behaviour of labelled cells. Contrary to the earlier morphological findings, fluorescent marker studies demonstrated that RG cells retain the entire pial process throughout division (Miyata et al 2001, Noctor et al 2001, 2002 2004, Tamamaki et al 2001, Gotz et al 2002, Weissman et al 2003; Figs 1 and 5). This finding demonstrates that retention of the pial or radial process is a common feature in cortical structures: both Bergmann glia in the postnatal cerebellum (Basco et al 1977, Noctor et al 2002), and radial astrocytes in the subgranular zone of the adult dentate gyrus (Seri et al 2004), retain radial processes during division. But, even if one assumed that dividing non-process bearing cells exist side by side with RG cells in the VZ, the demonstration that RG cells are numerous and retain a pial fibre during division raises the following question: why have EM studies failed to detect the pial process of dividing VZ cells? Several features of RG pial fibres may provide the explanation. Pial processes ascend from the VZ along a radial trajectory that is largely perpendicular with respect to the ventricular surface. The ventricular and pial contact points of RG cells appear to remain fixed during neurogenesis (Noctor et al 2004), but the pial fibre does not course along a direct path to the pia and often follows a curving course that changes across time (Fig. 2). Furthermore, the trajectory followed by pial fibres is related to the location of the RG cell along the medial–lateral and the anterior–posterior axes (Fig. 3). Thus, the orientation of a section of tissue along the medio–lateral and/or antero–posterior axis can influence whether the trajectory of a single pial fibre is contained within a single section of tissue, and pial fibres might therefore appear to be truncated, short or missing entirely. In addition, cell cycle specific changes in the pial fibre morphology may also make it difficult to trace the fibre in M-phase cells, as described below.

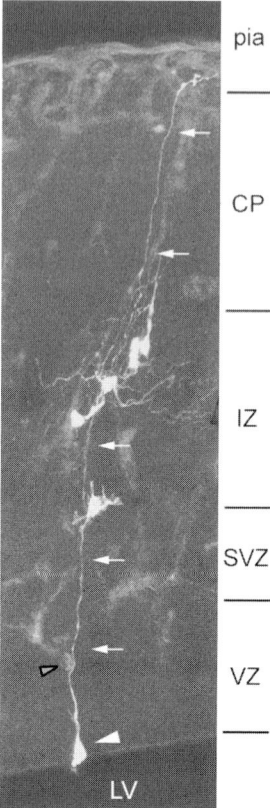

FIG. 1.   Radial glial cells (white arrowhead) undergo division at the edge of the lateral ventricle and retain their pial fiber. During M-phase the cell body rounds at the margin of the ventricle, and the pial fibre (small arrows) remains intact. Several daughter cells that were produced during previous divisions are located along the pial fibre in the subventricular zone (SVZ) and intermediate zone (IZ). Blood vessels are slightly autofluorescent and can be appreciated in outline in the embryonic brain. Pial fibres contact and course adjacent to blood vessels (grey arrowhead, black outline). LV, lateral ventricle; VZ, ventricular zone; CP, cortical plate.

We further examined the RG pial fibres with an eye towards identifying features that may correlate with neural stem cell properties. We labelled mitotic RG cells with *in utero* injections of a retrovirus that carries the eGFP reporter gene (for detailed methods see Noctor et al 2004), and prepared coronal sections of fixed tissue one to three days after retroviral injections. We imaged pial fibres in time-lapse movies and confirmed that pial fibres do not retract during M-phase as previously reported (see above). In addition, we found that the diameter of the pial fibre

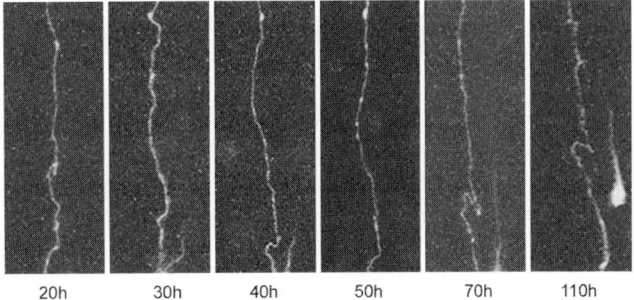

FIG. 2.   The pial fibre is a dynamic structure. A RG cell was labelled with eGFP and time-lapse imaged on a laser scanning confocal microscope. The series of images show that the pial fibre course changes regularly. In the right hand panel a daughter neuron generated by the same RG cell enters the field of view as it migrates toward the cortical plate. Time elapsed is shown below each panel. From Noctor et al (2004).

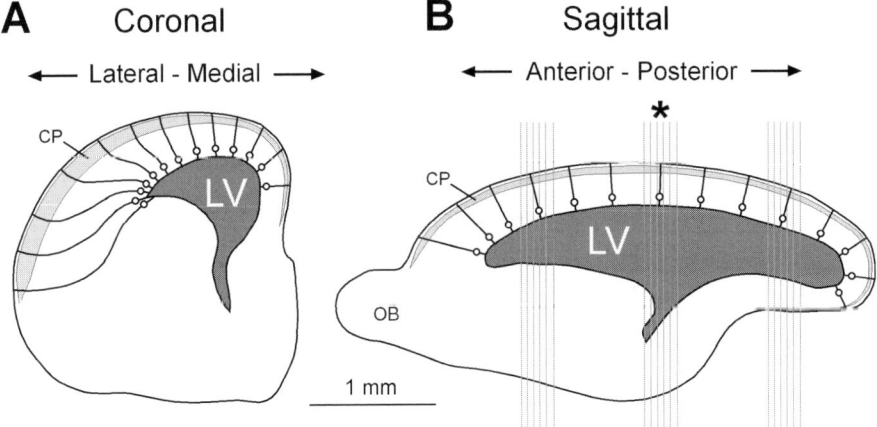

FIG. 3.   Schematic drawings of cut sections prepared from the embryonic day (E)18 rat brain in the coronal plane (A), and sagittal plane (B), indicating the trajectory of RG cells (black). (A) Representation of a coronal section at the level of the anterior commissure. RG cells radiate from the surface of the lateral ventricle (dark grey) through the cortical plate (light grey) to the pia. In the dorsal neocortex pial fibres course along a trajectory that is largely perpendicular to the ventricular surface. In lateral regions the pial fibres course along a curving 'S'-shaped trajectory to the pia. (B) The trajectory of RG cells also varies along the anterior–posterior axis as shown in this sagittal section drawn to scale from an E18 rat brain. A sampling of 50 μm thick sections (grey lines drawn to scale) shows that only a subset of sections in the dorsal telencephalon (asterisk) will contain the entire trajectory of the RG pial fiber. Sections obtained <500 μm more anterior or posterior will contain RG cell bodies in which the pial fibre appears truncated, short or entirely missing. The same principle holds true for cortical tissue cut in any plane: i.e. only a subset of sections will contain RG cells with the entire pial fiber. Analysis in thicker sections (i.e. >200 μm) yields a greater proportion of RG cells with long pial fibres. LV, lateral ventricle; CP, cortical plate; OB, olfactory bulb.

undergoes regular changes that are correlated with the RG cell cycle. The pial
process is approximately 1 µm in diameter during interphase, but becomes extremely
thin during M phase (Figs 4 and 5). We also noted the consistent appearance
of varicosities along the length of the thin pial fiber during M-phase (Fig. 4), and
found that many of the varicosities stream toward the RG soma prior to the initiation of cytokinesis (Fig. 5). Upon completion of cytokinesis the pial fibre rapidly
becomes thicker and most varicosities are no longer present or visible. The
extremely thin diameter, curving trajectory, and diaphanous bridging between the
fibre and the cell body during division (see Fig. 4 in Weissman et al 2003), may
have further contributed to the difficulty of process identification and thus led to
the conclusion that mitotic cells at the ventricular margin lack processes. The in-
ability of previous EM based studies to find any process bearing cells in the VZ
suggests that this approach is not best suited for all aspects of morphological
determination in the developing neocortex.

Determining whether RG cell divisions are self-renewing is dependent on the
reliable identification of both RG daughter cells. Time-lapse imaging studies have
shown that only one daughter cell inherits the pial process after RG divisions
(Miyata et al 2001, Noctor et al 2001, 2004, Tamamaki et al 2001), and that from

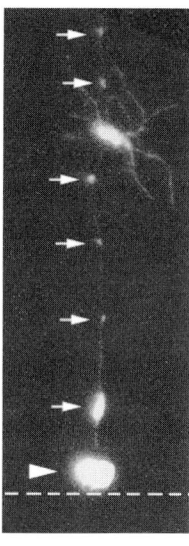

FIG. 4.   M-phase radial glial cells (arrowhead) divide at the surface of the lateral ventricle
(dashed line). The pial fibre becomes very thin during division but can be identified by
conspicuous varicosities (arrows) along the length of the pial fibre. Shown is a radial glial
cell in anaphase. A daughter cell generated at the previous cell cycle is positioned on the
parental pial fibre in the SVZ. Some pial fibre varicosities are large and measure up to 3µm
in diameter.

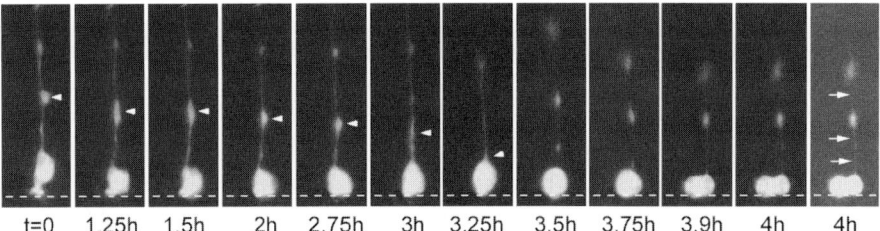

| t=0 | 1.25h | 1.5h | 2h | 2.75h | 3h | 3.25h | 3.5h | 3.75h | 3.9h | 4h | 4h |

FIG. 5. Pial fibre varicosities stream toward the radial glial cell body prior to cytokinesis. A radial glial cell at the G2/M phase transition is shown in the left panel at (t = 0). The dotted white line represents the ventricular surface, and time elapsed is shown below each panel. As the radial glial cell enters M phase the soma rounds up at the surface of the ventricle, the fibre becomes thinner, and varicosities (arrow) become apparent. The varicosities stream toward the cell body prior to cytokinesis. At 4 h the cell body is in telophase, and the thin pial fibre is difficult to detect. However, increased laser power or post-hoc image contrast demonstrate that the pial fibre (white arrowheads) is still present.

mid-stages of neurogenesis until the completion of cortical neurogenesis, some RG divisions produce daughter cells that inherit the pial process, detach from the ventricle, and translocate toward the cortical plate (Miyata et al 2001, Tamamaki et al 2001, Noctor et al 2004). Miyata et al (2001) reported that daughter neurons inherit the pial process from their parent RG cell. This interpretation might indicate that RG cells do not undergo self-renewing divisions, since mother and daughter cells do not share a ventricular contacting process. However, the translocating cells described in these studies match previous descriptions of RG cells that transform into GFAP expressing astrocytes at the end of cortical neurogenesis (Schmechel & Rakic 1979, Voigt 1989). In addition, electrophysiological recordings obtained from translocating daughter cells demonstrate that they lack the inward voltage gated currents that are expressed by immature neurons (Noctor et al 2004). Furthermore, studies in the developing human neocortex have reported that translocating cells do not express neuronal specific markers (deAzevedo et al 2003). The conclusion that neurons inherit the pial process in rodents may have been flawed since it relied on the Hu antibody to identify neurons (Miyata et al 2001). This antibody is restricted to neuronal lineages, but it labels both progenitor cells and immature neurons (Miyata et al 2004). Thus, translocating cells identified as neurons on the basis of Hu immunoreactivity in previous publications (Miyata et al 2001), may have been mitotically active translocating astrocytes that are known to be present at the end of cortical neurogenesis, and that may retain the potential to produce neurons (Merkle et al 2004). GFAP positive astrocytes produce neurons in the postnatal SVZ (Doetsch et al 1999), and therefore this interpretation could link the later stages of embryonic neurogenesis with postnatal neurogenesis in the cortical SVZ. Furthermore, this interpretation suggests that translocating cells are produced at the end of the RG lifespan, and does not rule out the hypothesis that RG cells undergo self-renewing divisions during cortical neurogenesis. In fact, live imaging studies have reported that mitotic RG cells produce daughter cells that retain the long pial process after division, resume interkinetic nuclear migration and divide at the ventricular surface, produce daughter cells with neuronal membrane properties and also support neuronal migration (Noctor et al 2004). Thus, these experiments demonstrate examples where both parent and daughter cells are RG cells based on at least three identifying criteria. These data indicate that RG cells undergo self-renewing divisions *in situ*, but do not rule out the possibility that the mother and daughter RG cells may differ in the expression of transcription factors or other genes.

The finding that embryonic neural stem cells exhibit characteristics that are similar to those described for RG cells further strengthens the claim that RG cells are neural stem cells. For example, during embryonic stem cell derivation of mouse and human neurons, the neurogenic cells appear to pass through an obligatory RG-like phase that is characterized by the formation of epithelial rosettes, the development of tight junctions, the expression of a variety of markers that are expressed by RG cells *in vivo*, and movements that resemble interkinetic nuclear migration (Perrier et al 2004, Conti et al 2005, Glaser & Brustle 2005). RG cells share additional characteristics with neural stem cells identified in other regions. For example, the RG pial fibre not only supports neuronal migration, but also makes frequent contact with blood vessels in the developing cortex (Misson et al 1988b, Noctor et al 2001; see Fig. 1, grey arrowhead). Recent work has demonstrated that neural stem cells in other regions of the brain share this feature. For example, stem cells in the adult dentate gyrus are located in clusters that are adjacent to blood vessels (Palmer et al 2000, Seri et al 2004), leading to the suggestion that endothelial cells or circulating factors influence neurogenesis in neural structures such as the embryonic neocortex (Shen et al 2004).

Temporal changes in gene expression increase cellular diversity

Neurogenesis in *Drosophila* involves the sequential expression of different genes during production of distinct neuroblasts (Brody & Odenwald 2005). Previous work in the developing neocortex has provided evidence to support this molecular model in the developing cerebral cortex. For example, the transcription factor Otx1 is expressed in the VZ only during the generation of deep layer cortical neurons (Frantz et al 1994). Additional genes and transcription factors such as M-fng (Ishii et al 2000), Svet1 (Tarabykin 2001), Cux1 and 2 (Nieto et al 2004), and Er81 (Yoneshima et al 2006), have been identified that are expressed first by proliferating cells in the VZ or SVZ, and later by cortical neurons in specific layers of the cortex. Temporal changes in gene expression by RG and IP cells during development may reflect changes in the cell generation potential. Thus, RG cells may progress along a maturational pathway of changing molecular expression patterns that might correlate with the known RG transitional cell forms: neuroepithelial cell → neurogenic RG cell → IP-genic RG cell → translocating astroglial cell → neurogenic astrocyte (see Alvarez-Buylla et al 2001). The progressive increase in GFAP expression by VZ cells that has been reported in the developing primate neocortex (Schmechel & Rakic 1979, Levitt et al 1983), may thus reflect the transition of some RG cells from the neurogenic phase to the translocating astroglial phase that is common at the end of cortical neurogenesis.

IP cells as restricted neural progenitors

At the onset of cortical neurogenesis RG cells begin producing IP cells (Haubensak et al 2004). The first generated IP cells reside within the VZ, but at later stages IP cells migrate away from the ventricle and form the SVZ (Noctor et al 2007). At this point both RG and IP cells produce cortical neurons (Haubensak et al 2004, Noctor et al 2004). However, unlike RG cells that largely divide asymmetrically during cortical neurogenesis, most embryonic IP cells primarily undergo symmetrical terminal divisions to produce paired daughter neurons (Haubensak et al 2004, Noctor et al 2004). It has not been determined whether IP cells undergo asymmetric self-renewing divisions during embryonic development, but a minority have been shown to undergo symmetrically self-renewing divisions based on the proliferative characteristics of both daughter cells (Noctor et al 2004). Embryonic IP cells are thus analogous in function to transit amplifying 'C' cells that have been identified in the postnatal and adult SVZ, since both cell types can rapidly amplify the number of specific neuronal cell types in developing and adult cortical structures (Doetsch et al 1997, Alvarez-Buylla et al 2001). In addition to neurogenic IP cells, additional distinct subsets of progenitor cells likely reside in the embryonic SVZ. For example, Olig2-expressing oligodendrocyte precursor cells are generated in the ventral forebrain, and migrate tangentially into the overlying dorsal cortex, taking a position in several structures including the embryonic SVZ (Takebayashi et al 2000). These Olig2 cells may represent the precursor cells that generate oligodendrocytes in the postnatal SVZ (Levison & Goldman 1993). Nonetheless, the relationship between IP cells and oligodendrocyte precursor cells in the SVZ remains to be determined.

Summary

RG cells undergo self-renewing divisions and generate multiple cell types including neurons, which supports the idea that RG cells are a form of neural stem cell in the developing cortex. Embryonic IP cells, on the other hand, are more restricted in their potential and appear to be dedicated to the production of cortical neurons. Pluripotency is often used to define stem cells, but another requisite for stem cell classification is self-renewal. Neural stem cells derived from various CNS structures have the potential to self-renew indefinitely under controlled conditions (Weiss et al 1996). Neural stem cells in the adult SVZ or dentate gyrus have limited potential to generate diverse cell types *in situ*, but retain the capacity to undergo self-renewing divisions throughout adulthood. In contrast, embryonic RG cells have a greater potential to generate diverse cell types, but are only present in the neocortex during development. However, many RG cells transform into astroglial cells (Schmechel & Rakic 1979, Voigt 1989, Noctor et al 2004) after neurogenesis, and some of these transitional cell forms retain the capacity for self-renewing divisions and appear to seed the neurogenic niche in the postnatal SVZ (Merkle et al 2004). Additional transformed RG cells may persist in other regions of the cortex and reveal their neurogenic potential only under special conditions (Magavi et al 2000). Thus, RG cells may represent the embryonic form of a self-renewing lineage of cortical neural stem cells that persist in the neocortex from early development through adulthood, while embryonic IP cells and adult C cells in the SVZ may represent the unipotent progeny of the RG lineage that amplify cell numbers, but have a limited potential for self-renewal.

References

Alvarez-Buylla A, García-Verdugo JM, Tramontin AD 2001 A unified hypothesis on the lineage of neural stem cells. Nat Rev Neurosci 2:287–293

Anthony TE, Klein C, Fishell G, Heintz N 2004 Radial glia serve as neuronal progenitors in all regions of the central nervous system. Neuron 41:881–890

Basco E, Hajos F, Fulop Z 1977 Proliferation of Bergmann-glia in the developing rat cerebellum. Anat Embryol (Berl) 151:219–222

Bentivoglio M, Mazzarello P 1999 The history of radial glia. Brain Res Bull 49:305–315

Bhardwaj RD, Curtis MA, Spalding KL et al 2006 Neocortical neurogenesis in humans is restricted to development. Proc Natl Acad Sci USA 103:12564–12568

Boulder Committee 1970 Embryonic vertebrate central nervous system: revised terminology. Anat Rec 166:257–261

Brody T, Odenwald WF 2005 Regulation of temporal identities during Drosophila neuroblast lineage development. Curr Opin Cell Biol 17:672–675

Chamberlin NL, Du B, de Lacalle S, Saper CB 1998 Recombinant adeno-associated virus vector: use for transgene expression and anterograde tract tracing in the CNS. Brain Res 793:169–175

Conti L, Pollard SM, Gorba T et al 2005 Niche-independent symmetrical self-renewal of a mammalian tissue stem cell. PLoS Biol 3:e283

Dahl D, Rueger DC, Bignami A, Weber K, Osborn M 1981 Vimentin, the 57 000 molecular weight protein of fibroblast filaments, is the major cytoskeletal component in immature glia. Eur J Cell Biol 24:191–196

Davis AA, Temple S 1994 A self-renewing multipotential stem cell in embryonic rat cerebral cortex. Nature 372:263–266

deAzevedo LC, Fallet C, Moura-Neto V, Daumas-Duport C, Hedin-Pereira C, Lent R 2003 Cortical radial glial cells in human fetuses: depth-correlated transformation into astrocytes. J Neurobiol 55:288–298

Doetsch F, García-Verdugo JM, Alvarez-Buylla A 1997 Cellular composition and three-dimensional organization of the subventricular germinal zone in the adult mammalian brain. J Neurosci 17:5046–5061

Doetsch F, Caillé I, Lim DA, García-Verdugo JM, Alvarez-Buylla A 1999 Subventricular zone astrocytes are neural stem cells in the adult mammalian brain. Cell 97:703–716

Feng L, Hatten ME, Heintz N 1994 Brain lipid-binding protein (BLBP): a novel signaling system in the developing mammalian CNS. Neuron 12:895–908

Fogarty M, Richardson WD, Kessaris N 2005 A subset of oligodendrocytes generated from radial glia in the dorsal spinal cord. Development 132:1951–1959

Frantz GD, Weimann JM, Levin ME, McConnell SK 1994 Otx1 and Otx2 define layers and regions in developing cerebral cortex and cerebellum. J Neurosci 14:5725–5740

Glaser T, Brustle O 2005 Retinoic acid induction of ES-cell-derived neurons: the radial glia connection. Trends Neurosci 28:397–400

Gotz M, Stoykova A, Gruss P 1998 Pax6 controls radial glia differentiation in the cerebral cortex. Neuron 21:1031–1044

Gotz M, Hartfuss E, Malatesta P 2002 Radial glial cells as neuronal precursors: a new perspective on the correlation of morphology and lineage restriction in the developing cerebral cortex of mice. Brain Res Bull 57:777–788

Haubensak W, Attardo A, Denk W, Huttner WB 2004 Neurons arise in the basal neuroepithelium of the early mammalian telencephalon: a major site of neurogenesis. Proc Natl Acad Sci USA 101:3196–3201

Hinds JW, Ruffett TL 1971 Cell proliferation in the neural tube: an electron microscopic and Golgi analysis in the mouse cerebral vesicle. Z Zellforsch Mikrosk Anat 115:226–264

Hockfield S, McKay RD 1985 Identification of major cell classes in the developing mammalian nervous system. J Neurosci 5:3310–3328

Ishii Y, Nakamura S, Osumi N 2000 Demarcation of early mammalian cortical development by differential expression of fringe genes. Brain Res Dev Brain Res 119:307–320

Levison SW, Goldman JE 1993 Both oligodendrocytes and astrocytes develop from progenitors in the subventricular zone of postnatal rat forebrain. Neuron 10:201–212

Levitt P, Cooper ML, Rakic P 1983 Early divergence and changing proportions of neuronal and glial precursor cells in the primate cerebral ventricular zone. Dev Biol 96:472–484

Magavi SS, Leavitt BR, Macklis JD 2000 Induction of neurogenesis in the neocortex of adult mice. Nature 405:951–955

Malatesta P, Hartfuss E, Gotz M 2000 Isolation of radial glial cells by fluorescent-activated cell sorting reveals a neuronal lineage. Development 127:5253–5263

McConnell SK, Kaznowski CE 1991 Cell cycle dependence of laminar determination in developing neocortex. Science 254:282–285

Merkle FT, Tramontin AD, Garcia-Verdugo JM, Alvarez-Buylla A 2004 Radial glia give rise to adult neural stem cells in the subventricular zone. Proc Natl Acad Sci USA 101:17528–17532

Misson JP, Edwards MA, Yamamoto M, Caviness VS Jr 1988a Mitotic cycling of radial glial cells of the fetal murine cerebral wall: a combined autoradiographic and immunohistochemical study. Brain Res 466:183–190

Misson JP, Edwards MA, Yamamoto M, Caviness VS Jr 1988b Identification of radial glial cells within the developing murine central nervous system: studies based upon a new immunohistochemical marker. Brain Res Dev Brain Res 44:95–108

Miyata T, Kawaguchi A, Okano H, Ogawa M 2001 Asymmetric inheritance of radial glial fibers by cortical neurons. Neuron 31:727–741

Miyata T, Kawaguchi A, Saito K, Kawano M, Muto T, Ogawa M 2004 Asymmetric production of surface-dividing and non-surface-dividing cortical progenitor cells. Development 131:3133–3145

Nadarajah B, Alifragis P, Wong RO, Parnavelas JG 2002 Ventricle-directed migration in the developing cerebral cortex. Nat Neurosci 5:218–224

Nieto M, Monuki ES, Tang H et al 2004 Expression of Cux-1 and Cux-2 in the subventricular zone and upper layers II-IV of the cerebral cortex. J Comp Neurol 479:168–180

Noctor SC, Flint AC, Weissman TA, Dammerman RS, Kriegstein AR 2001 Neurons derived from radial glial cells establish radial units in neocortex. Nature 409:714–720

Noctor SC, Flint AC, Weissman TA, Wong WS, Clinton BK, Kriegstein AR 2002 Dividing precursor cells of the embryonic cortical ventricular zone have morphological and molecular characteristics of radial glia. J Neurosci 22:3161–3173

Noctor SC, Martinez-Cerdeño V, Ivic L, Kriegstein AR 2004 Cortical neurons arise in symmetric and asymmetric division zones and migrate through specific phases. Nat Neurosci 7:136–144

Noctor SC, Martinez-Cerdeno V, Kriegstein AR 2007 Contribution of intermediate progenitor cells to cortical histogenesis. Arch Neurol 64:639–642

Pakkenberg B, Gundersen HJ 1997 Neocortical neuron number in humans: effect of sex and age. J Comp Neurol 384:312–320

Pakkenberg B, Pelvig D, Marner L et al 2003 Aging and the human neocortex. Exp Gerontol 38:95–99

Palmer TD, Willhoite AR, Gage FH 2000 Vascular niche for adult hippocampal neurogenesis. J Comp Neurol 425:479–494

Parnavelas JG 2000 The origin and migration of cortical neurones: new vistas. Trends Neurosci 23:126–131

Perrier AL, Tabar V, Barberi T et al 2004 Derivation of midbrain dopamine neurons from human embryonic stem cells. Proc Natl Acad Sci USA 101:12543–12548

Rakic P 1972 Mode of cell migration to the superficial layers of fetal monkey neocortex. J Comp Neurol 145:61–83

Ramón y Cajal S 1995 Histology of the nervous system of man and vertebrates. New York: Oxford University Press

Sauer FC 1935 Mitosis in the neural tube. J Comp Neurol 62:377–405

Schmechel DE, Rakic P 1979 A Golgi study of radial glial cells in developing monkey telencephalon: morphogenesis and transformation into astrocytes. Anat Embryol (Berl) 156:115–152

Seri B, Garcia-Verdugo JM, Collado-Morente L, McEwen BS, Alvarez-Buylla A 2004 Cell types, lineage, and architecture of the germinal zone in the adult dentate gyrus. J Comp Neurol 478:359–378

Shen Q, Goderie SK, Jin L et al 2004 Endothelial cells stimulate self-renewal and expand neurogenesis of neural stem cells. Science 304:1338–1340

Shen Q, Wang Y, Dimos JT et al 2006 The timing of cortical neurogenesis is encoded within lineages of individual progenitor cells. Nat Neurosci 9:743–751

Spassky N, Merkle FT, Flames N, Tramontin AD, Garcia-Verdugo JM, Alvarez-Buylla A 2005 Adult ependymal cells are postmitotic and are derived from radial glial cells during embryogenesis. J Neurosci 25:10–18

Takahashi T, Nowakowski RS, Caviness V Jr 1995 Early ontogeny of the secondary proliferative population of the embryonic murine cerebral wall. J Neurosci 15:6058–6068

Takebayashi H, Yoshida S, Sugimori M et al 2000 Dynamic expression of basic helix-loop-helix Olig family members: implication of Olig2 in neuron and oligodendrocyte differentiation and identification of a new member, Olig3. Mech Dev 99:143–148

Tamamaki N, Nakamura K, Okamoto K, Kaneko T 2001 Radial glia is a progenitor of neocortical neurons in the developing cerebral cortex. Neurosci Res 41:51–60

Tarabykin V, Stoykova A, Usman N, Gruss P 2001 Cortical upper layer neurons derive from the subventricular zone as indicated by Svet1 gene expression. Development 128:1983–1993

Voigt T 1989 Development of glial cells in the cerebral wall of ferrets: direct tracing of their transformation from radial glia into astrocytes. J Comp Neurol 289:74–88

Weiss S, Reynolds BA, Vescovi AL, Morshead C, Craig CG, van der Kooy D 1996 Is there a neural stem cell in the mammalian forebrain? Trends Neurosci 19:387–393

Weissman T, Noctor SC, Clinton BK, Honig LS, Kriegstein AR 2003 Neurogenic radial glial cells in reptile, rodent and human: from mitosis to migration. Cereb Cortex 13:550–559

Yoneshima H, Yamasaki S, Voelker CC et al 2006 Er81 is expressed in a subpopulation of layer 5 neurons in rodent and primate neocortices. Neuroscience 137:401–412

DISCUSSION

Mallamaci: I have a question about the sizing of the pool of intermediate progenitors (IPs). Do we have any evidence that the size of this pool is areally regulated? For example, we know that the paleocortex originates from a very small ventricular region, whereas the neocortex and the hippocampus originate from proportionately larger and larger ventricular zones. In relationship to this, is there any evidence that, for example, in paleocortical regions, the basal proliferative compartment is over-sized?

Kriegstein: It is true that there are variations in the sizes of the subventricular zones (SVZs). In ferret or primate, there are differences in the thickness in the SVZ that seem to presage whether they are going to produce areas of sulci or gyrii. In areas where there will be a gyrus, the SVZ is usually thicker, and where there is going to be a sulcus it is thinner. In areas like the ventral telencephalon, the MGE has a huge SVZ. In some of our time-lapse images it looks as if they routinely undergo multiple rounds of IP symmetrical divisions to expand their numbers. The MGE is essentially a factory for interneuron production, so it has a large SVZ with multiple cells undergoing symmetrical divisions to expand the number of neurons produced.

Tan: Do you see much flipping of bipolar to multipolar morphology as the cells migrate up?

Kriegstein: Yes. We see it in the opposite direction. When a cell is generated at the ventricular surface it rapidly moves away into the SVZ and becomes multipolar. Cells stay in the SVZ for about 24 h or so. Bayer & Altman (1991) saw this years ago when they examined the migration of cortical cells using tritiated thymidine labelling. Cells are initially labelled at the surface of the ventricle. They then move away, and arrest migration for around 24 h before resuming. We observe that when cells move to the SVZ, they become multipolar, and often have the ability to move tangentially. Then, in most cases, they descend to the ventricle again. As soon as a cell process reaches the ventricle, they pull in their other processes, become bipolar, and in the final phase of migration they move radially up along the glial fibre (Noctor et al 2004).

Molnár: What is the largest number of divisions of one particular IP that you and your colleagues have ever seen?

Kriegstein: We are limited by the amount of time we can visualize cells in slice-culture. We have seen the generation of four cells probably half a dozen times.

Molnár: It makes a huge difference whether you see them twice, three times or four times. This can double, quadruple or multiply eight times the numbers of neurons respectively. This is why the regulation of the IP division is such a fundamental question. How can one regulate the divisions of IPs? Colette Dehay and Henry Kennedy propose that peripheral input could modulate them (Lukaszevicz et al 2006).

Kriegstein: I have no idea how they are regulated. They are in the SVZ where they could be regulated by a range of different signalling molecules, including GABA. There is a predominant band of interneuron migration just in the area where the cells sojourn, and GABA or other factors produced by interneurons could influence IPs in this zone.

Molnár: This mechanism would solve the problem of regulating the GABA/glutamate ratio. Interneurons were not observed to divide in large numbers as they migrate towards the cortex. It is difficult to imagine that two relatively distant germinal zones can produce these cells in exactly the right proportions without minor adjustments within the target region. You could compensate for the numbers by regulating the generation of pyramidal cells by migrating interneurons.

Rakic: I have a comment on the question of whether all cells in the SVZ are radial glia (RG) in all species. This could be one of the major differences between rodents and human or non-human primates. There is very strong GFAP expression in RG cells from the beginning of cortical development in primates. RG have GFAP positive soma and a shaft that you can't miss using EM or immunohisto-chemistry. Next to RG are other cells that are not GFAP positive and are dividing (Levitt et al 1981, Zecevic 2004). So, at least in humans and non-human primates there are progenitors that are not RG. However, in mouse there are fewer of these cells and they may be missed. It's possible that small processes could be missed in individual EM photomicrographs because the sections are thin, but if you look at them in the serial sections you can't miss them. If RG are the only progenitors in the VZ in mouse, this would be the only part of the brain, because other progenitors that are not RG exist in the spinal cord, brain stem and cerebellum, ganglionic eminence, etc. Everywhere, there are dedicated neural progenitors that are not glial in nature. However, we recently found that with a battery of powerful methods including gapless serial EM sections that even in the mouse ventricular zone there are progenitors with short processes that are not RG (Gal et al 2006). Use of retroviral labelling can be selective and an entire population can be missed.

Kriegstein: There are several different ways that could be explained. I am talking about rodents because I have not worked in primates. The only cells that we have seen generate neurons are RG and IPs. But in the VZ at any given time there will be a variety of cell types including RG cells, IPs, young RG cells (which may have short growing processes), as well as newborn neurons. So at any given moment you may find some short-process-bearing cells that are not GFAP-positive, which could be young neurons. The evidence in your recent paper (Gal et al 2006) for short-process-bearing BrDU-positive cells is not adequate to rule out the possibility that these are just RG or IPs.

Fishell: One of the most striking features of migratory cells is their saltatory mode of migration. What happens to cells migrating on the RG when they are emasculated by having the cytoskeleton stripped out of it? Do the cells start migrating? Do they switch to other fibres?

Kriegstein: The time resolution of our imaging is not fine enough to say whether they stop migrating during RG cell cytokinesis.

Fishell: Did you notice that before the cell division the cell was migrating on the fibre, but after the cell division it seemed to have moved away and was migrating on another fibre?

Kriegstein: No. Quite the opposite. If a cell is migrating along a parental fibre and the cell then divides, it continues along the same fibre. It does not switch fibres.

Macklis: Most of these slice cultures were done around E16. Do you think these sets of mechanisms span from the earliest cortical development of deep layer projection neurons through superficial layers, or is this a temporally shifting set of mechanisms?

Kriegstein: That's a good question. I think there is a shift. Caviness and colleagues did kinetic studies and looked at the output from the proliferative zone (Takahashi et al 1996). They concluded that symmetrical progenitor divisions predominate early, then asymmetrical divisions occur together with symmetrical progenitor divisions, and toward the end of neurogenesis symmetrical terminal divisions occur. The symmetrical terminal divisions come to dominate at the end of neurogenesis. I think this scheme is generally true, but the cell populations involved are a bit different from what the authors were aware of when this work was done. The SVZ where most of the IPs reside is not the only place where they are found. Before there is a well defined SVZ there are a relatively large number of IPs. They are mixed together with RG in the VZ early on. The role for IPs in neurogenesis does not start late in neurogenesis; they are actually there are the very beginning. This was shown by Weiland Huttner using the Tis21 mouse that reports on cells that are entering the neurogenetic cell cycle (Haubensak et al 2004). There are quite a number of abventricular-located IP cells early on in corticogenesis. The idea that the SVZ appears only late is not entirely correct. The zone appears late but the IP cells are there from early on.

Macklis: Have you done the experiments to see whether the same kinds of RG morphology cells act as progenitors at E12?

Kriegstein: We haven't looked as carefully in the early stages as we have later on. This is a technical limitation. By eye it is only possible to do intraventricular injections in the rat at E14, but not earlier, and E12 in the mouse.

Walsh: When I was looking back at the meeting here 12 years ago—in terms of the talk about non-radial migration—there was a lot of beautiful work showing that many of these widespread clones are GABAergic. But right from the start when we were doing our retroviral work we noted that some of the neurons in these widespread clones were clearly pyramidal in their morphology. One early picture we showed in *Science* 20 years ago (Walsh & Cepko 1988) was almost like a static photocopy of that cell you showed migrating horizontally in the SVZ. It looks like there are widespread clones for pyramidal neurons as well. I have a question. In a paper we had in *Neuron* in 1995 (Reid et al 1995) we noticed that some of these widespread clones looked like they had subunits. We saw horizontal clusters that looked like they represented multiple cell divisions of some sort of IP. But then some of the clones we saw had more than one of these little clusters, spaced about 2 mm apart. Have you seen anything in your work that might suggest a mechanism for this? We saw them in the rat but not in the ferret.

Kriegstein: We do not watch our cells long enough to answer this question in our slice cultures. But if it's true that the IP cells undergo symmetrical divisions in the SVZ, you can imagine how an IP cell can move to the SVZ, undergo a symmetrical division, and then both identical daughter cells could migrate away. If they divide again and migrate radially you would have clusters of cells that are separated but would behave the same way. They'd look like stereotyped units at different distances.

Walsh: They weren't always directed so that the lower layer neuron was caudal and the upper layer neurons were far rostral, or vice versa. This could be consistent with a model where there was a progenitor in the middle which sent two daughters in opposite directions.

Yamamoto: Do you mean that the asymmetrical division is dependent on the plane of orientation of the cell?

Kriegstein: This has to do with a concept from the fly literature, where it is clear that plane of orientation is important for determining whether the daughter cells will have the same or different fates. This is because of the asymmetrical segregation of fate-determining substances. They are segregated to one site in the cytoplasm, so if the cleavage plane bisects this site and both daughter cells get equal amounts, they will be symmetrical cells. If the division is at right angles, however, one of the cells gets all of the determinants and the other gets none. This dictates an asymmetrical daughter cell fate.

Yamamoto: Did you see any unequal distribution of proteins or mRNAs?

Kriegstein: Some of you may know the work of Wieland Huettner (Kosodo et al 2004). He has evidence that there is a little bit of apical membrane that is key to determining whether cell divisions are symmetrical or not. The fate-determining substances could be anchored in just this small spot. Even a slightly oblique vertical cleavage could result in all of the membrane patch being inherited by one of the daughter cells. The other possibility is that the CNS of a fly (the most widely cited model) where there is apical–basal distribution of fate determinants is the wrong model, and we need to consider the planar dimension. This part of the discussion relates to neuroepithelial cells and RG, the IP cells divide symmetrically, and I don't believe that they have asymmetrical segregation of fate determinants. So the cleavage plane angle does not dictate fate in the IP cells.

Molnár: You wrote a nice review on the evolutionary effect on the IP cells (Kriegstein et al 2006). We also looked into these cells in embryonic turtle and chick forebrain (Molnár et al 2006a,b, Cheung et al 2007). They seem to be absent from the dorsal cortex. In turtle we have never seen this kind of amplification system at any stage studied. There are some abventricularly proliferating cells scattered in the cortex, but they never line up as an SVZ. The same is true for the chick dorsal cortex. The embryonic chick has some IP cells in subcortical regions (in subpallium and pallidum), which line up as an SVZ, but never in the dorsal cortex (Cheung et al 2007). This must be a fundamental amplification mechanism to produce a cortex in mammals.

Kriegstein: In the reptile DVR, an area that looks like the MGE where interneurons seem to come from, there is a large SVZ.

Molnár: But they never line up into a zone from a defined distance from VZ, as they do in embryonic chick. Also, in mammals abventricular divisions don't line up in the ganglionic eminence. The SVZs in the cortical and subcortical regions seems to be different. The key question is, why do abventricular divisions line up from a particular distance away from ventricular zone in cortex? Is it related to the transcriptional signalling which does show similar layers? You have a defined distance between the ventricular (apical) and SVZ (basal) divisions. In mammals the IPs are usually at a defined distance from VZ, and in turtle they seem to be scattered.

Hevner: We have been using anti-Tbr2 antibodies and Tbr2-GFP mice to look across ages in mouse embryos. We find that IPs are present at the onset of neurogenesis. Even in E10 mice we can see Tbr2-positive mitotic figures at the basal edge of the ventricular zone, where the SVZ will form. We also see many of these cells in the VZ, and a small percentage of them dividing at the apical surface. Interestingly, we also see many of them dividing adjacent to blood vessels. Some of them have the morphology of the short neural precursors (by GFP expression). In terms of their regulation, several papers have been published showing increase or decrease of Tbr2 (reviewed in Pontious et al 2007), which may be an indicator of IPs. For example, Sam Pleasure's group recently published a paper (Zhou et al 2006) knocking out the canonical Wnt signalling with the LRP receptor. All these things seem to be able to regulate at least Tbr2. It is hard to say if the basal progenitors themselves are increased or decreased, or if such changes are purely molecular.

References

Bayer SA, Altman J 1991 Neocortical development. Raven Press, New York

Cheung AFP, Pollen AA, Tavare A, DeProto J, Molnár Z 2007 Comparative aspects of cortical neurogenesis in vertebrates. J Anat 211:164–176

Gal JS, Morozov YM, Ayoub AE, Chatterjee M, Rakic P, Haydar TF 2006 Molecular and morphological heterogeneity of neural precursors in the mouse neocortical proliferative zones. J Neurosci 26:1045–1056

Haubensak W, Attardo A, Denk W, Huttner WB 2004 Neurons arise in the basal neuroepithelium of the early mammalian telencephalon: a major site of neurogenesis. Proc Natl Acad Sci USA 101:3196–3201

Kosodo Y, Roper K, Haubensak W, Marzesco AM, Corbeil D, Huttner WB 2004 Asymmetric distribution of the apical plasma membrane during neurogenic divisions of mammalian neuroepithelial cells. EMBO J 23:2314–2324

Kriegstein A, Noctor S, Martinez-Cerdeno V 2006 Patterns of neural stem and progenitor cell division may underlie evolutionary cortical expansion. Nat Rev Neurosci 7:883–90

Levitt P, Cooper ML, Rakic P 1981 Coexistence of neuronal and glial precursor cells in the cerebral ventricular zone of the fetal monkey: an ultrastructural immunoperoxidase analysis. J Neurosci l:27–39

Lukaszewicz A, Cortay V, Giroud P et al 2006 The concerted modulation of proliferation and migration contributes to the specification of the cytoarchitecture and dimensions of cortical areas. Cereb Cortex 16 Suppl 1:i26–34

Molnár Z, Métin C, Stoykova A et al 2006a Comparative aspects of cerebral cortical development. Eur J Neurosci 23:921–934

Molnár Z, Tavare A, Cheung AFP 2006b The origin of neocortex: lessons from comparative embryology. In: Kaas JH, Krubitzer LA (eds) Evolution of nervous system vol. 3 The evolution of nervous systems in mammals. Elsevier, Oxford, UK p 13–26

Noctor SC, Martinez-Cerdeno V, Ivic L, Kriegstein AR 2004 Cortical neurons arise in symmetric and asymmetric division zones and migrate through specific phases. Nat Neurosci 7:136–44

Pontious A, Kowalczyk T, Englund C, Hevner RF 2007 Role of intermediate progenitor cells in cerebral cortex development. Dev Neurosci, in press

Reid C, Liang I, Walsh C 1995 Systematic widespread clonal organization in cerebral cortex. Neuron 15:1–20

Takahashi T, Nowakowski RS, Caviness VS Jr 1996 The leaving or Q fraction of the murine cerebral proliferative epithelium: a general model of neocortical neuronogenesis. J Neurosci 16:6183–6196

Walsh C, Cepko CL 1988 Clonally related cortical cells show several migration patterns. Science 241:1342–1345

Zecevic N 2004 Specific characteristic of radial glia in the human fetal telencephalon. Glia 48:27–35

Zhou C-J, Borello U, Rubenstein JLR, Pleasure SJ 2006 Neuronal production and precursor proliferation defects in the neocortex of mice with loss of function in the canonical Wnt signaling pathway. Neuroscience 142:1119–1131

Genes that control the size of the cerebral cortex

Teresa H. Chae and Christopher A. Walsh

Howard Hughes Medical Institute, Beth Israel Deaconess Medical Center; Division of Genetics, Children's Hospital, Boston; and Program in Neuroscience, Harvard Medical School, 77 Avenue Louis Pasteur, Boston, MA 02115, USA

Abstract. Study of the mechanisms that control growth of the cerebral cortex has largely followed by analogy from work in invertebrate systems such as fly and worm. However, the identification of several genes that cause human microcephaly has provided new avenues of investigation into the mechanisms that control cell identity during cerebral cortical development. *In vivo* studies suggest that many forms of microcephaly result from defects in the control of cell fate: precocious formation of neurons during early developmental stages produces deficiencies in progenitor cells at later stages of neurogenesis, resulting in an overall small cerebral cortex. Also, some of the genes that are mutated in human microcephaly seem to have been targets in the evolution of humans from distant primate ancestors.

2007 Cortical development: genes and genetic abnormalities. Wiley, Chichester (Novartis Foundation Symposium 288) p 79–95

During the course of mammalian evolution, the cerebral cortex has undergone a remarkable increase in size disproportionate to the rest of the central nervous system. The surface area of the human cerebral cortex is 10 times that of the macaque monkey and 1000 times that of the mouse (Rakic 1995). Most would agree that this increase in cerebral cortical size somehow enabled humans to develop higher functions such as language and culture. However, despite the dramatic increase in size and function, the basic structure of the cerebral cortex is highly conserved in mammals as disparate as the human and mouse (Caviness et al 1995). And not surprisingly, many of the molecular mechanisms regulating cortical development appear to be conserved across mammalian species as well, making the genetic basis of cerebral cortical expansion all the more intriguing.

Development of the cerebral cortex is a complex process involving a balance between neural progenitor proliferation and differentiation. Neuronal progenitors proliferate in the pseudostratified epithelium adjacent to the ventricular lumen referred to as the ventricular zone (Fig. 1). In general terms, growth of the cerebral cortex can be divided into three overlapping phases. The first phase is composed primarily of symmetric cell divisions that produce a population of progenitor cells that doubles with each cell cycle (Takahashi et al 1996). The second phase is composed of both symmetric and asymmetric cell divisions. Asymmetric divisions produce one progenitor that will continue to divide and one cell that exits the cell cycle, migrates out of the ventricular zone, and differentiates into a neuron. The shifting balance from symmetric divisions to asymmetric divisions results in a

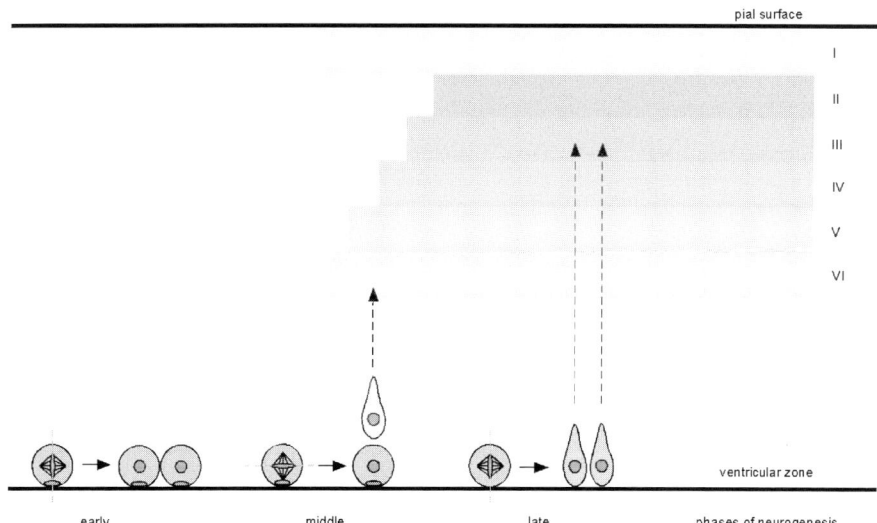

FIG. 1.   Development of the cerebral cortex. Neural progenitors (grey) proliferate in the ventricular zone adjacent to the ventricular lumen. As neurons are produced they migrate out of the ventricular zone (VZ) toward the pial surface. Later born neurons migrate past earlier born neurons to form six distinct cell layers (I–VI) in an 'inside-out' fashion. Neural progenitors can divide in a symmetric or asymmetric fashion producing daughter cells of identical or different fates respectively. Two factors that determine the mode of cell division are the polarized localization of cell fate determinants (oval) along the apical–basal axis and mitotic spindle. Symmetric divisions are characterized by the equal partitioning of cell fate determinants between the two daughter cells with the plane of cell division (vertical dashed line) perpendicular to the ventricular surface. Asymmetric divisions are characterized by unequal partitioning of cell fate determinants between two daughter cells with the plane of division parallel to the ventricular surface (horizontal dashed line), resulting in one progenitor cell that remains in the VZ and one neuron that migrates toward the pial surface. Symmetric divisions producing two progenitor cells predominate during early stages of neurogenesis, expanding the progenitor pool. There is a subsequent shift toward more asymmetric divisions that maintain the progenitor pool and produce post-mitotic neurons. During late states of neurogenesis, there is a shift back to symmetric divisions but progenitors divide to produce two neurons, depleting the progenitor pool while rapidly increasing neuronal production.

slowing of cortical expansion and increasing numbers of differentiated neurons and cortical thickness. Late in neurogenesis, cell divisions are again largely symmetric but now produce two post-mitotic cells such that the final, last-born neurons are generated rapidly and in large numbers, the progenitor pool is depleted, and neurogenesis ceases (Caviness et al 2003, Kriegstein et al 2006, Noctor et al 2002).

Modelling of cerebral cortical growth has suggested that rather minor alterations in the balance between proliferation and differentiation, or symmetric and asymmetric divisions, can result in large differences in brain size (Chenn & Walsh 2002, Rakic 1995). One key mechanism in generating an asymmetric division seems to be the differential inheritance of cell fate determinants, a process best studied in invertebrates (Bardin et al 2004, Lu et al 2000). In order for cell fate determinants to be preferentially inherited by one daughter cell, these determinants must be asymmetrically localized within the dividing cell and the mitotic spindle must be oriented along the same axis. In the *Drosophila* neuroblast, an apical complex of proteins consisting of Bazooka (Baz, *Drosophila* homologue of Par3), Par6, *Drosophila* atypical proteins kinase C (DaPKC), Inscuteable (Insc), Partner of Inscuteable (Pins) and a subunit of the heterotrimeric G protein complex (Gαi) directs the basal localization of the cell fate determinants Prospero, Numb, and Neuralized (Neur). This apical complex also directs the orientation of the mitotic spindle along the apical–basal axis (Kaltschmidt et al 2000) such that Pros, Numb and Neur are preferentially inherited by the basal daughter cell, the ganglion mother cell (GMC), while the apical daughter retains a neuroblast (NB), or progenitor, identity. Pros, Numb and Neur, in different ways, then influence cell identity. Pros, a homeodomain transcription factor, activates GMC-specific gene expression and represses NB-specific gene expression (Lu et al 2000). Numb and Neur act in subsequent GMC divisions allowing daughter cells to adopt different neuronal fates by inhibiting and promoting Notch signalling, respectively (Bardin et al 2004).

Whether unequal segregation of cell fate determinants plays an analogous role in the vertebrate nervous system is unclear. Imaging studies have reached a variety of conclusions about the relationship between mitotic spindle orientation and the fates of the daughter cells. A simple model suggests that mitoses oriented along the planar axis of the neuroepithelium results in symmetric divisions, whereas mitotic spindle orientation along the apical–basal axis results in an asymmetric division producing one progenitor cell and one neuron. While this is a popular model and some observations are consistent with it, the complex patterns of mitotic spindle orientation make it difficult to conclude that a very simple relationship is present (Gotz & Huttner 2005, Huttner & Kosodo 2005, Kriegstein et al 2006, Noctor et al 2004). Moreover, study of vertebrate homologues of Pros, Numb and Neur has not clearly shown a conservation of function across species.

The vertebrate Pros homologues have not been shown to asymmetrically localize during mitosis and vertebrate Neur homologues have not yet been studied in detail. Studies of vertebrate Numb have been hard to interpret. In contrast to *Drosophila*, mouse Numb has been reported to localize to the apical surface of the neuroepithelium (Zhong et al 1996), though a basal pattern of localization has been reported in other vertebrate species (Wakamatsu et al 1999). Mouse Numb has been suggested to segregate preferentially into the neuronal, presumably basal, daughter cell (Shen et al 2002). Consistent with invertebrate studies, Numb does appear to inhibit Notch signalling (McGill & McGlade 2003, Sestan et al 1999). However, whether mouse Numb promotes a progenitor or neuronal cell fate in the developing cerebral cortex has been controversial (Li et al 2003, Petersen et al 2002).

Whereas these and other studies in mammals have been based upon homology to genes studied in flies, recent 'forward genetics' approaches have identified regulators of cortical size in mammals. Microcephaly technically refers to a decrease in head circumference but is indicative of a decrease in brain size and primarily a decrease in cerebral cortical size. Microcephaly encompasses a heterogeneous group of disorders ranging from primary, or 'true' microcephaly in which the deficit in brain development is an isolated abnormality, to syndromes in which multiple tissue abnormalities are present (Mochida & Walsh 2001, Woods 2004). While there are environmental causes of microcephaly such as *in utero* infection or exposure to teratogens, there are many disorders associated with microcephaly that are thought to have a genetic basis. In this review we will discuss several genes recently identified to cause human microcephaly and the likely mechanisms by which they affect cortical growth, drawing upon complementary work in model systems.

Microcephaly and vesicle trafficking

Two genes potentially involved in vesicle trafficking have been identified to cause human microcephaly. One is *COH1*, which causes Cohen syndrome, a rare autosomal recessive disorder characterized by microcephaly, psychomotor retardation, dysmorphic facial features, childhood hypotonia, joint laxity, progressive retinochoroidal dystrophy, myopia and intermittent neutropaenia. *COH1* is a newly identified gene whose protein product shares homology to the *Saccharomyces cerevisiae* protein VPS13, a protein known to be involved in trafficking membrane proteins from the *trans*-Golgi network (TGN) (Kolehmainen et al 2003, Mochida et al 2004). Based on this homology, COH1 is postulated to be involved in intracellular vesicle-mediated sorting and transport or proteins. COH1 is expressed in multiple tissues with expression in brain down-regulated in adulthood, consistent with a key role in neurogenesis (Mochida et al 2004). However, whether the small brain seen in Cohen syndrome reflects defective neurogenesis, or some later defect such as neuronal process outgrowth, has not yet been demonstrated.

ARFGEF2 (ADP-ribosylation factor guanine nucleotide-exchange factor-2), which encodes a Sec7 homologue implicated in vesicle trafficking (Jones et al 2005), is mutated in a form of human microcephaly associated with periventricular heterotopia (Sheen et al 2004). Periventricular heterotopia are collections of neurons and/or glia abnormally located adjacent to the ventricular cavity and may represent a failure of neuronal migration, or abnormalities of proliferation or cell fate that have also led to microcephaly. ARFGEF2 is required for vesicle trafficking from the TGN. ARFs function in a GTP-dependent manner and are required for the assembly of coat proteins around intracellular vesicles. ARFGEFs accelerate the exchange of GDP for GTP, thereby activating ARFs. ARFGEF2 encodes the protein brefeldin A (BFA)-inhibited GEF2 (BIG2) that localizes to the TGN where proteins are sorted for various cellular destinations. Inhibition of BIG2-mediated vesicle trafficking in MDCK cells, a commonly used *in vitro* model for epithelia, results in the mislocalization of E-cadherin and β-catenin to the Golgi instead of forming adherens junctions at the apical–lateral cell surface (Sheen et al 2004).

Expression and functional studies also support a role for ARFGEF2 in neural proliferation. Within the CNS expression is most striking in proliferative regions such as the cerebral cortical ventricular zone (Lu et al 2006, Sheen et al 2004). Inhibition of BIG2 in neural progenitors and HEK293 cells using BFA and dominant negative BIG2, respectively, resulted in decreased proliferation *in vitro*. How might defects in vesicle trafficking regulate proliferation? One potential mechanism might involve the regulation of polarized protein expression in epithelial cells.

The establishment of cell polarity is critical for many aspects of embryonic development including proliferation and cell fate determination. Central to the establishment of cell polarity are the highly conserved PAR proteins, aPKC and Cdc42, a small G protein (Ahringer 2003, Macara 2004). As previously discussed, PAR3, PAR6 and aPKC are part of the apical complex in the *Drosophila* neuroblast, directing the basal localization of cell fate determinants and mitotic spindle orientation. Disruption of aPKC, Par6 or Par3 produces abnormalities of asymmetric division in *Drosophila* neuroblasts (Petronczki & Knoblich 2001, Schober et al 1999, Wodarz et al 2000). Analogous to the apical localization of these proteins in *Drosophila*, aPKC, Par6 and Par3 are localized to the apical aspect of mammalian neural progenitors, associating with adherens junctions (Manabe et al 2002), suggesting that the same molecular machinery is in place to establish cell polarity and to generate asymmetric divisions in the mammalian cerebral cortex.

Study of the *hyh* ('hydrocephalus with hop gait') mouse suggests that regulation of vesicle trafficking is required for the apical targeting of proteins that control cell polarity and cell fate. The *hyh* mouse is a spontaneous mutant with a CNS-specific phenotype that includes a small cerebral cortex (Chae et al 2004). *hyh* is caused by a missense mutation in the soluble NSF attachment protein α (α*Snap* or *Napa*) gene resulting in the Met105Ile substitution (Chae et al 2004, Hong et al 2004). The mutation does not appear to destabilize the protein but does reduce abundance of the mRNA and the level of the protein, while null αSnap mutations are early embryonic lethal (Chae et al 2004). αSnap is an integral component of SNARE-mediated vesicle trafficking and intracellular membrane fusion. SNAREs comprise a large family of proteins localized to distinct intracellular compartments and the interaction between SNAREs on vesicle and target membranes has been thought to provide the specificity of protein targeting within the cell (Xue & Zhang 2002). Specifically, the SNARE Vamp7 has been shown to form complexes with syntaxin 3, an apically localized protein in epithelial cells (Xue & Zhang 2002), and to be involved in the delivery of proteins to the apical membrane of MDCK cells (Lafont et al 1999). SNARE-mediated vesicle fusion events are dependent upon the ATPase NSF and the small family of SNAP proteins that together mediate the disassembly of SNARE complexes, making the components of the complex available for subsequent membrane fusion events. Apical protein targeting was found to require αSnap *in vitro*, being disrupted by dominant negative αSnap protein and antibodies to αSnap but independent of NSF. However, the role of these proteins in apical protein targeting *in vivo* remained to be demonstrated.

Reduction in αSnap activity in *hyh* results in abnormal apical vesicle localization as marked by Vamp7 and by abnormal localization of many apical proteins implicated in the regulation of neural cell fate, including E-cadherin, β-catenin, atypical protein kinase C (aPKC) and Pals1 (Chae et al 2004). These abnormalities in apical vesicle and protein localization are accompanied by alterations in cell fate, with premature withdrawal from the cell cycle of many neural progenitor cells, producing increased numbers of early-born, deep-layer cerebral cortical neurons and depleting the cortical progenitor pool consistent with an increase in asymmetric cell divisions or symmetric divisions producing neurons early in neurogenesis (Fig. 2). Consequently, late-born, upper-layer cortical neurons are severely underproduced, producing a small cortex.

One can hypothesize that a similar disturbance in cell polarity and cell fate is likely to underlie microcephaly caused by mutations in *ARFGEF2*. *In vitro* experiments of *ARFGEF2* dominant negative and antibodies in MDCK cells show abnormalities in E-cadherin and β-catenin, both integral components of adherens junctions that are also disturbed in the *hyh* mouse (Chae et al 2004, Sheen et al 2004). Work in *Drosophila* neuroblasts has suggested a role of adherens junctions in orientation of the mitotic spindle for symmetric cell divisions (Lu et al 2001). The adherens junction, located in the apical–lateral cell membrane, appears to

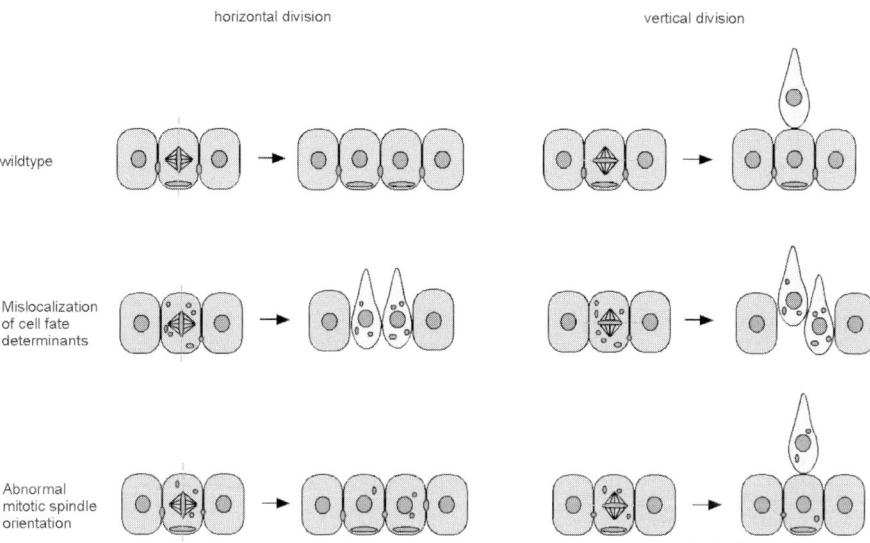

FIG. 2. Potential effects of vesicle trafficking defects on neural progenitor proliferation. Defects in the vesicle trafficking proteins αSnap and ARFGEF2/BIG2 have been shown to cause abnormalities in localization of proteins normally targeted to the apical surface. Potential ways in which abnormalities of apical protein targeting could cause a failure of cortical growth are illustrated here. Cell fate determinants (ovals) are normally asymmetrically localized within the cell. In this illustration, proteins that confer a progenitor cell fate are localized to the apical surface (drawn by convention as the lower surface) such that a horizontal division results in equal segregation into the daughter cells, producing more progenitors, and a vertical division results in unequal segregation such that the basal daughter (oblong cell, above the others) is able to adopt a neuronal fate while the apical (lower) daughter retains a progenitor identity. In the case of vesicle trafficking defects, these cell fate determinants are abnormally localized, trapped in vesicular compartments, and potentially degraded. Neither daughter cell, regardless of the plane of division, inherits sufficient active protein to confer a progenitor fate. A symmetric division producing two neurons results and the progenitor pool is not maintained. Alternatively, disruption of adherens junction (small dots between cells) localization to the apical–lateral surface caused by vesicle trafficking defects could cause abnormalities in mitotic spindle orientation. In *Drosophila*, adherens junctions have been suggested to maintain mitotic spindle orientation parallel to the epithelial surface to produce symmetric divisions. In the mammalian cerebral cortex, disruption of adherens junctions could lead to a premature shift to asymmetric divisions, increasing early neuronal production and decreasing progenitor expansion.

override cues that orient the mitotic spindle along the apical–basal axis. In keeping with this role in *Drosophila* neuroblasts, expression of a constitutively active form of β-catenin in mouse has been shown to cause a massively expanded cerebral cortex with fewer cells quitting the cell cycle (Chenn & Walsh 2002), suggesting an increase in symmetric divisions and expansion of the progenitor pool. Whether aberrant targeting of adherens junction proteins in ARFGEF2 mutant cells in fact leads to abnormal mitotic spindle orientation and fewer symmetric cell divisions to result in microcephaly will require further study.

Microcephaly and the mitotic spindle

In addition to *ARFGEF2* and *COH1*, four loci that cause primary microcephaly, called MCPH, have been cloned. The genes encode microcephalin (Jackson et al 2002), abnormal spindle-like microcephaly associated (ASPM) (Bond et al 2002), and most recently CENPJ and CDK5RAP2 (Bond et al 2005). Microcephalin, also known as BRIT1, encodes three BRCA1 C-terminal (BRCT) domains and plays essential roles in chromosome condensation as well as in the DNA repair pathway in response to DNA damage (Bartek 2006, Chaplet et al 2006, Jackson et al 2002, Lin et al 2005, Rai et al 2006). ASPM is the most common cause of MCPH and is the only locus affected in multiple ethnic groups (Bond et al 2003). ASPM is the human orthologue of the previously described abnormal spindle gene (asp) in *Drosophila* and is preferentially expressed in the cortical ventricular zone during neurogenesis, with its expression in adult brain limited to regions undergoing persistent neurogenesis (Bond et al 2002). Asp is a microtubule-associated protein (MAP) that localizes to the poles of mitotic spindles and is required in meiosis and mitosis in *Drosophila* (Riparbelli et al 2002, Saunders et al 1997). Asp is needed for the normal organization of microtubules at spindle poles and for the proper formation of the central mitotic spindle (do Carmo Avides & Glover 1999, Gonzalez et al 1990). Disruption of the central spindle results in failure of the contractile ring machinery and consequently failure of cytokinesis in *Drosophila* (Riparbelli et al 2002). This likely underlies the arrest of *Drosophila* neuroblasts in metaphase leading to reduced nervous system development analogous to MCPH. *CENPJ* and *CDK5RAP2* also encode potential proteins involved in spindle function, suggesting a potential general spindle mechanism for these forms of microcephaly (Bond et al 2005).

Does microcephaly resulting from *ASPM* mutations simply reflect a failure of cell division or could *ASPM* be serving other functions in mammalian neural progenitor cells? It has been proposed that ASPM could be involved in regulating mitotic spindle orientation, influencing the mode of cell division and cortical growth. Recent RNAi data suggests a potential role for ASPM in maintaining symmetrical cell divisions, and in orienting the mitotic spindle so that daughter cells divide next to one another in the plane of the epithelium (Fish et al 2006). Moreover, previous analysis of a mouse model with microcephaly, Nde1, which also encodes a centrosomal protein, shows that microcephaly is associated with abnormal mitotic spindle orientation (Feng & Walsh 2004), suggesting a potential link between spindle orientation and cell fate determination in vertebrates. While this link between mitotic spindle orientation and cell fate is strengthening however, there is still much to be learned about just how spindle orientation might relate to cell fate determination.

Evolutionary analysis of *ASPM* genes across species suggests that changes in *ASPM* may somehow correlate with the massive expansion of the cerebral cortex in humans. One of the most striking changes in the *ASPM* locus is the increased numbers of IQ (isoleucine-glutamine) repeats, which encode calmodulin-binding domains, with increasing brain size. *Caenorhabditis elegans*, *Drosophila*, mouse and human ASPM orthologues contain 2, 24, 61, and 74 IQ repeats respectively (Bond et al 2002). However, further analysis has suggested that the expansion of IQ repeats is common to most placental mammals and was deleted in rodents (Evans et al 2004, Zhang, 2003), arguing against a relationship between IQ repeat number and brain size in mammals. However, analysis of *ASPM* sequence in several primate species reveals an excess of non-synonymous amino acid substitutions in ape lineages leading to humans, suggesting that human *ASPM* has been subject to positive selection during the course of human evolution (Evans et al 2004, Kouprina et al 2004, Zhang, 2003). Functional studies of these amino acid substitutions may yield greater insight into *ASPM* function and its role in nervous system development and evolution.

More recently, analysis of human population data led to the suggestion that there might be ongoing evolutionary selection of *ASPM*, since the appearance of modern humans, and potentially as recently as 5000–6000 years ago (Mekel-Bobrov et al 2005). Following the appearance of this paper, a systematic analysis of human genetic variation, called the HapMap, was released (The International HapMap Consortium 2005), and analysis of this catalogue does not confirm evidence for recent evolutionary selection at *ASPM* (Yu et al 2007). These data suggest then, that while the evidence for ancient selection on *ASPM* during the evolution of humans from other primates is strong, the evidence for very recent selection is not as strong. This makes general sense with the functional role of ASPM in humans, since cortical size varies greatly between humans and other primates, but does not differ appreciably between modern human populations.

The regulation of the balance between symmetric and asymmetric divisions, or neural progenitor proliferation and differentiation, is a potentially powerful mechanism through which to expand the size of the cerebral cortex during the course of evolution. Even small shifts can have a profound effect on the normal growth of the cerebral cortex. For instance, allowing just three additional rounds of symmetric divisions during the course of neurogenesis would result in eight times as many neuronal precursors, which alone could account for the difference in cortical size between the macaque monkey and human (Rakic 1995). Insight into those mechanisms that underlie the expansion of the cerebral cortex in humans will likely require not only an extension of work from invertebrate systems but also identification of genes that cause human cortical defects such as microcephaly. As the functions of these genes are investigated we will see whether these genes indeed fit into the models of neurogenesis that have been developed in model systems until this point, or offer new molecular mechanisms by which cortical growth is regulated and perhaps even by which cortical growth has evolved.

Acknowledgements

THC was supported by the Medical Scientist Training Program training grant (T32 GM07753) to Harvard Medical School, and was an Adams-Quan fellow. CAW is an Investigator of the HHMI and is also supported by grants from the NINDS.

References

Ahringer J 2003 Control of cell polarity and mitotic spindle positioning in animal cells. Curr Opin Cell Biol 15:73–81

Bardin AJ, Le Borgne R, Schweisguth F 2004 Asymmetric localization and function of cell-fate determinants: a fly's view. Curr Opin Neurobiol 14:6–14

Bartek J 2006 Microcephalin guards against small brains, genetic instability, and cancer. Cancer Cell 10:91–93

Bond J, Roberts E, Mochida GH et al 2002 ASPM is a major determinant of cerebral cortical size. Nat Genet 32:316–320

Bond J, Scott S, Hampshire DJ et al 2003 Protein-truncating mutations in ASPM cause variable reduction in brain size. Am J Hum Genet 73:1170–1177

Bond J, Roberts E, Springell K et al 2005 A centrosomal mechanism involving CDK5RAP2 and CENPJ controls brain size. Nat Genet 37:353–355

Caviness VS Jr, Takahashi T, Nowakowski RS 1995 Numbers, time and neocortical neurono-genesis: a general developmental and evolutionary model. Trends Neurosci 18:379–383

Caviness VS Jr, Goto T, Tarui T, Takahashi T, Bhide PG, Nowakowski RS 2003 Cell output, cell cycle duration and neuronal specification: a model of integrated mechanisms of the neocortical proliferative process. Cereb Cortex 13:592–598

Chae TH, Kim S, Marz KE, Hanson PI, Walsh CA 2004 The hyh mutation uncovers roles for alpha Snap in apical protein localization and control of neural cell fate. Nat Genet 36: 264–270

Chaplet M, Rai R, Jackson-Bernitsas D, Li K, Lin SY 2006 BRIT1/MCPH1: a guardian of genome and an enemy of tumors. Cell Cycle 5:2579–2583

Chenn A, Walsh CA 2002 Regulation of cerebral cortical size by control of cell cycle exit in neural precursors. Science 297:365–369

do Carmo Avides M, Glover DM 1999 Abnormal spindle protein, Asp, and the integrity of mitotic centrosomal microtubule organizing centers. Science 283:1733–1735

Evans PD, Anderson JR, Vallender EJ et al 2004 Adaptive evolution of ASPM, a major deter-minant of cerebral cortical size in humans. Hum Mol Genet 13:489–494

Feng Y, Walsh CA 2004 Mitotic spindle regulation by Nde1 controls cerebral cortical size. Neuron 44:279–293

Fish JL, Kosodo Y, Enard W, Paabo S, Huttner WB 2006 ASPM specifically maintains symmetric proliferative divisions of neuroepithelial cells. Proc Natl Acad Sci USA 103:10438–10443

Gonzalez C, Saunders RD, Casal J et al 1990 Mutations at the asp locus of Drosophila lead to multiple free centrosomes in syncytial embryos, but restrict centrosome duplication in larval neuroblasts. J Cell Sci 96:605–616

Gotz M, Huttner WB 2005 The cell biology of neurogenesis. Nat Rev Mol Cell Biol 6:777–788

Hong HK, Chakravarti A, Takahashi JS 2004 The gene for soluble N-ethylmaleimide sensitive factor attachment protein alpha is mutated in hydrocephaly with hop gait (hyh) mice. Proc Natl Acad Sci USA 101:1748–1753

Huttner WB, Kosodo Y 2005 Symmetric versus asymmetric cell division during neurogenesis in the developing vertebrate central nervous system. Curr Opin Cell Biol 17:648–657

Jackson AP, Eastwood H, Bell SM et al 2002 Identification of microcephalin, a protein implicated in determining the size of the human brain. Am J Hum Genet 71:136–142

Jones HD, Moss J, Vaughan M 2005 BIG1 and BIG2, brefeldin A-inhibited guanine nucleotide-exchange factors for ADP-ribosylation factors. Methods Enzymol 404:174–184

Kaltschmidt JA, Davidson CM, Brown NH, Brand AH 2000 Rotation and asymmetry of the mitotic spindle direct asymmetric cell division in the developing central nervous system. Nat Cell Biol 2:7–12

Kolehmainen J, Black GC, Saarinen A et al 2003 Cohen syndrome is caused by mutations in a novel gene, COH1, encoding a transmembrane protein with a presumed role in vesicle-mediated sorting and intracellular protein transport. Am J Hum Genet 72:1359–1369

Kouprina N, Pavlicek A, Mochida GH et al 2004 Accelerated evolution of the ASPM gene controlling brain size begins prior to human brain expansion. PLoS Biol 2:e126

Kriegstein A, Noctor S, Martinez-Cerdeno V 2006 Patterns of neural stem and progenitor cell division may underlie evolutionary cortical expansion. Nat Rev Neurosci 7:883–890

Lafont F, Verkade P, Galli T, Wimmer C, Louvard D, Simons K 1999 Raft association of SNAP receptors acting in apical trafficking in Madin-Darby canine kidney cells. Proc Natl Acad Sci USA 96:3734–3738

Li HS, Wang D, Shen Q et al 2003 Inactivation of Numb and Numblike in embryonic dorsal forebrain impairs neurogenesis and disrupts cortical morphogenesis. Neuron 40:1105–1118

Lin SY, Rai R, Li K, Xu ZX, Elledge SJ 2005 BRIT1/MCPH1 is a DNA damage responsive protein that regulates the Brca1-Chk1 pathway, implicating checkpoint dysfunction in microcephaly. Proc Natl Acad Sci USA 102:15105–15109

Lu B, Jan L, Jan YN 2000 Control of cell divisions in the nervous system: symmetry and asymmetry. Annu Rev Neurosci 23:531–556

Lu B, Roegiers F, Jan LY, Jan YN 2001 Adherens junctions inhibit asymmetric division in the Drosophila epithelium. Nature 409:522–525

Lu J, Tiao G, Folkerth R, Hecht J, Walsh C, Sheen V 2006 Overlapping expression of ARFGEF2 and Filamin A in the neuroependymal lining of the lateral ventricles: insights into the cause of periventricular heterotopia. J Comp Neurol 494:476–484

Macara IG 2004 Parsing the polarity code. Nat Rev Mol Cell Biol 5:220–231

Manabe N, Hirai S, Imai F, Nakanishi H, Takai Y, Ohno S 2002 Association of ASIP/mPAR-3 with adherens junctions of mouse neuroepithelial cells. Dev Dyn 225:61–69

McGill MA, McGlade CJ 2003 Mammalian numb proteins promote Notch1 receptor ubiquitination and degradation of the Notch1 intracellular domain. J Biol Chem 278:23196–23203

Mekel-Bobrov N, Gilbert SL, Evans PD et al 2005 Ongoing adaptive evolution of ASPM, a brain size determinant in Homo sapiens. Science 309:1720–1722

Mochida GH, Walsh CA 2001 Molecular genetics of human microcephaly. Curr Opin Neurol 14:151–156

Mochida GH, Rajab A, Eyaid W et al 2004 Broader geographical spectrum of Cohen syndrome due to COH1 mutations. J Med Genet 41:e87

Noctor SC, Flint AC, Weissman TA, Wong WS, Clinton BK, Kriegstein AR 2002 Dividing precursor cells of the embryonic cortical ventricular zone have morphological and molecular characteristics of radial glia. J Neurosci 22:3161–3173

Noctor SC, Martinez-Cerdeno V, Ivic L, Kriegstein AR 2004 Cortical neurons arise in symmetric and asymmetric division zones and migrate through specific phases. Nat Neurosci 7:136–144

Petersen PH, Zou K, Hwang JK, Jan YN, Zhong W 2002 Progenitor cell maintenance requires numb and numblike during mouse neurogenesis. Nature 419:929–934

Petronczki M, Knoblich JA 2001 DmPAR-6 directs epithelial polarity and asymmetric cell division of neuroblasts in Drosophila. Nat Cell Biol 3:43–49

Rai R, Dai H, Multani AS et al 2006 BRIT1 regulates early DNA damage response, chromosomal integrity, and cancer. Cancer Cell 10:145–157

Rakic P 1995 A small step for the cell, a giant leap for mankind: a hypothesis of neocortical expansion during evolution. Trends Neurosci 18:383–388

Riparbelli MG, Callaini G, Glover DM, Avides Mdo C 2002 A requirement for the Abnormal Spindle protein to organise microtubules of the central spindle for cytokinesis in Drosophila. J Cell Sci 115:913–922

Saunders RD, Avides MC, Howard T, Gonzalez C, Glover DM 1997 The Drosophila gene abnormal spindle encodes a novel microtubule-associated protein that associates with the polar regions of the mitotic spindle. J Cell Biol 137:881–890

Schober M, Schaefer M, Knoblich JA 1999 Bazooka recruits Inscuteable to orient asymmetric cell divisions in Drosophila neuroblasts. Nature 402:548–551

Sestan N, Artavanis-Tsakonas S, Rakic P 1999 Contact-dependent inhibition of cortical neurite growth mediated by Notch signaling. Science 286:741–746

Sheen VL, Ganesh VS, Topcu M et al 2004 Mutations in ARFGEF2 implicate vesicle trafficking in neural progenitor proliferation and migration in the human cerebral cortex. Nat Genet 36:69–76

Shen Q, Zhong W, Jan YN, Temple S 2002 Asymmetric Numb distribution is critical for asymmetric cell division of mouse cerebral cortical stem cells and neuroblasts. Development 129:4843–4853

Takahashi T, Nowakowski RS, Caviness VS Jr 1996 The leaving or Q fraction of the murine cerebral proliferative epithelium: a general model of neocortical neuronogenesis. J Neurosci 16:6183–6196

The International HapMap Consortium 2005 A haplotype map of the human genome. Nature 437:1299–1320

Wakamatsu Y, Maynard TM, Jones SU, Weston JA 1999 NUMB localizes in the basal cortex of mitotic avian neuroepithelial cells and modulates neuronal differentiation by binding to NOTCH-1. Neuron 23:71–81

Wodarz A, Ramrath A, Grimm A, Knust E 2000 Drosophila atypical protein kinase C associates with Bazooka and controls polarity of epithelia and neuroblasts. J Cell Biol 150:1361–1374

Woods CG 2004 Human microcephaly. Curr Opin Neurobiol 14:112–117

Xue M, Zhang B 2002 Do SNARE proteins confer specificity for vesicle fusion? Proc Natl Acad Sci USA 99:13359–13361

Yu F, Hill RS, Schaffner SF et al 2007 Comment on 'Ongoing adaptive evolution of ASPM, a brain size determinant in Homo sapiens'. Science 316:370

Zhang J 2003 Evolution of the human ASPM gene, a major determinant of brain size. Genetics 165:2063–2070

Zhong W, Feder JN, Jiang MM, Jan LY, Jan YN 1996 Asymmetric localization of a mammalian numb homolog during mouse cortical neurogenesis. Neuron 17:43–53

DISCUSSION

Macklis: How much is known about the large-brained dolphins and whales? How evolutionarily different are they?

Walsh: I don't know much about them. They have a very different histology, such that their upper layers are not like the upper layers of primates. They don't have the same kind of elaboration of layers II and III. It is hard to map the upper layers of dolphins and whales on to primates.

Rakic: The dolphin brain has a bigger surface than humans, but the cortex is thinner. They have more radial units but fewer cells per unit. No one has looked at baby dolphins to see whether they have a longer period of ventricular zone formation before neurogenesis starts and a shorter total period, as could be predicted from the radial unit hypothesis.

Walsh: I wasn't aware that it was bigger but thinner. This could suggest that there are other ways of independently regulating thickness versus surface area of the cortex.

Molnár: You mentioned that during the evolution of the cerebral cortex there is an increase in cell number and elaboration in the supragranular layers. Indeed, if you lined up sections prepared from the brains of different species, this is the impression you would get. However, almost 30 years ago Tom Powell published the cell numbers in different species (Rockel et al 1980). This study claimed that in a 30 μm unit column of any mammalian cortex, exactly the same number of neurons can be found, 110, in every species in every cortical regions with the exception of the primate primary visual cortex, which is 2.5 times more.

Walsh: Did he break this out into layers?

Molnár: No, these are just overall number of neurons in an arbitrary cortical unit column. When we discussed this paper at our lab meeting, we didn't believe it at first. But then Dr Amanda Cheung, a postdoctoral fellow in my laboratory, started to look at these numbers in primates, echidna, platypus and many other species in collaboration with Patrick Hof (Mount Sinai Medical School, NY). She found that Powell was more or less right (Rockel et al 1980). The unit columns have the same numbers of neurons with the exception of the primate primary visual cortex (Cheung et al 2007).

Walsh: The primates do have the elaboration of the upper layers. If you look at this on a layer-specific basis you find that they have the same number of layer II cells in a column and the same number of layer III cells and so on.

Molnár: If the supragranular layers indeed contain more neurons, then they have to expand at the expense of the infragranular layer cells.

Walsh: In these mutations you basically swap lower layer cells for the upper layer cells. Many of these genetic mutations alter the balance. You might imagine that if you had a milder overexpression of β-catenin, you would see the reverse, with fewer of the lower layer neurons and elaboration of the upper layer ones. None of these mutations seem to change the overall time-course of neurogenesis. For all the mice it starts at E11 and ends E18. The mutations just seem to change what the brain does with that time-course. It makes me think that a lot of these shifts will shift the balance of the lower layers versus upper layers.

Molnár: Although in Amanda's quantification we did find similar number of neurons in a unit column in mouse and monkey, there was a big difference in the glial cell number (Cheung et al 2007).

Rakic: I would like to point out that Rockel et al (1980) didn't look at the width of the columns. They took arbitrary sized columns, which is on the micrometer on their microscope, which is artificial and not biological.

Walsh: Again, I am not sure what to do with those numbers. How do we define the size of the unit given that the cell size differs so much?

Tan: Takahashi et al (1997) and ourselves have pointed out that a mouse on average undergoes 11 cell cycles and produces about 150 neurons from a founder cell at E11.5 (Tan et al 1998). I like your idea about how the operative developmental unit is the column rather than a single cell. Do you think that your cells undergo 11 cell cycles even though they are generating a smaller number of neurons?

Walsh: I don't know, but I would guess that some progenitors are. If we stain an *hyh* mutant or *Nde1* mutant for Ki67 at different ages, they all have progenitors left at E17, but fewer than normal. I don't think all progenitors are going through 11 cell cycles. I am not so sure that it is as simple as every progenitor going through 11 cycles. I suspect that in most of these microcephalic mutants the population average will be shifted to 9. However, we still see some progenitors all the way out to E17 or 18, just fewer than normal.

Harrison: Some genes, when knocked out or mutated, produce these gross changes in brain size. Is there any evidence that polymorphic variation or more subtle changes in expression are associated with more minor individual differences or disease-associated changes in brain size?

Walsh: We have been trying to look at this. We have been collaborating with Jay Giedd at the NIMH. They have a fantastic database from which they have done a volumetric analysis of the brains of people with ADHD and people with dyslexia. We started by sequencing ASPM in the two shoulders of the distribution for normals and ADHD. We found lots of polymorphisms. Some of the early ones we found were in African-Americans and not Europeans. Then we found a couple of patients with heterozygous stop codons right in the middle of the gene. We are puzzled about how to interpret this because it is a huge gene with many polymorphisms. Bruce Lahn had another paper in which they studied people of differing intelligence. They found no correlation between any of the single nucleotide polymorphisms (SNPs) they checked and intelligence (Mekel-Bobrov et al

2007). There was another paper (Woods et al 2006) showing that within the normal range there was no obvious correlation of SNPs in ASPM and head size.

Hevner: One of my favourite papers on cortical size and thickness was from Kyuson Yun (Yun et al 2004) on Id4 mutant mice. This is an anti-differentiation gene. In this mutant they found that the surface area decreased, but despite the reduced surface area, the cortical thickness actually increased. The Tbr2 expression was increased as well. This is an example of where cortical surface area and thickness can be dissociated in some situations. I have a question. In one of your mutants there was a shift to the horizontal plane. What age was this?

Walsh: E14.

Hevner: If it was that early, and there was a shift to basal progenitors, this could explain the precocious production of deep layer cells.

Walsh: I guess that would explain it. There must be mutants that do differentiate surface area from thickness. The ones we have don't: perhaps this is because they are part of a generalized cell fate determination pathway, not a specific one.

Keays: With your *hyh* mouse, how do you know that the altered cell fate you observe isn't just a product of the increased pressure as a result of the hydrocephalus?

Walsh: The hydrocephalus is not obviously present until after birth. At the earliest stages the ventricle isn't enlarged. They seem to have a cell fate defect early on, and the hydrocephalus really gets going postnatally.

Macklis: You mentioned in passing that you saw no aneuploidy in the *Nde1*. Did you see no difference? When you do the same kinds of analysis as Gerald Chun, do you see that dramatic aneuploidy generally?

Walsh: We didn't look for aneuploidy in the brain because we didn't apply Gerald's methods. We grew fibroblasts in culture and they showed aneuploidy somewhat more commonly than wild-type. There wasn't enough to account for this sort of microcephaly. We don't see a lot of chromosome missegregation or cell death in the ventricular zone that might suggest that aneuploidy is causing the microcephaly.

Rubenstein: Is ASPM expressed in other places besides the brain?

Walsh: Yes.

Rubenstein: Do you have any idea what the IQ domains are binding to? Have you replaced the mouse gene with a human one to see whether this is the mechanism?

Walsh: The expression looks like it is pretty widespread in neural stem cells throughout the brain. It would be a good marker for neural stem cells. We can see a few resident ASPM positive cells in the ventricular zone of the cortex until P9, then they are gone. This suggests that there might be a couple of neural stem cells in the ventricular zone.

Rubenstein: Is there a distortion in the birth dating in different parts of CNS besides the cerebral cortex?

Walsh: We haven't looked in ASPM carefully enough to know. It is a gross defect in the cortex. It is expressed in several other proliferating tissues, but there doesn't seem to be a phenotype in non-CNS tissues. There have been no reports of frequent tumours in children with ASPM mutants. ASPM mutation is the most common cause of primary microcephaly.

Rubenstein: It seems to have something to do with stem cell biology. Does it have any effect on longevity?

Walsh: I am not aware of any. We might suspect that those carrying the mutation would have some sort of progeria. We should look at this.

Mallamaci: In the microcephalic models, did you find any change in the relative size of the cortical areas targeted by specific sensory afferents, such as visual?

Walsh: This would probably require functional magnetic resonance imaging (fMRI) and we haven't done this. Once we have these genetic mutations, what happens to the functional architecture of the cortex? This is a fascinating question. We haven't found a family with ASPM mutations in the Boston area. There are other human disorders that seem to affect the patterning of the cortex. One of these is a mutation in GPR56. There is a recent one that is not published yet.

Wilkins: Loss-of-function mutations are often uninformative about the normal developmental function. If you overexpress any of these genes in mice, do you get bigger brains?

Walsh: We haven't tried hard, but we haven't succeeded in overexpressing these genes. Now that we have an ASPM mutant we are working hard to put the human into the mouse.

Wilkins: Do rats have the same number of IQ units as mice?

Walsh: I think so. Rat, hamster and mouse all seem to have the same deletion of 12 IQ domains. In proteins that have only one or two IQ domains, they usually bind calmodulin. I don't know whether they still bind calmodulin if you have 72 of them.

Yamamoto: The time required for cell cycle is also an important aspect of this. In early development the cell cycle is very short. Gradually the time required gets longer. How about these mutant mice?

Walsh: We haven't measured cell cycle time *per se*. We have looked at the quitting fraction and then labelling index. The labelling indices are not obviously changed. If we administer a pulse of BrDU we see the same number of labelled cells, except at the end of neurogenesis when the mutants generally run out of progenitors.

References

Cheung AFP, Pollen AA, Tavare A, DeProto J, Molnár Z 2007 Comparative aspects of cortical neurogenesis in vertebrates. J Anat 211:164–176

Mekel-Bobrov N, Posthuma D, Gilbert SL et al 2007 The ongoing adaptive evolution of ASPM and Microcephalin is not explained by increased intelligence. Hum Mol Genet 1:600–608

Rockel AJ, Hiorns RW, Powell TP 1980 The basic uniformity in structure of the neocortex. Brain 103:221–244

Takahashi T, Nowakowski RS, Caviness VS Jr 1997 The mathematics of neocortical neuronogenesis. Dev Neurosci 19:17–22

Tan SS, Kalloniatis M, Sturm K, Tam PP, Reese BE, Faulkner-Jones B 1998 Separate progenitors for radial and tangential cell dispersion during development of the cerebral neocortex. Neuron 21:295–304

Woods RP, Freimer NB, De Young JA et al 2006 Normal variants of Microcephalin and ASPM do not account for brain size variability. Hum Mol Genet 15:2025–2029

Yun K, Mantani A, Garel S, Rubenstein J, Israel MA 2004 Id4 regulates neural progenitor proliferation and differentiation in vivo. Development 131:5441–5448

General Discussion I

Rubenstein: We haven't heard anything about the production of ependymal cells. As the neuroepitheium matures and there is a progressive transition from neural progenitor cells to neurons, this would deplete the ventricular surface of radial glial-type cells. As radial glia are depleted, how are the potential gaps filled in? Or is this simultaneously coupled with the production of ependymal cells that then fill in these loci?

Rakic: I think that ependymal cells are modified radial glial cells. All descendents of the radial glial cells have cilia including the ependyma. As radial glia begin to disappear at late stages of cortical development, some of them transform into ependymal cells which, like neurons, do not divide anymore. Additional cells that are dividing above the ependyma are now often called subventricular cells, though I think that this term should be reserved for the cells in the subventricular zone during development. Once the ependyma develops, at least in the field of neuropathology of the adult human brain, these cells are well known as the subependymal glia and are considered to be the source of brain tumours (gliomas) as well as a potential source of astrocytes that migrate to the site of lesion such as stroke. They, at least *in vivo*, never give origin to neurons or neuronal cancers.

Kriegstein: Arturo Alvarez-Buylla has a recent paper (Spassky et al 2005) in which he looked at the lineage relationship. He showed that radial glia generate ependyma.

Walsh: I'm thinking about Arnold Kriegstein's comment about the difference between the β-catenin overexpressing mouse and the ferret. A lot of the human congenital genetic malformations have enlarged ventricles. It is often referred to as hydrocephalus, but it is not: the ventricles just seem to remain in an enlarged state. This has been interpreted as a persistence of the fetal state, where the ventricles normally become reduced in size. This may represent some sort of ependymal remodelling where the ventricles have to actively reduce in size. In the β-catenin overexpressing mice, all these progenitors are formed, but then don't seem to go through the remodelling/contraction phase. The ventricle remains large as it does in many other developmental disorders.

Tan: Chris Walsh, you stated that malformations can come from somatic mutations. Can you explain more?

Walsh: I have heard a suggestion that every person has a somatic mutation somewhere in their body for every gene in their genome. If you run the calculations for the number of cells and rate of mutation of DNA polymerase and number of cell divisions it is clear that we carry a fair number of somatic mutations. As long as cancer genes aren't involved they don't generally cause disease.

Chelly: I think it depends on the stage when these mutations occur. If they occur early during development there can be knock-on effects. There could be focal abnormal neuron migration that could lead to epilepsy.

Walsh: You can get estimates from tuberous sclerosis, because in this you carry a germline mutation and multiple lesions in the body each represent somatic mutations just in that one gene.

Crino: Tuberous sclerosis is an autosomal dominant disorder in which a patient inherits one germline mutation, and focal manifestations of the disease, for example in the case of brain lesions, only occur when there is a second somatic hit. In theory, the same patient could have 12 malformations in the brain each with a common genotype but with 12 distinct somatic mutational events. Somatic mutational events have been described in hypothalamic hamartoma and a variety of other brain neoplasms. Many of the focal malformations associated with epilepsy are not genome events, and we will need to extract genomic DNA from the brain to find the mutation.

Walsh: If you use the same sort of statistic that people typically use when screening a library or a forward mutagenesis screen to calculate the saturation, you say that if you have three independent mutations in the same gene, then this means that 95% of the genome must have mutated. A tuberous sclerosis patient with 12 or 20 lesions means they have 12 or 20 mutations in the same gene, which means they have some other somatic clone somewhere in their body that is mutated for every other gene in the genome.

Tan: Because there is no turnover of neurons once they are made, they will stay there forever.

Chelly: I have a comment about brain size. Is the number of synapses also normal in the mutants with smaller brains?

Walsh: That's a great question. I would guess that the answer is no.

Parnavelas: It will be a difficult to answer this question.

Goffinet: One way to do it is to run a fluorodeoxyglucose (FDG) scan and see whether the density of metabolism is the same.

Parnavelas: You could use immunohistochemistry with antibodies against synaptic proteins in order to get a rough idea.

Goffinet: Does anyone know the relationship between the position of the centromere in ventricular zone cells in relation to when they divide? I was told that the centromere is on the ventricular side in ependymal cells, for example. When the cell migrates, the centromere is on the leading process side. Is its position important? Is it passive or does it reorganize the cytoskeleton in relation to asymmetric/symmetric divisions?

Walsh: We have done some pericentrin staining but it is a little hard to come up with a simple rule. There is a lot of centrosome staining along the ventricular surface. We haven't looked in great detail.

Parnavelas: The prevailing thought is that centrosomes are contained within the leading processes of migrating neurons. I was recently visiting Dr M. Kengaku at the Riken Institute in Tokyo, who has been looking at migration of granule cells in the cerebellum. She finds that quite often the centrosome is in the trailing process. I do not know whether this represents neuronal type differences, but it seems that these organelles are not always within the leading processes of migrating neurons.

Murakami: She has been using slices of the cerebellum, looking at tangential migration. At least for this case, she thinks that the centrosome is not necessary.

Kriegstein: In radial glia at interphase the centrosome is right at the base of the cilium. Radial glia withdraw their cilia when they enter mitosis, and at the G1–S phase transition the centrosome replicates and moves to either side laterally.

Goffinet: And then the centrosome relocates to the other side when it has to migrate.

Walsh: Bill Harris (Zolessi et al 2006) was also looking at migration of retinal ganglion cells. He suggested that the centrosome was sometimes in the trailing process, not the leading process.

Goffinet: The idea that it is on the side where the axon elongates is not a general rule then.

Parnavelas: It doesn't sound like it, but I say this with some caution.

Goffinet: Every time you have a cilium there is a centriole. The centriole is not necessarily the centromere of the cell. It is not completely clear.

Murakami: Apoptosis plays an important role in relation to the size of the brain. Do β-catenin knockout mice have changes in apoptosis?

Walsh: They have a little more apoptosis than wild-type mice, probably because lots of things are messed up. It is probably twofold increased cell death.

References

Spassky N, Merkle FT, Flames N, Tramontin AD, Garcia-Verdugo JM, Alvarez-Buylla A 2005 Adult ependymal cells are postmitotic and are derived from radial glial cells during embryogenesis. J Neurosci 25:10–18

Zolessi FR, Poggi L, Wilkinson CJ, Chien CB, Harris WA 2006 Polarization and orientation of retinal ganglion cells in vivo. Neural Develop 1:2

Control of cortical neuron layering: lessons from mouse chimeras

Vicki Hammond, Joanne Britto, Eva So, Holly Cate and Seong-Seng Tan

Howard Florey Institute, The University of Melbourne, Parkville 3010, Victoria, Australia

Abstract. How is the activation of Reelin signalling within neurons translated into the layering of cortical neurons? To address this question, we made mouse chimeras to test the reciprocal effects of neurons possessing different genotypes but sharing a common cortical environment during development. In chimeras composed of wild-type and mutant neurons (for either *Reelin*, *Dab1* or *p35* genes), a common observation was the formation of a second set of cortical layers on top of an inverted mutant cortex. The secondary cortex was invariably layered in the correct order, and in *Dab1* and *p35* chimeras, they were principally composed of wild-type neurons. In contrast to these cell-autonomous effects, *Reelin* chimeras displayed non cell-autonomous effects. In these chimeras, only a small number of wild-type neurons were required to be present in order for a secondary cortex to be formed. Interestingly, the principal constituents of the secondary cortex are not wild-type but mutant neurons, suggesting non cell-autonomous signalling by low levels of Reelin. Overall, these results suggest that information for the generation of cortical layers is vested within neuroepithelial progenitors even before the first neurons have been born, but the guidance of successive generations of daughter neurons to their proper locations requires the activation of Reelin and p35.

2007 Cortical development: genes and genetic abnormalities. Wiley, Chichester (Novartis Foundation Symposium 288) p 99–115

In the mammalian brain, the arrangement of neurons into morphologically distinct layers is a feature of many structures including the neocortex, hippocampus, cerebellum, thalamus and colliculus. This arrangement of neurons into discrete layers reflects both architectural and functional economy of brain function. Inherently, this requires genetic adaptations to co-ordinate the order of neurogenesis with their migration. The list of genes that are responsible for the control of this process is still incomplete but a number have been identified. To date, the best known and studied have been members of the Reelin signalling pathway (Rice & Curran 2001, Tissir & Goffinet 2003). Indeed, this biochemical pathway is so tightly linked to the generation of neuronal layers that its perturbation is known to result in layer disruptions in multiple brain structures, including the cerebellum, hippocampus and cortex (D'Arcangelo 2006). This would suggest

99

Reelin signalling to be a highly-successful biochemical mechanism for generating layered structures in the brain and probably an important means for the evolution of the vertebrate brain.

The Reelin signalling pathway has been most extensively characterized in cortical layering of rodents, principally because of the availability of *reeler* mice resulting from *Reelin* gene mutations. When this gene was first characterized and cloned in 1995, it rapidly opened up the field, leading to the subsequent identification of other members of this signalling pathway. These members include the transmembrane Reelin receptors, *very low-density lipoprotein receptor* (*VLDLR*) and *apolipoprotein E receptor 2* (*ApoER2*), and the cytoplasmic signal effector *disabled 1* (*Dab1*) (D'Arcangelo et al 1995, Howell et al 1997, Sheldon et al 1997, Trommsdorff et al 1999). Mutations in any of these genes are known to cause migration abnormalities leading to inversion of cortical layers in the neocortex. Reelin binding leads to receptor clustering and phosphorylation of Dab1 through the activation of Src family kinases. Phosphorylated Dab1 in turn binds to a range of proteins that potentially activate cytoskeletal dynamics for neuronal migration. Principal among these is Lis1, a subunit of the PAFAH1b complex (platelet-activating factor acetyl hydrolase 1b) that is capable of binding to dynein and mNudE, proteins important for microtubule structure and function (Feng & Walsh 2001). However, the core question is: how does Reelin act upon its target cells to ensure that migrating neurons settle sequentially in distinct layers?

There are some clues, however, and one of these is that Reelin protein is most concentrated in the marginal zone (MZ) away from where neurons are generated in the ventricular zone (VZ), suggesting that Reelin acts by long-range signalling on migrating neurons situated some distance away (Hartfuss et al 2003, Perez-Garcia et al 2004). However, there appears to be no requirement for Reelin to be produced locally in the MZ, because the addition of exogenous Reelin to *reeler* slice cultures is capable of rescuing the mutant phenotype (Jossin et al 2004). Furthermore, the ectopic expression of *Reelin* cDNA under the control of the nestin promoter is able to effect partial rescue of transgenic cortices (Magdaleno et al 2002). In addition, it would appear that proper cortical layering can proceed quite adequately without very much Reelin (Meyer et al 2002, Takiguchi-Hayashi et al 2004, Yoshida et al 2006).

The above issues highlight the need for employing additional models to study Reelin function in the developing neocortex. Given the highly conserved role for Reelin in cortical layering, one important question concerns the intrinsic and extrinsic consequences of Reelin activation in all mammals, including humans. One would like to know whether a neuron's capacity to migrate to a particular layer is fully cell-autonomous or does it require additional input from, or interaction with, its environment? We have tried to address this question using mouse chimeras that offer a means of studying, in a shared developmental space, the behaviour of cells with different genotypes (wild-type and mutant) and the interactions of these cells with one another.

Materials and methods

We constructed chimeras by using one of two methods as previously described (Hammond et al 2001, 2004) (Fig. 1). To produce strong unbalanced chimeras, one to five wild-type embryonic stem (ES) cells carrying the *lacZ* marker were injected into blastocysts with either $Reelin^{-/-}$, $Dab1^{-/-}$ or $p35^{-/-}$ genotypes (Fig. 1A). These produced chimeras with the following genotypic compositions: $Reelin^{+/+} \leftrightarrow Reelin^{-/-}$ (*Reelin* chimeras); $Dab1^{+/+} \leftrightarrow Dab1^{-/-}$ (*Dab1* chimeras), and $p35^{+/+} \leftrightarrow p35^{-/-}$ (*p35* chimeras). Injected blastocysts were implanted into pseudopregnant foster females, and chimeric pups reared for 14–23 days (Fig. 1A) before genotype analysis and processing for X-gal histochemistry and immunocytochemistry.

Other chimeras were constructed by morulae aggregation (Mintz 1970). Briefly, eight-cell stage morulae from H253 mice carrying the *lacZ* gene (Tan et al 1993) were fused with morulae carrying homozygous mutant alleles (Fig. 1B). Fused morulae were cultured overnight to the blastocyst stage before implantation into pseudopregnant foster females. Chimeric pups were reared until 14–23 d before genotype analysis, X-gal histochemistry and immunocytochemical analysis.

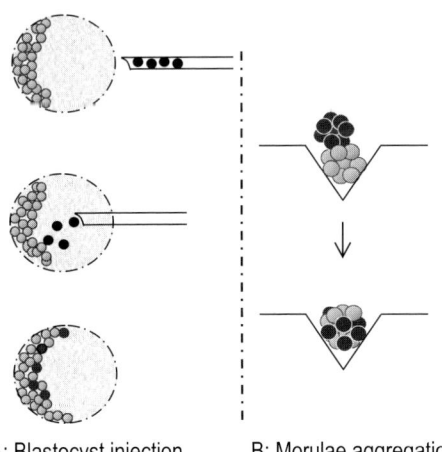

A: Blastocyst injection   B: Morulae aggregation

FIG. 1.   Generation of mouse chimeras by blastocyst injection or morulae aggregation. Two methods were used (A) the microinjection of *lacZ*-tagged embryonic stem (ES) cells into 3.5 day blastocysts homozygous for mutations in *Dab1*, *Reelin* or *p35* genes; (B) the aggregation of 2.5 day morulae from wild-type mice (carrying the *lacZ* reporter gene) and mice lacking Dab1, Reelin or p35.

Results and discussion

Being a fairly large glycoprotein, Reelin is not known to diffuse over long distances (Tissir & Goffinet 2003). Following secretion into the MZ, Reelin is believed to exert non cell-autonomous functions on its target cells. On the other hand, Dab1 is a cytoplasmic effector protein and therefore its activation is expected to result in cell-autonomous consequences. In the present work, chimeras were constructed using cells that lack either *Reelin* or *Dab1*, allowing a direct comparison of cell with non cell-autonomous gene functions. The results are indeed instructive. In *Dab1* chimeras, $Dab1^{+/+}$ neurons (identified by staining for β-galactosidase) (β-gal) appeared to be concentrated under the pial surface in a densely-packed arrangement (Fig. 2A). Deeper into the cortex, a minor fraction of β-gal-positive cells was

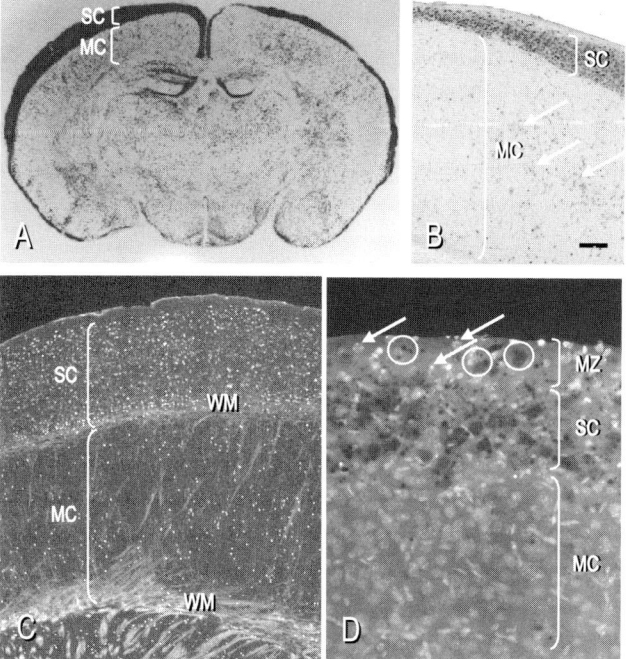

FIG. 2. $Dab1^{+/+}$ cells in chimeras form a secondary cortex (SC) beneath the pial surface. (A) In *Dab1* chimeras the wild-type cells (β-gal positive) formed a SC underneath the pial surface above the mutant cortex (MC). (B) Higher power magnification showing densely packed wild-type cells forming a SC, with a smaller number of wild-type cells (arrows) underneath in the MC intermingled with the *Dab1* mutant cells. (C) The SC is supported by a secondary white matter (WM) underneath as revealed by CNPase staining for myelin. (D) The normally cell free marginal zone (MZ) is invaded by both wild-type (β-gal positive, circles) and mutant (bis-benzamide positive, arrows) cells. Scale bar: A, 500 μm; B, 100 μm; C, 90 μm; D, 20 μm.

diffusely intermingled with $Dab1^{-/-}$ cells that form the rest of the cortical tissue underneath the densely packed $Dab1^{+/+}$ cells (Fig. 2A). All *Dab1* chimeras displayed the intense packing of β-gal-positive $Dab1^{+/+}$ cells into a dense superficial band, forming a secondary neocortex (Fig. 2B). Neurons in the secondary cortex are correctly arranged in the inside-out manner, with layer II neurons closest to the pial surface and layer VI neurons the furthest away (Hammond et al 2001). Indeed, the notion of a secondary cortex is fully supported by the existence of a secondary white matter beneath layer VI, and revealed by staining for myelin using CNPase antibodies (Fig. 2C). Despite their ectopic location, neurons of the secondary cortex are functionally wired; for example, backfilling of brainstem with fluoro-gold reveals fluorescent cells bodies in layer V pyramidals (Hammond et al 2004). Underneath this, there is a second set of cortical layers composed principally of $Dab1^{-/-}$ neurons but in the reversed order. Thus, layer II/III neurons of the mutant cortex are located closest to the ventricle, and layers V/VI are directly adjacent to the secondary white matter. These results would indicate a high degree of cell-autonomy with regard to Dab1 function. Since Reelin protein is expressed in these *Dab1* chimeras, it is reasonable to assume that layer patterning in these chimeras is reflective of Reelin-signalling among responding $Dab1^{+/+}$ neurons. Conversely, layer patterning of $Dab1^{-/-}$ neurons is reversed in the absence of Reelin signalling, and does not appear to be rescued even if Reelin-signalling is activated in a large number of $Dab1^{+/+}$ cells present in the neighbourhood. In addition, $Dab1^{-/-}$ neurons are seldom integrated into the correctly-layered secondary cortex, leading to the conclusion that although layering is a community effort, the process is quite dis-criminative and neurons incapable of Reelin signalling are excluded. However, it would appear that this cell-autonomy in cortical layering is not absolute. In the MZ situated above the secondary cortex, the normally cell-poor structure is invaded by a large number of cell bodies (Fig. 2D), similar to mutant mice lacking either Reelin, Dab1 or ApoER2/VLDLR. In the chimeric MZ, cells of both geno-types are present, suggesting lack of rescue of this mutant phenotype. One may conclude that the tendency of neurons to over-migrate into the MZ is non cell-autonomous, and both neurons with or without Reelin signalling may be embed-ded in the MZ. In extremely weak *Dab1* chimeras, the formation of a secondary cortex appears to be quite local (Fig. 3A). Although many of the $Dab1^{+/+}$ neurons have migrated into the secondary cortex in these weak chimeras, there are still significant numbers of $Dab1^{+/+}$ neurons trailing within the mutant cortical space (Fig. 3A). In fact, trailing neurons can be observed in both strong and weak chi-meras (Fig. 2A, 2B, 3A), suggesting that the ratio of mutant versus wild-type cells within a hemisphere is not the reason some $Dab1^{+/+}$ neurons fail to make it to the secondary cortex. Taken together, the above results would point to a strong rela-tionship between Dab1 function and cell-autonomy, but this relationship is far from being unconditional.

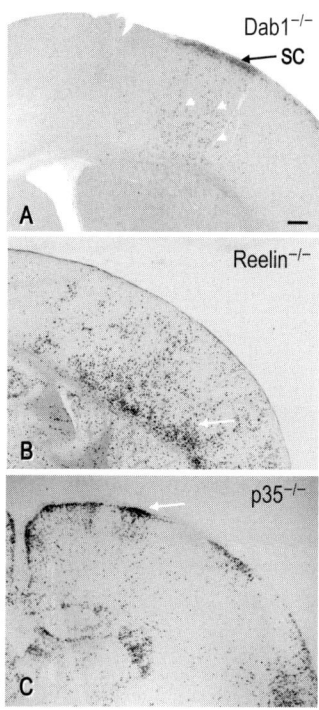

FIG. 3.   A secondary cortex is present in chimeras with all three genotypes (*Dab1*, *Reelin* and *p35*). (A) In a *Dab1* chimera, there is tight packing of wild-type cells which form a SC beneath the pial surface but other wild-type cells may also be found trailing underneath in the mutant cortex (arrowheads). (B) In the *Reelin* chimera, the majority of wild-type cells are located in the lower layers (arrow). (C) In the *p35* chimera, the vast majority of wild-type cells are localized in the upper layers of the cortex (arrow). Scale bar: 150 μm.

In contrast, *Reelin* chimeras present with a different phenotype (Fig. 3B). In these cortices, two separate cortices are also present, with a secondary cortex situated above the inverted cortex (Fig. 4). The secondary cortex is also correctly layered in the inside-out fashion but unlike *Dab1* chimeras, the neurons of the secondary cortex are principally composed of $Reelin^{-/-}$ neurons. Underneath this, the cortex consists of inverted layers, with layer II closest to the ventricle. This has the net effect of producing a mirror symmetry between the two sets of cortical layers in a single hemisphere. Apparently, the proportion of $Reelin^{+/+}$ neurons present in these cortices is not a deciding factor in the ability to form a secondary cortex because a secondary cortex was also found in *Reelin* chimeras with varying numbers of $Reelin^{+/+}$ neurons. In other words, a secondary cortex was formed even when the proportion of β-gal-positive cells in the cortex was low, suggesting non

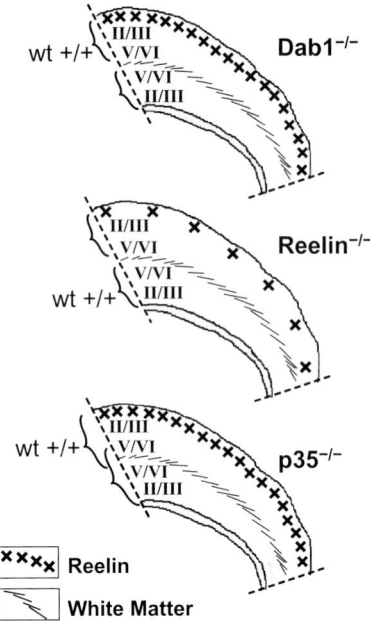

FIG. 4.   Schematic representation of chimeric cortices. (Upper) *Dab1* chimera showing a correctly layered wild-type cortex (wt$^{+/+}$) located above a secondary white matter with an inverted *Dab1* mutant cortex underneath. (Middle) *Reelin* chimera consisting of an essentially mutant cortex situated above a secondary white matter with a correctly layered wild-type cortex underneath. (Lower) *p35* chimera showing a wild-type cortex above a mutant cortex with slight overlap of wild-type and mutant layers V/VI.

cell-autonomous effects. On the other hand, when a large number of wild-type neurons were present, the entire cortex was rescued, with six layers present in the inside-out order and without a secondary cortex. In *Reelin* chimeras with two cortices, the *Reelin*$^{+/+}$ neurons appear to be preferentially localized in the lower half of the chimeric cortex (Fig. 3B, 4). This is an interesting finding given the *a priori* expectation that both *Reelin*$^{+/+}$ and *Reelin*$^{-/-}$ neurons should be capable of activating Dab1. The consistency of forming a secondary cortex in these chimeras leads to the question of whether or not a secondary cortex is specific only to the Reelin signalling pathway. It would be interesting to examine what happens in chimeras containing wild-type and mutant cells for *Cdk5* or *p35*; both of these genes do not participate directly in Reelin signalling but are known to be important for cortical layering (Gilmore et al 1998, Kwon & Tsai 1998). To test this, we also generated chimeras containing *p35*$^{+/+}$ and *p35*$^{-/-}$ neurons (Fig. 3C, 4). The results demonstrated that a secondary cortex was also present in these chimeras, and the layers

in this structure are properly layered (Fig. 4) (Hammond et al 2004). Underneath this, a mutant cortex with inverted layers was also found.

The above observations highlight a number of interesting points. First, data from knock-out mice suggest that when either *Reelin* or *Dab1* genes are disrupted, cortical layers are inverted. This phenotype is recessive, suggesting that the loss of one mutant allele *per se* is insufficient to result in layer inversion, and this notion is borne out by the normal layering observed in heterozygous mutants. Thus, cortical layering is able to proceed quite normally with only half the levels of Reelin or Dab1. However, when the cortical space contains a mixture of $Dab1^{+/+}$ and $Dab1^{-/-}$ genotypes, the predominant result is the generation of a second set of cortical layers by the $Dab1^{+/+}$ neurons. This effect was observed even if only a small minority of $Dab1^{+/+}$ neurons were present within the $Dab1^{-/-}$ cortex (Fig. 3A). These results suggest that the process of cortical layering is a community effect between neurons of similar genotype, and this effect is mutually exclusive between neurons of different genotypes. In support of this, $Dab1^{+/+}$ neurons were never seen to integrate fully (as detected with BrdU) with $Dab1^{-/-}$ neurons to form a particular layer, and vice versa. In the case of *Reelin* chimeras, the situation appears to be slightly different. A properly-layered secondary cortex was formed in response to the presence of $Reelin^{+/+}$ cells but the main constituents of this cortex were $Reelin^{-/-}$ cells. It would appear that the effect of Reelin signalling would be to attempt building a properly-layered cortex but the recruitment of neurons to this secondary cortex appears to be non genotype-specific. This leads to the second observation—the secondary cortex (in all three different chimeras) invariably contained layers II–VI of the same genotype. Thus, it would appear that cortical layering is the result of a 'domino effect', i.e. once layer VI neurons are in place, the process continues on within the same community until the last layer (layer II) is generated. One could reasonably conclude that information for generating the complete set of cortical layers (but not layering order) is already contained within the neuroepithelial progenitors at E11.5, and Reelin and its signalling associates do not influence the generation of layers, but have effects on their destinations over the next 11 cell cycles. An open question for future investigations is whether or not Reelin proteins need to be present in the MZ for the entire duration of 11 cell cycles (over 6 days) for correct cortical layering (Caviness et al 1995).

In conclusion, Reelin signalling is indispensable for the formation of cortical layers but the mechanisms remain unclear. Mouse chimeras provide a tool for testing the reciprocal effects of neurons with or without Reelin signalling within the same cortical environment. The results suggest that cortical layering is the result of co-operative interactions between neurons of the same genotype. Given a choice, neurons with the same genotype are more likely to form a second set of cortical layers, rather than integrate with neurons of a different genotype.

Acknowledgements

This work was supported by Program and Project Grants from the NH&MRC. We wish to thank Andre Goffinet for the gift of *Reeler* mice, Brian Howell for the gift of *Dab1* mutant mice, Li-Huei Tsai for the *p35* mutant mice, Frank Weissenborn for expert technical assistance and Fiona Christensen for making the chimeras.

References

Caviness VS, Takahashi T, Nowakowski RS 1995 Numbers, time and neocortical neuronogenesis: a general developmental and evolutionary model. Trends Neurosci 18:379–383

D'Arcangelo G 2006 Reelin mouse mutants as models of cortical development disorders. Epilepsy Behav 8:81–90

D'Arcangelo G, Miao GG, Chen SC, Soares HD, Morgan JI, Curran T 1995 A protein related to extracellular matrix proteins deleted in the mouse mutant reeler. Nature 374:719–723

Feng Y, Walsh CA 2001 Protein-protein interactions, cytoskeletal regulation and neuronal migration. Nat Rev Neurosci 2:408–416

Gilmore EC, Ohshima T, Goffinet AM, Kulkarni AB, Herrup K 1998 Cyclin-dependent kinase 5-deficient mice demonstrate novel developmental arrest in cerebral cortex. J Neurosci 18: 6370–6377

Hammond V, Howell B, Godinho L, Tan SS 2001 disabled-1 functions cell autonomously during radial migration and cortical layering of pyramidal neurons. J Neurosci 21: 8798–8808

Hammond V, Tsai LH, Tan SS 2004 Control of cortical neuron migration and layering: cell and non cell-autonomous effects of p35. J Neurosci 24:576–587

Hartfuss E, Forster E, Bock HH et al 2003 Reelin signaling directly affects radial glia morphology and biochemical maturation. Development 130:4597–4609

Howell BW, Hawkes R, Soriano P, Cooper JA 1997 Neuronal position in the developing brain is regulated by mouse disabled-1. Nature 389:733–737

Jossin Y, Ignatova N, Hiesberger T, Herz J, Lambert de Rouvroit C, Goffinet AM 2004 The central fragment of Reelin, generated by proteolytic processing in vivo, is critical to its function during cortical plate development. J Neurosci 24:514–521

Kwon YT, Tsai LH 1998 A novel disruption of cortical development in p35(−/−) mice distinct from reeler. J Comp Neurol 395:510–522

Magdaleno S, Keshvara L, Curran T 2002 Rescue of ataxia and preplate splitting by ectopic expression of Reelin in reeler mice. Neuron 33:573–586

Meyer G, Perez-Garcia CG, Abraham H, Caput D 2002 Expression of p73 and Reelin in the developing human cortex. J Neurosci 22:4973–4986

Mintz B 1970 Gene expression in allophenic mice. In: Padykula HA (ed) Control mechanisms in the expression of cellular phenotypes. Academic Press, New York & London p 15–42

Perez-Garcia CG, Tissir F, Goffinet AM, Meyer G 2004 Reelin receptors in developing laminated brain structures of mouse and human. Eur J Neurosci 20:2827–2832

Rice DS, Curran T 2001 Role of the reelin signaling pathway in central nervous system development. Annu Rev Neurosci 24:1005–1039

Sheldon M, Rice DS, D'Arcangelo G et al 1997 Scrambler and yotari disrupt the disabled gene and produce a reeler-like phenotype in mice. Nature 389:730–733

Takiguchi-Hayashi K, Sekiguchi M, Ashigaki S et al 2004 Generation of reelin-positive marginal zone cells from the caudomedial wall of telencephalic vesicles. J Neurosci 24:2286–2295

Tan S-S, Williams EA, Tam PPL 1993 X-chromosome inactivation occurs at different times in different tissues of the post-implantation mouse embryo. Nat Genet 3:170–174

Tissir F, Goffinet AM 2003 Reelin and brain development. Nat Rev Neurosci 4:496–505

Trommsdorff M, Gotthardt M, Hiesberger T et al 1999 Reeler/Disabled-like disruption of neuronal migration in knockout mice lacking the VLDL receptor and ApoE receptor 2. Cell 97:689–701

Yoshida M, Assimacopoulos S, Jones KR, Grove EA 2006 Massive loss of Cajal-Retzius cells does not disrupt neocortical layer order. Development 133:537–545

DISCUSSION

Macklis: You pointed out that the effect of Reelin is cell autonomous in the chimeric neurons. Is it really an effect in the neurons, or could it be secondary to the fact that when you make your chimeras you are also bringing along mutant glia, radial glia, progenitors and substrates of migration? How strong are the genetic data with specific expression or knockout, showing that this is a long distance repulsive signal? I find it hard to understand how this works when the cortex gets thicker during development. Could there be effects through other cell types than the actively migrating neuron?

Tan: Do you mean that there could be intermediate cell types that perhaps respond to Reelin and thereby pass on the signal to the neurons?

Macklis: Yes, the radial glial substrates, for example.

Tan: Magdalena Gotz (Hartfuss et al 2003) has evidence that radial glia are also able to respond to Reelin signalling. In this light, what you are saying is not too far-fetched. Perhaps the radial glia that generate the neurons are responding to Reelin and in turn influence the neurons that migrate on top of the glia.

Goffinet: When you rescue the Reelin mutant embryos with normal cells, is the rescue proportional to the number of normal Cajal-Retzius cells you get? Have you tried to look at Cajal-Retzius cells and quantify the numbers of Reelin mutant and normal ones?

Tan: We have tried these experiments by transplanting a small number of cortical hem cells where the Cajal-Retzius cells originate to see if these small number of cells can produce a properly layered cortex. We haven't been successful yet, but it would be nice to be able to titrate the critical dose of Reelin needed for full rescue.

Goffinet: This should happen in your chimeras.

Tan: Well not exactly because we analyse our animals at adult stages when the Cajal-Retzius cells have disappeared. We infer that in chimeras where there is clearly a large proportion of wild-type cells there is rescue because of large numbers of Cajal-Retzius cells present. Conversely where the contribution of wild-type cells is miniscule, we see a mutant phenotype. We conclude that a secondary cortex is produced in Reelin chimeras where there is just enough Reelin present to initiate proper layering, but not enough to effect total rescue of the resident mutant cortex.

Molnár: How certain are you about the normal splitting of the marginal zone and the subplate in the p35 knockout cortex?

Tan: We have used calretinin as a marker, and this is our way of saying that they have split properly. Normally, p35 mutants do split. In the adult the subplate in the p35 mutant is localized all the way down underneath layer VI neurons (Kwon et al 1998). But Cdk5 is the interesting one: even though they split, the subplate remains marooned up between layer V and VI (Gilmore et al 1998). The two genes are in the same pathway but seem to have different effects on the subplate.

Molnár: In John Parnavelas's group, Sonia Rakić has looked at this quantitatively, birth dating and then looking at the split. There is a quantitative difference. The proportion of split is altered. This has implications for the thalamocortical path finding and fibre pattern which we studied in collaboration with them (Rakić et al 2006).

Tan: Where is the split localized?

Molnár: The proportions between the marginal zone and subplate are different in the p35 knockout. More early-generated neurons are located in MZ at the expense of the subplate. It is not as bad as in *reeler* (Molnár et al 1998) or Shaking Rat Kawasaki (Higashi et al 2005) where most subplate neurons end up in superplate, but there is a significant quantitative difference.

Parnavelas: Calretinin is not a good marker for studying splitting. The birthdating is much more reliable.

Tan: We have not looked carefully at splitting in the Reelin chimeras.

Fishell: It seems that your model of lack of repulsion might be easier to test with a grafting experiment rather than a chimera, where you have a particular cohort of mutant cells and you are encountering a completely wild-type environment, rather than the fine 'salt and pepper' mix you get in a chimera.

Tan: Ideally I'd like to image the neurons in real time to see whether they actually respond according to the repulsion model.

Price: In any given chimera, does the global chimerism predict the proportions of the mutant and wild-type cells that you see in the cortex? Is there any evidence that the mutant and wild-type cells are contributing differentially because of their expression or lack of it?

Tan: No, global chimerism in the animal is not a solid indicator of brain chimerism. This is because there appears to be no relationship between the number of wild-type cells present in the neural tube, compared to the rest of the body.

Rakic: In your model, why do you have to invoke repulsion? Some cells have the ability to come on top of previously generated cells. This is possibly the only rule they have to obey. If they lose that ability, they don't reach their proper destination and are randomly distributed among previously generated neurons.

Tan: That is precisely what I am saying. In the repulsion model, neurons normally have no trouble in migrating towards the pia, but once they come close to the MZ they activate the Reelin signal, and become repelled away from the MZ. Successive waves of repulsion lead to compaction of later-arriving neurons on top of earlier-arriving ones. In the absence of Reelin, earlier neurons infiltrate the MZ because they are not repelled. Later-arriving neurons become log-jammed underneath, leading to inversion of layers. What is more, the lack of compaction leads to diffused positioning of neurons, which is what you observe in the *reeler* mice. The Curran lab made a Reelin transgenic where Reelin is expressed close to the VZ under the nestin promoter. My interpretation here is that neurons are repelled by the malpositioned Reelin, but because the source of repulsion is not in the MZ, they are repelled haphazardly and you fail to get a properly-layered cortex, which is what they observe.

Stoykova: What might be the role of Reelin in layer V neurons?

Tan: There are no clear data on this. Perhaps they are there during late neurogenesis (Alcantara et al 1998, Meyer et al 2002), when the neurons need a bit of a kick to repel them upwards towards the cortical plate. Reelin might help them move away from their place of birth. Some people in the oligodendrocyte field suggest that Reelin is secreted there to help attract oligodendrocytes coming in. The data just are not strong enough yet.

Keays: With your repulsion experiment you have a slice. Do you then take another slice which you say is secreting Reelin, and put it on top of that slice?

Tan: Yes, we have done that, but most of our experiments are with H293 cells that have been transfected with cDNA and made into an agarose block of cells which we put on top of the slice.

Harrison: You talked about whether this phenotype is reproduced in diseases associated with decreased expression of Reelin, and that schizophrenia is probably the best example. There are two or three issues one has to take into account regarding this. First of all, the magnitude of the reduction and the cells of origin that are expressing less Reelin is probably more subtle and localized in the disease than it is in this model. Secondly, the timing of the Reelin reduction in schizophrenia isn't clear. Third, the changes in Reelin are but one of a panoply of expression changes, including effects on related genes, such as the semaphorins. The net effect of expressing less Reelin in schizophrenia should be taken in this broader context.

Tan: I agree whole heartedly, but clearly levels of Reelin are important considerations. Consistently, schizophrenic brains have lower levels of Reelin due to promoter methylation (Abdolmaleky et al 2005). I know the data on that point aren't very clear and we don't known whether or not there is malpositioning of neurons in humans with schizophrenia.

Fishell: Experiments by Tom Curran showing that the Reeler phenotype can be rescued in mutant mice carrying a transgene where Reelin expression is directed from a Nestin promoter argue that source of Reelin doesn't seem all that important (Magdaleno et al 2002).

Tan: The question is not how much Reelin is there but that it should be present at least at some threshold level in a local source, preferably the MZ. In mice there is an excess of Reelin in the MZ compared with primates, which is why if you take away most of that Reelin, but keep the position of Reelin intact, then layering is still possible. That is how I would interpret the results of the ablation experiments that remove most of the Cajal-Retzius cells (Bielle et al 2005, Yoshida et al 2006).

Goffinet: I don't think it's clear yet. The idea is that you have to anchor it in the marginal zone, then cleave it in order to make it available.

Macklis: I'm confused by this. I would think that in the chimeras, if you could have a much lower source in Cajal-Retzius neurons, then all the normal ones that get up there would be enough. I would have thought that there needs to be some other cellular mechanism.

Tan: That is what we see in the weak chimeras, where despite a lower number of Cajal-Retzius cells (and by implication, Reelin), we still get a layering effect in the secondary cortex. But as you say, this does not appear to be enough because the rest of the mutant cortex still remains inverted. Clearly another mechanism needs to be invoked.

Macklis: How late is Reelin expressed? Is it expressed in adults?

Tan: Yes, but not at the same levels.

Macklis: Paul Harrison, you made a comment on the levels of Reelin expressed in schizophrenic brains.

Harrison: Yes, Reelin goes on being expressed throughout life in identifiable cell populations in humans, albeit at somewhat lower levels.

Rubenstein: Since you can remove the Cajal-Retzius cells and get normal lamination, then perhaps in the case of Reelin mutants, the Cajal-Retzius cells, which I believe are still present, may be producing another molecule that is causing the lamination phenotype. Reelin could be an inhibitor of this other molecule, and allows the lamination phenomena to occur normally. If that is the case it might be worth thinking about the other molecules made by the Cajal-Retzius cells that might act dominantly to invert the cortex.

Tan: That is a very interesting idea and worth keeping in mind. Of course Reelin is not the only molecule whose perturbation results in layer inversion. Mutant mice for Brn1 and Brn2 also exhibit cortical layer inversion (McEvilly et al 2002) and it would be instructive to look for a possible inhibitor of Reelin in the MZ.

Walsh: We have looked at the role of Dab in migrating neurons. Last year Eric Olson (Olson et al 2006) published a paper showing that cell autonomous suppression of Dab with RNAi blocks branching. As the cells hit the marginal zone, migrating neurons have this branch point. We proposed that this branch point might be the point at which the nucleus stops. Reelin stimulates branching, and therefore the branch might be what causes the cell to stop. This might explain why Reelin's effects are indirect. The other thing we noticed is that although the VLDL mRNA is expressed in the migrating neuron, the receptor itself localizes to the MZ, so the cells that express it seem to localize receptor to the MZ. This may explain why even when Reelin is expressed all over the place, it may still only function in the MZ. They are now expressing Reelin in the radial glial cells, but the Reelin may float around, and would only actually function in the MZ.

Tan: You have an interesting point there but you still need to explain how the lack of branching in the Reeler physically causes layer inversion.

Walsh: Would this kind of branching model, where Reelin stimulates branching, be consistent with your overlaying experiments? Do you see more branching? Is it consistent as an interpretation of your chimeras?

Tan: We'll need to look into this, i.e. to see if our time-lapse experiments reveal more branching of cells following exposure to Reelin.

Rakic: I like that idea very much. If Reelin acts through Notch, Notch has an effect on branching patterns.

Walsh: The branching also provides a mechanism for why the cells could then release from the radial glial cells: when they branch, they have to let go of the radial glia. This allows the next cohort of cells to migrate past the radial glia cells, if this gives a mechanism for release.

Tan: Are you suggesting that in the Reelin or Dab mutants, you get failure of branching?

Walsh: As the cells are migrating through the cortical plate they are in a simple bipolar state. Then, just as it hits the MZ, the leading edge branches. The same happens with radial glia cells: their process in the MZ branches. This is consistent with the idea that both cell types branch in the MZ and that Reelin has a branching activity. As the cells branch, the nucleus comes up and hits a branch point. Where is it going to go? It can't migrate any further. The branch is what causes the cell to stop.

O'Leary: I wonder how this mechanism relates to the fact that a certain population of interneurons that migrate into cortex express Reelin. They migrate in and are found relatively early during development, during formation of the cortical plate and during radial migration of later generated neurons, primarily in anterior cortex. How would this source of Reelin impact on your model? Migrating cells are going to sense or come into contact with the Reelin source before they get to the MZ and perhaps before the leading process reaches the MZ.

Do you see anterior–posterior defects or differences in your chimeras that might be explained on the basis of this extra Reelin source? At these developmental stages, this source of Reelin expression is higher in anterior cortex than in posterior cortex.

Tan: My understanding is that if these early interneurons make Reelin, they would be concentrated in the MZ where they would settle first, as demonstrated by Pasko's lab (Ang et al 2003). However, your point about AP analysis is interesting: and we have not looked at this in our chimeras.

O'Leary: If the spatial localization of Reelin is not critical, as suggested, then the interneuronal source could replace the Cajal-Retzius neuron source. In addition, I suspect that this alternate source of Reelin would have a major impact on the branching model presented by Dr Walsh.

Goffinet: The interneurons start making Reelin a bit later. Do you find Reelin in interneurons prior to E17?

O'Leary: I am not certain when Reelin from this source can first be detected in mouse. However, in newborn mice a fairly dense band of Reelin-expressing interneurons is already present within the cortical plate across most of the anterior–posterior cortical axis. Based on this observation, I would assume that Reelin-expressing interneurons enter the cortex several days prior to birth, possibly during the migration of supragranular layer neurons.

Parnavelas: We have localized Reelin in interneurons arising in the ganglionic eminence at E15 in the rat. There is a substantial number of such cells in the MZ.

Goffinet: We only see it later.

Murakami: We have been studying the nuclear formation using the pontine mutant cells in hindbrain. By *in situ* hybridization experiments these cells express Reelin. There are no Cajal-Retzius cells in hindbrain. There is clear birth date-dependent layer-like structure in the pontine nucleus. Is the Reelin function different in different parts of the brain?

Tan: Perhaps here you are seeing an autocrine effect, with the layering dependent on the Reelin secreted by the neurons themselves. This is not incompatible with what I am saying. As long as there a bit of Reelin somewhere, it appears to provide an impetus for cortical layering. This still doesn't explain the Tom Curran experiment, where Reelin in the wrong position appears insufficient to correct the layer inversion.

Walsh: The Curran experiments never visualized where the Reelin protein was. It was expressed in a transgenic and they were able to show expression of mRNA, but were unsuccessful in showing Reelin protein by immunohistochemistry.

Tan: I would agree with you. Indeed they do get partial rescue in the subplate.

Goffinet: If you take a reeler embryonic brain slice *in vitro* and add Reelin in the medium, it cures it. It doesn't matter where Reelin comes from. On the other hand, I think it must be important in the MZ and there is perhaps a mechanism to concentrate Reelin in the MZ, such as a binding partner in the matrix.

Chelly: How do you explain the formation of the two cortices?

Tan: In the case of the Dab mutant it is comparatively easy. The cells respond because they are cell autonomous, they respond to Reelin signalling and then form a separate community of cortical neurons in proper layers. In the Reelin chimeras, it is hard to understand. There appears to be a threshold dosage to kick off the layering process. For both Dab1 and Reelin chimeras, it would appear that once a few layer VI neurons begin to settle in the correct positions, there is a 'domino' effect that results in sequential layering until you get layers II to VI in the secondary cortex. In other words, you never get a layered cortex with only lower layers present and upper layers missing, or vice versa. Of course it remains unclear whether the later stages of this process is also dependent upon Reelin signalling; for that, one would require a means of studying deep and superficial layering separately.

References

Abdolmaleky HM, Cheng KH, Russo A et al 2005 Hypermethylation of the reelin (RELN) promoter in the brain of schizophrenic patients: a preliminary report. Am J Med Genet B Neuropsychiatr Genet 134:60–66

Alcantara S, Ruiz M, D'Arcangelo G et al 1998 Regional and cellular patterns of reelin mRNA expression in the forebrain of the developing and adult mouse. J Neurosci 18:7779–7799

Ang ES Jr, Haydar TF, Gluncic V, Rakic P 2003 Four-dimensional migratory coordinates of GABAergic interneurons in the developing mouse cortex. J Neurosci 23:5805–5815

Bielle F, Griveau A, Narboux-Neme N et al 2005 Multiple origins of Cajal-Retzius cells at the borders of the developing pallium. Nat Neurosci 8:1002–1012

Gilmore EC, Ohshima T, Goffinet AM, Kulkarni AB, Herrup K 1998 Cyclin-dependent kinase 5-deficient mice demonstrate novel developmental arrest in cerebral cortex. J Neurosci 18:6370–6377

Hartfuss E, Forster E, Bock HH et al 2003 Reelin signaling directly affects radial glia morphology and biochemical maturation. Development 130:4597–4609

Higashi S, Hioki K, Kurotani T, Molnár Z 2005 Functional thalamocortical synapse reorganization from subplate to layer IV during postnatal development in the reeler-like mutant rat (Shaking rat Kawasaki): an optical recording study. J Neurosci 25:1395–1406

Kwon YT, Tsai LH 1998 A novel disruption of cortical development in p35(−/−) mice distinct from reeler. J Comp Neurol 395:510–522

Magdaleno S, Keshvara L, Curran T 2002 Rescue of ataxia and preplate splitting by ectopic expression of Reelin in reeler mice. Neuron 33:573–586

McEvilly RJ, de Diaz MO, Schonemann MD, Hooshmand F, Rosenfeld MG 2002 Transcriptional regulation of cortical neuron migration by POU domain factors. Science 295:1528–1532

Meyer G, Perez-Garcia CG, Abraham H, Caput D 2002 Expression of p73 and Reelin in the developing human cortex. J Neurosci 22:4973–4986

Molnár Z, Adams R, Blakemore C 1998 Mechanisms underlying the establishment of topographically ordered early thalamo-cortical connections in the rat. J Neurosci 18:5723–5745

Olson EC, Kim S, Walsh CA 2006 Impaired neuronal positioning and dendritogenesis in the neocortex after cell-autonomous Dab1 suppression. J Neurosci 26:1767–1775

Rakić S, Davis C, Molnár Z, Nikolić M, Parnavelas JG 2006 Role of p35/Cdk5 in preplate splitting. Cereb Cortex 16(Suppl 1):i35–i45

Yoshida M, Assimacopoulos S, Jones KR, Grove EA 2006 Massive loss of Cajal-Retzius cells does not disrupt neocortical layer order. Development 133:537–545

Intracortical multidirectional migration of cortical interneurons

Fujio Murakami, Daisuke Tanaka, Mitsutoshi Yanagida and Emi Yamazaki

Graduate School of Frontier Biosciences, Osaka University, Yamadaoka 1-3, Suita, Osaka 565-0871, Japan

Abstract. It is well documented that most cortical interneurons originate from the basal forebrain and migrate tangentially to the cortex. However, relatively little is known about their migration after their arrival at the cortex. To elucidate the route and mode of intracortical migration of the interneurons, we performed real-time analysis by utilizing glutamate decarboxylase (GAD)67/green fluorescence protein (GFP) knock-in mice and an electroporation-based gene transfer of DsRed into the ganglionic eminence (GE) of a mouse embryo. Cortical interneurons show a diverse mode of migration. In coronal slices, ventrolateral-to-dorsomedial migration predominantly occurs in the lower-intermediate zone. However, a substantial number of interneurons migrate radially either towards the pial or ventricular surface. There are also quiescent neurons. Observations of the marginal zone or the ventricular zone in flat-mounted cortex from the pial or the ventricular surface, respectively, revealed that the interneurons tangentially migrate in all directions. Medial GE-derived interneurons visualized by DsRed electroporation show similar migratory behaviours. Thus, final settlement of cortical interneurons in their destinations may be a result of successive migratory process of different modes within the cortex.

2007 Cortical development: genes and genetic abnormalities. Wiley, Chichester (Novartis Foundation Symposium 288) p 116–129

Postmitotic neurons migrate from the site of their origin to final positions. While some of them migrate for only a short distance, others migrate over long distances. The journey of the latter on some occasions takes several days, and involves intricate pathways. Their settlement in final positions is an outcome of successive migratory processes with distinct modes. Cortical interneurons are typical examples that undergo such long-distance migration. In this review, we describe recent progress on our understanding of their intracortical migration.

Dorsomedially directed tangential migration of cortical interneurons

Neurons in the cerebral cortex are composed of two major populations, excitatory projection neurons and inhibitory (local circuit) interneurons. Of these, excitatory neurons originate from the ventricular zone of the pallium and reach the cortical plate by way of radial migration (Tan et al 1998, Hatanaka & Murakami 2002). It has relatively recently been demonstrated that a major population of cortical interneurons originate from the medial ganglionic eminence in the basal forebrain and tangentially migrate towards the pallium (De Carlos et al 1996, Anderson et al 1997, Tamamaki et al 1997). Analysis of interneurons in fixed brains and *in vitro* preparations indicate that this mode of migration takes place along two major routes: marginal zone (MZ) and the lower intermediate/subventricular zones (Lavdas et al 1999, Del Rio et al 1992).

The glutamic acid decarboxylase (GAD)67/green fluorescent protein (GFP) knock-in mouse is a powerful tool for studying migration of GABAergic interneurons. In this mouse line, *Gfp* is introduced into the locus of *Gad67*, a rate-limiting enzyme for synthesis of γ-amino butyric acid (GABA). Thus, all GABAergic neurons can be visualized by GFP fluorescence in this mouse line (Tamamaki et al 2003). By making use of this transgenic line, we carried out real-time imaging of GABAergic interneurons in the developing cortex. In coronal slices, we observed lateral-to-medial tangential migration of GABAergic interneurons in agreement with previous studies, but mainly in the lower intermediate zone (Tanaka et al 2003). Although GFP-positive cells were also observed in the MZ, the motility of the neurons in this zone was very low. This suggests that lateral-to-medial tangential migration largely occurs in the lower intermediate/subventricular zones. Since many neurons migrated from the lower intermediate zone towards the pial surface (Fig. 1), we speculated that most neurons in the MZ, if not all, are supplied from these deeper zones of the cortex (Tanaka et al 2003).

Cortical interneurons show a diverse mode of migratory behaviour

Cortical interneurons show various modes of migration. Real-time imaging of E13.5 and E15.5 GAD67-GFP mouse coronal slices demonstrated that cortical interneurons migrate in many directions, although most prevalent was ventrolateral-to-mediodorsal tangential migration in the lower intermediate/subventricular zones. Some neurons migrate radially towards the pial surface, while others migrate obliquely to the radial axis (Tanaka et al 2003). There were also neurons migrating towards the ventricular surface from the MZ. Curiously, a small number of neurons migrate tangentially but in ventrolateral directions, namely, opposite to most tangentially migrating neurons (Tanaka et al 2003). Thus, real-time imaging of GABAergic interneurons in GAD67-GFP mice demonstrates that these neurons show a diverse mode and multidirectional migration in the coronal plane.

Visualization of cortical interneurons by other methods supports occurrence of a diverse mode of migration for cortical interneurons. In accordance with our

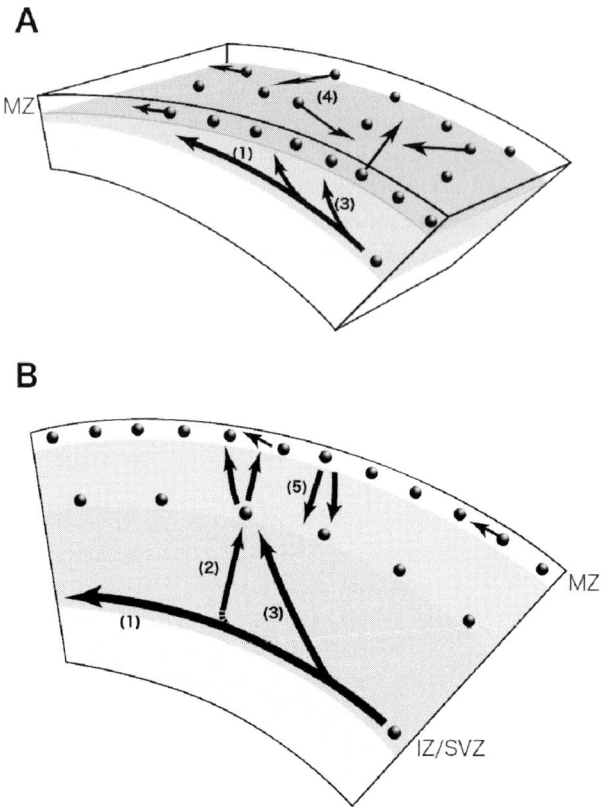

FIG. 1.   A model for intracortical migration of GABAergic neurons. (A) A three-dimensional view of the neocortex in an early stage of development, and (B) a coronal view of the cortex at a later stage. In the coronal plane, dorsomedially directed tangential migration is most prevalent in the lower intermediate zone (IZ)/subventricular zones (1), but some of the neurons deflect towards the marginal zone, migrating radially (2) or with some angle to the radial axis (3). Neurons in the MZ tangentially migrate in all directions (4), while many of them are stationary. MZ neurons descend from the MZ to the cortical plate (5).

observations, Nadarajah et al (2002) found that some interneurons show ventricle-directed migration, by labelling neurons with Oregon Green BAPTA 488 AM. These migrating neurons contact the ventricular surface and, after a pause of approximately 45 min, start to migrate towards the pial surface. Observation of exogenously supplied medial ganglionic eminence (MGE) neurons also showed a diverse mode of migration. Migration of ganglionic eminence-derived neurons were imaged in slice preparations in which a cortical slice was placed next to a slice of ganglionic eminence from a GFP transgenic mouse in which *Gfp* was driven by the β actin promoter (Polleux et al 2002). Migrating neurons in this preparation initially travelled within the marginal and intermediate zones, and then entered the cortical plate from either location. While most cells migrate coherently from the lateral-to-medial aspect of the cortex, there were cells that migrated medio-laterally, down into the cortical plate (CP) and from the CP to the MZ. Curiously, migratory cells displayed a prolonged pause (50–70 minutes) at the CP/MZ inter-face before crossing over into the new zone. Real-time imaging of CellTracker Green-labelled MZ neurons in E15 coronal slices also demonstrated downward migration of cortical interneurons from the MZ (Ang et al 2003).

Altogether, real-time imaging of GABAergic interneurons demonstrates that these neurons display a diverse mode of migration within the cortex, although ventrolateral-to-mediodorsal migration in the lower intermediate/subventricular zones form the predominant stream of migration (Fig. 1).

Cortical interneurons migrate in all directions in the marginal zone

As described above, cortical interneurons in the MZ appear to be quiescent in real-time imaging of coronal sections; the same was true in horizontal sections. This, however, does not necessarily mean that these neurons lack motility *in vivo*, because sectioning of the brain tissue might interfere with their motility: the neurons may be unable to move in the direction perpendicular to the plane of section; it would also be possible that surface of the cortex is specifically damaged during the course of preparing slices. To examine whether the apparent quiescence of MZ interneurons in slices was caused by brain sectioning, we carried out real-time imaging without sectioning the brain. This was accomplished by observing the mouse embryonic cortex from the dorsal surface. The cortex was removed, unfolded and flat mounted. We then observed the neurons in the surface, possibly of the MZ (Tanaka et al 2003, 2006). We found robust migration of neurons, although there were also quiescent neurons dispersed among them. Unexpectedly, these neurons do not show biased migration towards medial directions. Instead, they migrate in all directions including rostrocaudal directions, which is perpen-dicular to the coronal plane (Fig. 1). Multidirectional tangential migration was also observed for CellTracker Green-labelled MZ neurons in E15 (Ang et al 2003). Ang et al labelled the cells in the MZ by application of CellTracker Green after removing the meninges and observed the behaviour of cells in the flattened cortex. The labelled cells migrated in all directions similarly to GAD67-GFP cells in the MZ.

Thus, the apparent lack of interneuron motility in coronal slices is likely an artefact caused by sectioning of the brain and vigorous migration of cortical interneurons takes place not only in the subventricular/intermediate zones but also in the MZ.

Individual MGE-derived interneurons change their direction of migration in the MZ

Due to the high density of GFP-positive neurons in the MZ, it is not possible to track the migration of individual neurons for a long period of time in GAD-GFP mice. Therefore, multidirectional migration in the MZ observed in GAD67-GFP mice might be explained in two ways. (1) Individual neurons change the direction of migration and can migrate in all directions; (2) Individual neurons have their own particular direction of migration, but the direction is different among different neurons. To determine which idea is correct, we labelled ganglionic eminence-derived neurons by electroporation-based gene transfer (Tanaka et al 2006). DsRed plasmid was introduced into the lateral ventricle of E12 mouse embryos and electric pulses were then applied. Three and a half days after the electroporation, labelled neurons were observed in the MZ and the morphology and migrating behaviour of labelled neurons could be easily analysed. Real-time imaging of flatmount cortical preparations revealed that the labelled neurons in the MZ occasionally changed their direction of migration during several hours of imaging. Similar multidirectional migration was observed in animals injected with DsRed at E15 (Yanagida, Tanaka & Murakami, unpublished work, 2007). Thus, individual cortical neurons originating from ganglionic eminences tangentially migrate in many directions within the MZ.

Long-distance tangential migration of cortical interneurons in the MZ

Real-time imaging allows us to observe the behaviour of ganglionic eminence-derived neurons for up to a period of a few days. However, since the field of observation is confined to a small area, we cannot determine whether these neurons just drift locally or migrate over long distances. To obtain an answer to this question, we employed a different method, namely, labelling of MZ cells by implanting a lipophilic dye, 1,1′-dioctadecyl-3,3,3′,3′-tetramethylindodicarbocyanine (DiD) into the cortical surface. DiD was injected into the cortex of E15.5 embryos and they were allowed to survive for a few days after injection. Then the embryos were fixed and observed from the dorsal surface (Tanaka et al 2003). The result of this experiment indicated that DiD-labelled neurons were dispersed far from the injection site: some of the labelled cells were located as far as 2–3 mm away. These findings indicate that cortical interneuron migrate over long distances in the MZ.

Cortical interneurons may stay in the MZ for a prolonged period

Occurrence of long-distance tangential migration in the MZ indicates that cortical interneurons may stay within the MZ for a long period. If we assume that an interneuron migrates at a speed of 20 μm/h (see Tanaka et al 2003) and if it migrates in a straight line within the MZ, this would mean that it takes 100 h to travel over a distance of 2 mm. Do interneurons really stay in the MZ for such a long period? DiD labelling, unfortunately, does not allow us to determine the pathway of their migration and, therefore, does not directly answer this question. However, our preliminary long-time imaging of migrating cortical interneurons did indeed demonstrate that these neurons stay in the MZ for at least one day. Cortical interneurons, therefore, appear to spend a long period of time in the MZ during their journey, although the significance and mechanism for this phenomenon remain unknown.

Cortical interneurons actively migrate downwards toward the cortical plate

Cortical interneurons settle in an inside-out manner like excitatory neurons (Fairén et al 1986, Peduzzi 1988). Inside-out relations between excitatory neuron birthdate and laminar fate, in which early-born neurons settle in deep layers and late-born neurons settle in superficial layers, is thought to result from radial migration of later-born neurons past earlier-born neurons. Occurrence of radial migration for inhibitory interneurons, towards the pial surface raises the possibility that the inside-out settlement of these neurons takes place in a similar fashion to excitatory neurons. However, the presence of numerous interneurons in the MZ in early stages of development raises the possibility that cortical interneurons reach their final position by their downward migration via the MZ. In this scenario, later-born neurons may simply settle more superficially to earlier born neurons.

Bromodeoxyuridine (BrdU) birthdating studies indicated that MZ neurons shift downwards to the cortical plate during postnatal development. Hevner et al (2004) labelled embryonic cortical neurons with BrdU and compared layer distribution of cortical interneurons as identified by DLX expression at postnatal day (P) 0.5 and P7.5. While a large number of BrdU and DLX double-positive neurons were located in the MZ at P0.5, they were predominantly distributed in the cortical plate at P7.5. These findings were ascribed to downward radial migration of cortical interneurons. In accord with this idea, in our DiD labelling experiments we found, in the cortical plate, labelled neurons which leave their trailing process in the MZ (Tanaka et al 2006). A limitation of BrdU labelling, however, is that one can not determine the pathway of neuronal migration. It is therefore possible that they take a roundabout route to reach the cortical plate, not a simple radial route. To address whether MZ neurons migrate downwards at a late stage of development, we carried out a time-lapse analysis of GABAergic interneurons at late embryonic stages (E18) and observed their behaviour in the coronal plane. The result of preliminary experiments indicated that many GABAergic interneurons migrate downwards towards the cortical plate (Yamazaki, Tanaka and Murakami, unpublished work, 2007). Taken together, it is likely that cortical interneurons, once reaching the MZ, migrate downwards to the cortical plate during perinatal development.

Although there is no doubt that a subset of cortical interneurons migrate from the MZ to the CP, it is not possible to exclude the possibility that there is also a subset that reaches the CP without passing through the MZ. It would be interesting to examine whether there is a difference in the properties of MZ-passing and non-passing neurons.

Significance of long-distance tangential migration in the MZ

GABAergic interneurons originate from specific locations in the ganglionic eminences, although they are eventually dispersed throughout the entire area of the cerebral cortex. Therefore, some mechanism that allows dispersion is necessary for them to achieve the final distribution. Our findings that multidirectional, long-distance tangential migration of cortical interneurons widely occurs within the cortex suggest that this mode of migration may contribute to the dispersion of cortical interneurons in the cortex, although we can not exclude the possibility that the interneurons from distinct locations within the ganglionic eminence differentially contribute to different cortical areas (Yozu et al 2005).

MZ as a permissive corridor for long-distance tangential migration

Migrating neurons have to move through an environment filled with cellular processes. Therefore, migrating neurons must somehow create a space through which to penetrate. Although the mechanisms of how migrating cells slip through cell-filled milieu are not known, it seems quite efficient for translocation of neurons that vigorous migration occurs in cell-sparse MZ. It is interesting to note that tangential migration of neurons in many instances occurs underneath the pial surface such as the MZ. Granule cell precursors in the cerebellum and precerebellar neurons in the hindbrain are typical examples (Kawauchi et al 2006) in other brain regions. These neurons also migrate tangentially over long distances and it would be advantageous for them to pass underneath the pial surface.

Tangential migration of cortical interneurons in the ventricular zone

If the surface of the cortex provides a preferred milieu, one might expect that robust migration also should occur on the ventricular surface. To explore this possibility, we prepared flat mount preparations of GAD67-GFP mice cortex, placed the pial surface down and performed real-time analysis. We found that interneurons in the ventricular zone also robustly migrate in all directions on the tangential plane, similarly to those in the MZ (Tanaka et al 2006). Thus, dispersion of cortical interneurons also occurs in the ventricular zone.

Problems with data interpretation of real-time imaging

Progress of imaging instruments such as the laser confocal microscope and visualization methods such as introduction of genes coding for fluorescent proteins have enabled us to observe migrating neurons within tissues in real time. This has brought about a significant progress in understanding the migrating behaviour of neurons. It is to be noted, however, that most real-time imaging experiments have been done *in vitro* because of technical reasons. Therefore, contamination by *in vitro* artefacts should be carefully assessed. In many instances, brain slices are used to visualize migrating neurons. In such preparations, cutting procedures of brain slices might cause damage to the tissue. In this context, it would be reasonable to assume that preparations in which the motility of neurons is high are healthy and the behaviour of cells in such preparations is likely to recapitulate that *in vivo*. However, high motility of neurons in *in vitro* preparations does not guarantee that one is observing phenomena occurring *in vivo*. For example, certain factors that suppress neuronal motility *in vivo* might be inactivated. It would also be possible that cells in a slice are moving on its surface instead of migrating within the tissue. Another potential problem with slice preparations stems from tissue isolation: neuronal migration in directions perpendicular to the plane of section is restricted, so supply of neurons from external sources as well as their departure from the slice are impeded.

Although *in situ* analysis of neuronal migration in live animals is ideal to avoid the misinterpretation that derives from *in vitro* artefacts, it is technically very difficult. What needs to be done at this moment is to carefully compare the morphologies of neurons *in vitro* with those in fixed *in vivo* preparations. Morphological features that can only be seen *in vitro* should be considered to reflect some kind of artefact.

Future directions

A notable feature of intracortical migration of cortical interneurons is its disorderliness. Although there may be rough predetermined routes of migration common to all cortical neurons, the pathway of individual neurons might be different among different neurons. The diversity of migratory behaviour leads us to speculate that there might even be a stochastic process that governs their migration. There are many common mechanisms such as chemoattraction (Yee et al 1999) and chemorepulsion (Wu et al 1999) between axon guidance and cell migration. However, these mechanisms can not explain disordered migration of cortical interneurons.

It will be important to elucidate novel mechanisms that can explain the diverse migratory behaviour of individual neurons.

Another intriguing feature is that cortical interneurons do not migrate straight from their origin to their final destinations but take a roundabout route. This suggests that there are intermediate destinations for these neurons. It will therefore be important to elucidate mechanisms that operate at such points.

References

Anderson SA, Eisenstat DD, Shi L, Rubenstein JLR 1997 Interneuron migration from basal forebrain to neocortex: dependence on Dlx genes. Science 278:474–476

Ang ES Jr, Haydar TF, Gluncic V, Rakic P 2003 Four-dimensional migratory coordinates of GABAergic interneurons in the developing mouse cortex. J Neurosci 23:5805–5815

De Carlos JA, Lopez-Mascaraque L, Valverde F 1996 Dynamics of cell migration from the lateral ganglionic eminence in the rat. J Neurosci 16:6146–6156

Del Rio JA, Soriano E, Ferrer I 1992 Development of GABA-immunoreactivity in the neocortex of the mouse. J Comp Neurol 326:501–526

Hevner RF, Daza RA, Englund C, Kohtz J, Fink A 2004 Postnatal shifts of interneuron position in the neocortex of normal and reeler mice: evidence for inward radial migration. Neuroscience 124:605–618

Fairén A, Cobas A, Fonseca M 1986 Times of generation of glutamic acid decarboxylase immunoreactive neurons in mouse somatosensory cortex. J Comp Neurol 251:67–83

Hatanaka Y, Murakami F 2002 In vitro analysis of the origin, migratory behavior, and maturation of cortical pyramidal cells. J Comp Neurol 454:1–14

Kawauchi D, Taniguchi H, Watanabe H, Saito T, Murakami F 2006 Direct visualization of nucleogenesis by precerebellar neurons: involvement of ventricle-directed, radial fibre-associated migration. Development 133:1113–1123

Lavdas AA, Grigoriou M, Pachnis V, Parnavelas JG 1999 The medial ganglionic eminence gives rise to a population of early neurons in the developing cerebral cortex. J Neurosci 19:7881–7888

Nadarajah B, Alifragis P, Wong RO, Parnavelas JG 2002 Ventricle-directed migration in the developing cerebral cortex. Nat Neurosci 53:218–224

Peduzzi JD 1988 Genesis of GABA-immunoreactive neurons in the ferret visual cortex. J Neurosci 8:920–931

Pla R, Borrell V, Flames N, Marin O 2006 Layer acquisition by cortical GABAergic interneurons is independent of Reelin signaling J. Neurosci 26:6924–6934

Tamamaki N, Fujimori KE, Takauji R 1997 Origin and route of tangentially migrating neurons in the developing neocortical intermediate zone. J Neurosci 17:8313–8323

Tamamaki N, Yanagawa Y, Tomioka R, Miyazaki J, Obata K, Kaneko T 2003 Green fluorescent protein expression and colocalization with calretinin, parvalbumin, and somatostatin in the GAD67-GFP knock-in mouse. J Comp Neurol 467:60–79

Tan SS, Kalloniatis M, Sturm K, Tam PP, Reese BE, Faulkner-Jones B 1998. Separate progenitors for radial and tangential cell dispersion during development of the cerebral neocortex. Neuron 21:295–304

Tanaka D, Nakaya Y, Yanagawa Y, Obata K, Murakami F 2003 Multimodal tangential migration of neocortical GABAergic neurons independent of GPI-anchored proteins. Development 130:5803–5813

Tanaka DH, Maekawa K, Yanagawa Y, Obata K, Murakami F 2006 Multidirectional and multizonal tangential migration of GABAergic interneurons in the developing cerebral cortex. Development 133:2167–2176

Wu W, Wong K, Chen J et al 1999 Directional guidance of neuronal migration in the olfactory system by the protein Slit. Nature 400:331–336

Yee KT, Simon HH, Tessier-Lavigne M, O'Leary DM 1999 Extension of long leading processes and neuronal migration in the mammalian brain directed by the chemoattractant netrin-1. Neuron 24:607–622

Yozu M, Tabata H, Nakajima K 2005 The caudal migratory stream: a novel migratory stream of interneurons derived from the caudal ganglionic eminence in the developing mouse forebrain. J Neurosci 25:7268–7277

DISCUSSION

O'Leary: I imagine Gord Fishell is pleased to see your results, because a few years ago when he was a postdoc, Mary Beth Hatten and Gord published a paper (Fishell et al 1993) showing a tremendous amount of dispersion of cells labelled in the ventricular zone. He took a lot of grief over that finding, and you have now shown that it is real, and probably one of the likely sources of those moving cells described by Gord. John Parnavelas and others have shown, as you do, that cells are in the marginal zone and they then migrate radially into the cortical plate. It has been suggested that they migrate along radial glia guides. An alternate possibility is that they are migrating along the apical dendrites of glutamatergic protection neurons, as your model suggests.

Murakami: There are two possibilities, as you suggested: that they migrate along the radial glia, or they migrate along the dendrites. We are interested in this issue, and examined this by staining the cortex with an antibody against nestin. Some of them look to be closely attached to nestin-positive fibres, but not all of them. The second possibility is hard to investigate, because we don't have a tool for staining all the dendrites of the pyramidal neurons. Even if we could, it would be difficult to see their association, because there are so many dendrites.

Rakic: We all use slice preparations, but we should not forget that they are detached from the rest of the brain: if we think that cells are reading some type of cues from gradients that have a source and sink, we have disturbed this by making the slice preparation. There are perhaps fewer of these molecules, and therefore some cells could be confused and end up in the wrong place. In reality, if we can work *in vivo* without disturbing the gradient we might see more precision than in slice preparations.

Murakami: You are right. One of the problems we find is that these experiments can only be done in *in vitro* preparations. Some of the data we showed were from slice preparations, but others were done in flattened cortex. We have to be careful about *in vitro* artefacts. We always compare our results with fixed preparations. If we find the cell morphology or distribution different from what is seen in fixed preparations, we consider it to be an artefact. I understand the importance of thinking about the real physiological conditions. We are now trying to develop *in vivo* imaging of these neurons.

Fishell: Looking at your results, one of the things that strikes me is that we might be seeing random movement because we are looking at diverse populations. One observation we saw with our transplants suggests that individual populations really do know that they have preferential directions. When we did CG transplants, we originally caught this stream coming backwards from the MGA, which contained a lot of somatostatin cells. If we avoided that stream we could get rid of them. But we never saw parvalbumin cells coming there, suggesting that these cells go straight up, whereas somatostatin cells can migrate caudally. There are now a number of transgenics for parvalbumin, somatostatin and calretinin: it would be lovely to re-explore your studies using GFP lines where you can discern the different types.

Molnár: At the moment it is a bit like looking at pedestrians in Tokyo in rush hour! However, if you put different coloured shirts on different cells you'll have a clearer picture and might discover specific migratory principles for each categories.

Murakami: During the migration, particular neurons change their direction during the course of migration.

Fishell: I think they are doing some shopping, but perhaps some only visit certain stores.

Parnavelas: If they spend such a long time in the marginal zone, as you suggested (up to three days), this zone would be bulging with interneurons, given that they arrive there at a considerable rate.

Murakami: The marginal zone is the thickest at E18. After this it becomes thinner. During postnatal development there is downward movement and the size of the marginal zone is reduced.

Walsh: When you are watching the cells migrate do they appear to repel one another? If so, does the repelling require contact?

Murakami: When we began these experiments, we wondered why they behaved like this. This is what we thought at the beginning. We also observed the behaviour of these cells in dissociated conditions quite carefully, but we didn't see any signs of repellent activity.

Tan: We have analysed neurons in culture and they do repel each other, but not from contact of processes, rather from contact of soma. Only when the somas touch do they repel.

Walsh: Do they divide while they are migrating?

Murakami: We haven't seen this.

Parnavelas: We have given BrdU following DiI injections in the ganglionic eminence. In these experiments, we never observed DiI labelled cells to also stain for

BrdU, suggesting that cortical interneurons do not divide once they leave the ganglionic eminence.

Hevner: There is something I have always wondered: why do interneurons enter the cortical plate from both above and below, as opposed to projection neurons which always come from below? For many types of neurons, including cortical projection neurons and granule cells in the cerebellum, the trailing process becomes the axon. Many interneurons have axons that project into the marginal zone, the Martinotti cells. Could this mechanism be how the Martinotti cells are trailing their axons?

Murakami: It is possible, considering the behaviour of other types of cells, such as the excitatory neurons of the cortex. It is also possible that these cells extend axons irrespective of the direction of the migration. Now we can do long-term imaging experiments, and we can do imaging for later stage events. So, we can see the maturation process of the interneurons. I have seen just one example where a cell starts to extend axons after terminating migration.

Parnavelas: How long are you imaging for?

Murakami: In that case it was about 40 h.

Rakic: What kind of role would cell death have in correcting mistakes made by migrating neurons? For example, when we look at the position of neurons in a mouse sacrificed at P10, you see some the neurons out of proper position. If you then wait another two months you find that they are not there any more. In other words, the cells that have made a mistake and don't get trophic support eventually disappear. Since these are short-term experiments we don't see correction. Is this a possibility?

Murakami: You are asking me to do an experiment involving one month of live imaging!

Rakic: I don't expect you to do the live imaging, but you should consider the possibility that cells that don't migrate in slices may not survive *in vivo*. There could be a precision *in vivo* that is tighter than you see in your experiments.

Murakami: I can't exclude that possibility. We try to interpret our data on the basis of what we have seen, but we should consider such possibilities.

Macklis: Early in your paper you showed that at E12.5 the cells are down at ganglionic eminence, and then a day later at E13.5 the lower intermediate cells are in the marginal zone, which is when you showed us this migration. Has anyone in your lab looked at E13? It seems like some magic happens during this one day period. This could test your hypothesis that there is an initial migration here that then goes up to marginal zone.

Murakami: What we can say from our results is that there are such cells. We can't say that all the cells in the marginal zone originated from the lower part of the brain. There might be cells that have taken a pathway that we don't know of.

Wilkins: I was intrigued by your final statement about the possible function of this movement: that they are moving to get the right excitation. What does this mean in molecular and cellular terms?

Murakami: I don't know.

Stoykova: Are there any data on the para-olfactory neuronal epithelium, or the neuroepithelium of the olfactory bulb as a source of the interneurons for the rostral-most pallium?

Murakami: Not that I know of.

Macklis: I think there may be data from a couple of investigators that would argue against this.

Parnavelas: There is no evidence to suggest that interneurons arise in the neuro-epithelium of the olfactory bulb and migrate to the cortex.

Rubenstein: Alfonso Fairen has data that there are some cells from the pre-plate that arise from a para-olfactory bulb locus that might also contribute some of the GABAergic neurons. This is worth considering (Meyer et al 1998).

Parnavelas: As far as I know, those neurons do not make it to the neocortex.

Rubenstein: When you do your experiments, do you keep the meninges on?

Murakami: Yes.

Rubenstein: Can you comment on the role of SDF1 and CXCR4 in regulating the maintenance of cells in the marginal zone, or their regulation when interneurons dive from the marginal zone into the cortical plate?

Murakami: We have some evidence that SDF1 has an attractive activity towards these interneurons. We think that one of the mechanisms of such long-term stay of the interneurons is mediated by SDF1/CXCR4 signalling.

Parnavelas: SDF1-containing beads placed in the developing neocortex appear to attract strongly ganglionic eminence cells (Nadarajah et al, unpublished observations).

Tan: We have been looking at similar time-lapse images of migrating interneurons too. One thing that is puzzling and unsatisfying is that you can pick and choose what you want to see. The interneurons seem to have all kinds of modes of migration, and move all kinds of distances. In comparison, the projection neuron migration is stereotyped and clear. To pick up on Pasko Rakic's point, perhaps interneurons have more selection later after they have migrated, for example by apoptosis.

Chelly: What proportion of these cells migrate from the surface to the deep layer? If the proportion is minor, the physiological role of this migration is questionable.

Murakami: I don't have direct data to answer your question. Judging from the frequency of these downward movements, I don't think it is a minor population of cells.

Fishell: I agree: it isn't minor. We have done transplants looking at E16: the marginal zone is packed with cells. It winnows down extremely over the next few days, suggesting that what is seen here is a robust phenomenon.

References

Fishell G, Mason CA, Hatten ME 1993 Dispersion of neural progenitors within the germinal zones of the forebrain. Nature 362:636–638

Meyer G, Soria JM, Martinez-Galan JR, Martin-Clemente B, Fairen A 1998 Different origins and developmental histories of transient neurons in the marginal zone of the fetal and neonatal rat cortex. J Comp Neurol 397:493–518

The atypical cadherin Celsr3 regulates the development of the axonal blueprint

Libing Zhou, Fadel Tissir and André M. Goffinet

University of Louvain, Developmental Neurobiology Unit, 73, Avenue Mounier, Bte DENE7382, B1200 Brussels, Belgium

Abstract. Celsr3, the murine orthologue of *Drosophila* Flamingo/Starry night, is a brain-specific, atypical sevenpass cadherin that plays a key role during brain development. *Celsr3* mutant mice die at birth of central hypoventilation. They have major anomalies of major tracts, particularly absence of anterior commissure and of all components of the internal capsule. In addition, the medial lemniscus and several longitudinal bundles in the brainstem and spinal cord are defective. This phenotype is similar to that generated by inactivation of *Frizzled3* (*Fzd3*). As both *Flamingo* and *Frizzled* are two core planar cell polarity (PCP) genes in flies, our results indicate that *Celsr3* and *Fzd3* are part of a genetic network with similarity to the PCP network. We studied by *in situ* hybridization the expression patterns of other murine PCP genes *Dvl1–3*, *Vangl1,2* and *Prickle1,2*. Together with data from the literature, results suggest a mechanism whereby *Celsr1–3*, *Fzd3* and *6*, and *Vangl2* may interact to control neural tube closure and axonal development. In order to study this mechanism further, we generated a conditional *Celsr3* mutant mouse that allows inactivation of *Celsr3* in the forebrain and cerebral cortex, by crossing with mice that express Cre under the *Foxg1* and *Emx1* promoters, respectively.

2007 Cortical development: genes and genetic abnormalities. Wiley, Chichester (Novartis Foundation Symposium 288) p 130–140

Cadherins are a large family of several dozen proteins, most of which act via homophilic calcium-dependent interactions. In addition to classical cadherins and protocadherins that have one transmembrane segment, a subfamily of atypical cadherins are anchored to the membrane by seven hydrophobic segments and bear similarity to G protein-coupled receptors. In all cases, the N-terminus is extracellular and the C-terminus intracellular (Takeichi 2007). In *Drosophila* the atypical cadherin named Flamingo (Fmi) or Starry night (Stan) regulates three key features of development, namely: (i) epithelial planar cell polarity (PCP, also called 'tissue polarity'), a readout of which is the wing epithelium; (ii) formation of axonal bundles; (iii) development of dendrites in some neuronal types (Chae et al 1999, Lu et al 1999, Usui et al 1999).

Mammalian genomes contain three orthologues of *Fmi/Stan* that go by the poetic name of *Celsr1–3* (**C**adherin **EGF LAG S**even-pass G-type **R**eceptor) (Hadjantonakis et al 1997, Formstone et al 2000). Corresponding proteins have 8 or 9 cadherin repeats encoded by a large 5′ exon, followed by a series of motifs (EGF, LAG), seven transmembrane segments and a long C-terminal tail. Although there is high similarity between all three Celsr proteins in the ectodomain and in the transmembrane region, no sequence similarity was found in the cytoplasmic tails, indicating that they may activate different intracellular signalling pathways.

In mice, two ENU-induced mutants, *Spin cycle (Scy)* and *crash (Crsh)*, were shown to be point mutations in the *Celsr1* coding sequence, with substitution of Arg for Lys at codon 110 in *Scy*, and Gly for Asp at codon 1040 in *Crsh*. Rather surprisingly, both mutations are embryonic lethal in homozygotes, with failure of neural tube closure followed by degeneration of brain tissue. Both mutations are haploinsufficient, as heterozygous animals have an anomaly of the organization of stereocilia at the surface of inner and outer ciliated cells in the inner ear (Curtin et al 2003). Both the neural tube closure defect and the disorganization of cilia in the ear are typical of mutations in genes that control PCP. Very similar traits are present, for example, in looptail mice that have mutations in *Ltap* or *Van Gogh-like 2* (*Vangl2*) that codes for a tetraspanin (Kibar et al 2001, Murdoch et al 2001, Montcouquiol et al 2003), in *Scrb1* (mutated in 'circle' tail mutant mice; Murdoch et al 2003), in *Frizzled 3* and *6* (Wang et al 2002, 2006), and *Dishevelled 1* and *2* (Hamblet et al 2002) mutant mice. Thus, it is reasonable to propose that, like its orthologue *Fmi/Stan* in flies, *Celsr1* plays a central role in the control of epithelial PCP in mice, probably in all mammals, and possibly in all vertebrates.

In mice, *in situ* hybridization studies showed that *Celsr3* is heavily and specifically expressed in the developing brain, and that its expression is higher in postmitotic neurons than in precursors in ventricular zones (Tissir et al 2002, Tissir & Goffinet 2006). Furthermore, *Celsr3* expression peaks around birth and is progressively down-regulated to reach very weak levels after maturation, when expression is found only in the olfactory bulb, the dentate gyrus and cerebellar granule cells. Expression in postmigratory neurons and down-regulation in the adult indicate a role in neuronal maturation.

We inactivated the gene by homologous recombination in embryonic stem (ES) cells. As *Celsr3* is a complex gene with 35 exons, we targeted for germline deletion a part of the gene that contains exons 19–27. We postulated that, if any residual protein is made, it should be a truncated protein without transmembrane segments, and we now have evidence that no protein is present in brains from homozygous animals, confirming that this mutation is a complete loss of function (Tissir et al 2005, Price et al 2006). Homozygous *Celsr3*$^{-/-}$

mice die at most a few hours after birth in a cyanotic state. Lungs are histologically normal, but poorly inflated. Phrenic nerve endings are normally present in flat mount preparations of diaphragms, and mRNA corresponding to the four surfactant genes are normally expressed, thus suggesting that animals die of central ventilation failure.

In the brain, the overall architectonics is unremarkable, but there are prominent defects in several major axonal tracts. The most spectacular are the complete absence of anterior commissure (both arms), and internal capsule, including pyramidal tract, so that the cortex of mutant mice does not receive any thalamic input and does not send any efferent fibres to the thalamus, striatum or other subcortical target. The only connections of the mutant cortex are ipsilateral cortico–cortical connections and commissural connections through the corpus callosum. Other tracts are defective. For example, the medial lemniscus is absent and several bundles are abnormal in the spinal cord and brainstem. Detailed neuroanatomical studies have not yet been carried out and are particularly difficult in the newborn brain, when most tracts are still actively growing. In contrast, some axonal bundles, such as the fornix, the retroflexus and the mammillothalamic are normal, showing that the mutation does not simply generate a diffuse axonopathy, but likely hits a specific developmental target.

A main element in the mechanism of action of Cclsr3 is provided by the observation that inactivation of Fzd3, one of the ten mammalian orthologues of *Drosophila* Frizzled, generates an axonal phenotype that is almost identical to that in *Celsr3*$^{-/-}$ mice (Wang et al 2002). This observation suggests strongly that both proteins are two elements of the same molecular mechanism. Contrary to *Celsr3* mutants, homozygous *Fzd3* mutant mice have a looping tail, and some of them die during the second half of gestation with defective neural tube closure. Furthermore, mice with double inactivation of *Fzd3* and *Fzd6* all die prenatally with craniorachischisis (Wang et al 2006), an anomaly closely similar to that in *Celsr1*$^{-/-}$ and *Vangl2*$^{-/-}$. These genetic data indicate clearly that mammals possess at least two paralogue systems related to that which controls PCP in flies. A first set composed of *Celsr1*, *Fzd6* (to a lesser extent *Fzd3*) and *Vangl2* would control epithelial PCP, particularly in the skin, in the neural plate prior to and during its folding into the neural tube, and in the inner ear. A second paralogue genetic pathway, composed of *Celsr3*, *Fzd3* and possibly *Vangl2*, would then control the development of the early axonal blueprint. This schematic view raises the question of the role of *Celsr2*, expression of which is largely superposed to that of *Celsr1* and *Celsr3*, and remains high after maturation (Shima et al 2002, Tissir et al 2002, Tissir & Goffinet 2006). One obvious possibility would be that *Celsr2* is part of a third paralogue network, with a role in the regulation of dendrite development, a hypothesis supported by studies on *in vitro* models (Shima et al 2004). Mutant *Celsr2* mice recently became available. Mutant animals are viable and macroscopically normal, and studies on the role of *Celsr2* on the maturation of dendrites will soon be possible.

The perinatal lethality of *Celsr3* mutant mice prevents a detailed analysis of cortical maturation in the absence of connections, and hampers the understanding of the mechanisms by which *Celsr3* regulates development. We therefore produced a conditional *Celsr3* mutant in which the region deleted in the constitutive mutations is flanked with *loxP* sites (*Celsr3* 'floxed'). Floxed *Celsr3* mutant mice are viable and fertile, showing that the introduction of *loxP* sites did not perturb expression of the gene. We have set out to generate mutants in which *Celsr3* is inactivated in specific regions in the brain by crossing *Celsr3* floxed mice with transgenic mice that express the Cre recombinase regionally. We will first use two Cre mice. *Foxg1-Cre* animals express Cre in the telencephalon (cortex, hippocampus, basal forebrain, and olfactory bulb), with minimal expression in the diencephalon and particularly the dorsal thalamus from which the majority of thalamocortical afferents arise. These crosses should enable us to determine whether *Celsr3* acts in a cell autonomous fashion, in which case thalamocortical axons will run normally to the basal forebrain and cortex, or whether *Celsr3* is required cell non autonomously for their turning through the diencephalons-telencephalon junction (DTJ). The second strain is *Emx1-Cre*, which expresses Cre in the hippocampus, cortex and olfactory bulb, but not in basal forebrain nor in diencephalons. By crossing with floxed *Celsr3* mice, we should be able to find out whether thalamic axons can reach the cortex and whether cortical efferent fibres are able to leave the cortex and navigate towards their subcortical targets. These crosses will also enable us to assess the formation of the anterior commissure after ablating *Celsr3* specifically in the neurons of origin of its constituent axons and not in the intermediate territory. We also plan to use other Cre-expressing mouse lines, such as *Nkx2.1* and *Gsh2-Cre* that express Cre in parts of the basal forebrain and diencephalon (Kessaris et al 2006), and Ror-α-Cre that drive Cre in some thalamic nuclei such as dorsal thalamus and lateral geniculate (O'Leary et al unpublished).

In conclusion, *Celsr3* and *Fzd3* are two elements in a newly identified genetic network that regulates the development of the early axonal blueprint. Mutant mice should enable us to define better the mechanism of action of corresponding proteins, prior to biochemical and cell biological studies.

References

Chae J, Kim MJ, Goo JH et al 1999 The Drosophila tissue polarity gene starry night encodes a member of the protocadherin family. Development 126:5421–5429

Curtin JA, Quint E, Tsipouri V et al 2003 Mutation of Celsr1 disrupts planar polarity of inner ear hair cells and causes severe neural tube defects in the mouse. Curr Biol 13:1129–1133

Formstone CJ, Barclay J, Rees M, Little PF 2000 Chromosomal localization of Celsr2 and Celsr3 in the mouse; Celsr3 is a candidate for the tippy (tip) lethal mutant on chromosome 9. Mamm Genome 11:392–394

Hadjantonakis AK, Sheward WJ, Harmar AJ, de Galan L, Hoovers JM, Little PF 1997 Celsr1, a neural-specific gene encoding an unusual seven-pass transmembrane receptor, maps to mouse chromosome 15 and human chromosome 22qter. Genomics 45:97–104

Hamblet NS, Lijam N, Ruiz-Lozano P et al 2002 Dishevelled 2 is essential for cardiac outflow tract development, somite segmentation and neural tube closure. Development 129:5827–5838

Kessaris N, Fogarty M, Iannarelli P, Grist M, Wegner M, Richardson WD 2006 Competing waves of oligodendrocytes in the forebrain and postnatal elimination of an embryonic lineage. Nat Neurosci 9:173–179

Kibar Z, Vogan KJ, Groulx N, Justice MJ, Underhill DA, Gros P 2001 Ltap, a mammalian homolog of Drosophila Strabismus/Van Gogh, is altered in the mouse neural tube mutant Loop-tail. Nat Genet 28:251–255

Lu B, Usui T, Uemura T, Jan L, Jan YN 1999 Flamingo controls the planar polarity of sensory bristles and asymmetric division of sensory organ precursors in Drosophila. Curr Biol 9:1247–1250

Montcouquiol M, Rachel RA, Lanford PJ, Copeland NG, Jenkins NA, Kelley MW 2003 Identification of Vangl2 and Scrb1 as planar polarity genes in mammals. Nature 423:173–177

Murdoch JN, Doudney K, Paternotte C, Copp AJ, Stanier P 2001 Severe neural tube defects in the loop-tail mouse result from mutation of Lpp1, a novel gene involved in floor plate specification. Hum Mol Genet 10:2593–2601

Murdoch JN, Henderson DJ, Doudney K et al 2003 Disruption of scribble (Scrb1) causes severe neural tube defects in the circletail mouse. Hum Mol Genet 12:87–98

Price DJ, Kennedy H, Dehay C et al 2006 The development of cortical connections. Eur J Neurosci 23:910–920

Shima Y, Copeland NG, Gilbert DJ et al 2002 Differential expression of the seven-pass trans-membrane cadherin genes Celsr1-3 and distribution of the Celsr2 protein during mouse development. Dev Dyn 223:321–332

Shima Y, Kengaku M, Hirano T, Takeichi M, Uemura T 2004 Regulation of dendritic maintenance and growth by a mammalian 7-pass transmembrane cadherin. Dev Cell 7:205–216

Takeichi M 2007 The cadherin superfamily in neuronal connections and interactions. Nat Rev Neurosci 8:11–20

Tissir F, Goffinet AM 2006 Expression of planar cell polarity genes during development of the mouse central nervous system. Eur J Neurosci 23:597–607

Tissir F, De-Backer O, Goffinet AM, Lambert de Rouvroit C 2002 Developmental expression profiles of Celsr (Flamingo) genes in the mouse. Mech Dev 112:157–160

Tissir F, Bar I, Jossin Y, Goffinet AM 2005 Protocadherin Celsr3 is crucial in axonal tract development. Nat Neurosci 8:451–457

Usui T, Shima Y, Shimada Y et al 1999 Flamingo, a seven-pass transmembrane cadherin, regulates planar cell polarity under the control of Frizzled. Cell 98:585–595

Wang Y, Thekdi N, Smallwood PM, Macke JP, Nathans J 2002 Frizzled-3 is required for the development of major fiber tracts in the rostral CNS. J Neurosci 22:8563–8573

Wang Y, Guo N, Nathans J 2006 The role of Frizzled3 and Frizzled6 in neural tube closure and in the planar polarity of inner-ear sensory hair cells. J Neurosci 26:2147–2156

DISCUSSION

O'Leary: In mutants that lack corticothalamic, thalamocortical reciprocal connections, for example the *Emx1/Emx2* double mutant, interneurons fail to migrate into the cortex. Their numbers are dramatically diminished, leading to an axon-philic migration hypothesis. What about interneurons in these mice?

Goffinet: They do well. There is a big difference between the mutants you mention and these ones. These are transcription factors and mutations can modify the differentiation programme of the cells. Our protein is more of a terminal effector, and signalling by transcription factors is likely to be normal. It is less interesting in that cell fate isn't changed, but it is cleaner to look at specific phe-notypes such as anterior commissure or internal capsule, because the effect of the protein is likely to be direct, or at least more directly related to the phenotype that when transcriptional modulators are mutated.

O'Leary: *Emx1* and *Emx2* are expressed in cortex but not in the ganglionic eminence, germinal zones that gives rise to the majority of cortical interneurons. Thus, the mechanisms of reduced interneuron number in the cortex of the *Emx1/Emx2* double mutants do not appear to be fate dependent, but instead they appear to be dependent on the axonal pathways being intact.

Goffinet: Yes. Clearly, the interneurons can migrate normally in the absence of axons.

Price: In your mutants made using *Foxg1-Cre*, how much have you characterized the expression of genes encoding molecules such as the neuregulins and others, that make the corridor described by Lopez-Bendito et al (2006) as permissive for thalamocortical axonal growth?

Goffinet: We haven't looked at this yet.

Rubenstein: Did you find any phenotype with *Emx1-Cre* of axons within the cortex?

Goffinet: We didn't look closely within the cortex itself, but apparently this is not affected.

Rubenstein: So there is no phenotype of any kind of cortical axon?

Goffinet: It is difficult to tell, but we didn't see anything. With the *Foxg1*, when I saw the increased marginal zone my first thought was that fibres from olfactory bulb had extended beyond the piriform cortex, but this is not the case. The rest has to be corticocortical. But then why is there a huge marginal zone?

Rubenstein: What is the basis for the different effects of *Foxg1-Cre* versus *Emx1-Cre?* on the cortex.

Goffinet: *Emx1-Cre* receives the thalamic afferent. Everything looks normal in the cortex.

Rakic: Most of the forebrains are smaller in your mutants compared to controls. Could this explain the larger thickness of the marginal zone? If most of the marginal zone is made by the input from the lower part of the neuraxis, there may be normal amount of afferents that are distributed in the smaller forebrain, so they contribute to the larger relative volume of the marginal zone.

Goffinet: But there is no afferent here. What you say should apply to cortico–cortical fibres, which is a possibility.

Molnár: I might have an idea about what happen in the marginal zone. You mentioned that the cell numbers weren't increased. So what else is in the marginal zone? A major component is the apical tufts of different pyramidal neurons. There are at least three classes of neurons with transient apical tufts. The spiny stellates in layer IV (Peinado & Katz 1990), and layer V neurons both with callosal projections (Koster & O'Leary 1992, Kasper et al 1994), and also some of the subplate neurons have a transient apical tuft in the marginal zone which they retract during the first postnatal week (Hoerder et al 2006). These somatodendritic modifications might depend on afferent input into the cortex and perhaps the normal retraction of these does not occur in the isolated cortex.

Walsh: I share your concern with the *Bf1-Cre*. I think we are using the same mouse. We did an extensive analysis of a *FilaminA* conditional knock-out using this and ended up throwing out the best part of a year's work because we found that the *Bf1-Cre* itself had a microcephaly phenotype. I have now commiserated with other people who have had the same experience.

Goffinet: We use the heterozygote as a control. The heterozygote could be abnormal, so we control for that. The promoter is leaky: there is some blue in thalamus as well.

Walsh: We thought that *FilaminA* was involved in control of brain size, and this is what we were seeing with *Bf1-Cre*. But then we realized that it is the *Bf1-Cre* causing this.

Goffinet: What we saw fits in well with what is described by Hebert & McConnell (2000).

Hevner: Getting back to the handshake hypothesis, the *Emx1-Cre* result would suggest that *Flamingo* or *Celsr3* is not necessary on the subplate neurons. Have you carefully tested whether the subplate neurons, because they are born so early, might escape the Cre inactivation of *Celsr3*?

Goffinet: It's possible. They probably escape Cre inactivation, because *Emx1* inactivation starts medially at E12.

Stoykova: I would consider it possible that the formation of the subplate might also be regionally affected in the *Flamingo* knockout mice. According to Gorski et al (2002) and Li et al (2003), the *Emx1-Cre* line initiates *in vivo* recombination at stage E9.5 at the medial telencephalic wall, progressively including progenitors of the dorsal and lateral pallium at later stages and reaching full recombination around E12.5, but still affecting only early progenitors of the ventral pallium.

Walsh: You are quite right about the mediolateral gradient with *Emx1*. We have found that for the *Pals1* mutant, for example, there is a reserved part of cortex out laterally that doesn't really show the most severe phenotype. Did you have a similar effect?

Goffinet: It is expressed later in the lateral bud. But it deletes maybe one day later.

Price: Presumably the effect of the mutation will be different if there is another mutation in the same animal. If one *Foxg1* allele is mutant (*Foxg1-Cre*) and there is also a mutation in another gene, this may complicate interpretation. If the $Foxg1^{+/-}$ heterozygote has a phenotype, it is not necessarily simply additive. It may change qualitatively what happens.

Richards: Since the Floxed mice give different phenotypes depending on whether they are crossed to *Foxg1-Cre* or *Emx1-Cre*, is this due to the timing of the gene recombination or are you saying that the *Foxg1-Cre* mice are on a mixed background that influences the phenotype?

Goffinet: The *Foxg1-Cre* come from colleagues and I don't know their background. The *Celsr3* are on a mixed background because they carry some CD1 genes. We have now backcrossed for 10 generations, and the part of the phenotype I described here is consistent. We have variation in other bits, such as corpus callosum, which I disregard.

Richards: The *Fizzled 3* mice also have agenesis of the corpus callosum (Wang et al 2002).

Goffinet: There is some defect, but it is not agenesis. In some there is a small corpus callosum and in others there is none. This is a part of the phenotype that differs a bit between *Fizzled 3* and *Celsr3*. I think there is an overlap with *Fizzled 3* and *6*. The homozygote *Fizzled 3* mice have looping tails which we never see in *Celsr3*, and the double knockout of *Fizzled 3* and *6* has the same phenotype as the *Ceslr1*. There is probably redundancy between the *Ceslr3* and the two *Frizzled 3* and *6*.

Molnár: You mentioned that once the thalamic cortical projections get through the region of the internal capsule and pallial/subpallial boundary and extend to the cortex, they can sort themselves out. I agree with this suggestion based on our own experiments. We also looked at a couple of knockouts in which thalamocortical projections scramble up a bit on their way towards the cortex, such as the *L1* or *L1/Tag1* double and *Sema6A* knockouts. Although the fasciculation pattern and trajectories of some or all of thalamocortical fibres are changed in these mutants the overall laminar and areal targeting patterns are indistinguishable and they all have normal cytoarchitecture and periphery-related patterning of thalamocortical projections in the barrel field. It suggests that thalamocortical axons can sort themselves out when they accumulate within the white matter.

Chelly: To go back to the polarity, did you study the subcellular localization of these proteins?

Goffinet: Not yet. We have antibodies, but they don't work unless we overexpress. We see it in the plasma membrane. In neurons it seems to be on all processes, but this is with overexpression.

Fishell: Work by Dennis and John has shown that in animals that die at birth the arealization looks pretty good. Have you attempted to look at arealization in your P20 animals?

Goffinet: Not yet.

Fishell: You mentioned barrel fields.

Goffinet: In *Foxg1* mutants the barrel fields are completely lost. Most are markers that I know are dependent on thalamic innervation.

O'Leary: Prenatally, all the markers we have looked at exhibit a normal differential expression pattern in the absence of thalamocortical input, none of them show a difference. Another group has published that there is a slight difference in a few of the genes.

Goffinet: I agree that we need to study distribution of markers to try and define cortical fields better.

Fishell: Changes in the barrel field should be fairly obvious just with cytochrome oxidase staining.

Goffinet: In *Emx1* mutants we have a perfect barrel field. For *Foxg1* there are no thalamic inputs.

Fishell: Dennis O'Leary was arguing that thalamic inputs aren't needed.

O'Leary: Prenatally, thalamic input is not required for normal differential gene expression; postnatally, the few genes that we have examined develop normal patterns postnatally in the absence of thalamic input. However, in mice that lack thalamic input to S1 postnatally, the barrel field doesn't develop barrels, and markers that take on a barrel-like pattern fail to develop their appropriate expression pattern.

Molnár: To come back to this, it is important to distinguish that what is seen with Nissl stain and what is seen with thalamic markers is totally different. Nissl stain reveals the cell patterning and the cortical cytoarchitecture, while cytochrome oxidase, serotonin transporter immunohistochemistry and various fibre stains reveal the periphery related fibre clustering. Your point is a good one: we need to look at both. It is not enough just to look at the cytochrome oxidase stain, because thalamic projections can assume the periphery related pattern without the cytoarchitectonic differentiation that is dependent on receptor-mediated signal transduction (Hannan et al 2001).

Goffinet: Do you think we could see barrels in *Foxg1* mutants with Nissl staining?

Molnár: No way.

Goffinet: I used cytochrome oxidase because for me this is the best marker.

Molnár: If you do not have normal periphery related patterning of thalamocortical axons revealed by cytochrome oxidase, then there is no chance that they impose cytoarchitectonic differentiation which you would reveal with Nissl staining.

Yamamoto: The arealization might be changed. You showed that the size is decreased. Is this due to the lack of thalamocortical axons?

Goffinet: There are no afferents, and there are no efferents from cortex. The fact that nothing is leaving the cortex can also impact on the differentiation of the efferent a bit.

Yamamoto: Is Celsr itself expressed in the cortex?

Goffinet: Yes, until maturation.

Macklis: I have a simple technical question. When you said you did your back labelling to say that nothing is leaving the cortex, where did you do this?

Goffinet: In striatum and thalamus.

Mallamaci: And, what about brainstem or spinal cord?

Goffinet: We injected DiI in the stain brain stem, but I never saw back labelled neurons in cortex.

Mallamaci: Did you look at the proliferation and neuronal differentiation kinetics in these mutants?

Goffinet: BrdU gave normal results. We didn't look in detail at differentiation rates. We looked at BrdU at E10 and 12, and looked at the laminar differentiation. We did not proceed.

References

Gorski JA, Talley T, Qiu M, Puelles L, Rubenstein JL, Jones KR 2002 Cortical excitatory neurons and glia, but not GABAergic neurons, are produced in the Emx1-expressing lineage. J Neurosci 22:6309–6314

Hannan AJ, Blakemore C, Katsnelson A et al 2001 PLC-beta1, activated via mGluRs, mediates activity-dependent differentiation in cerebral cortex. Nat Neurosci 4:282–288

Hebert JM, McConnell SK 2000 Targeting of cre to the Foxg1 (BF-1) locus mediates loxP recombination in the telencephalon and other developing head structures. Dev Biol 222:296–306

Hoerder A, Paulsen O, Molnar Z 2006 Developmental changes in the dendritic morphology of subplate cells with known projections in the mouse cortex. FENS Abstracts

Kasper EM, Larkman AU, Lübke J, Blakemore C 1994 Pyramidal neurons in layer 5 of the rat visual cortex. III. Differential maturation of axon targeting, dendritic morphology and electrophysiological properties. J Comp Neurol 339:495–518

Koester SE, O'Leary DD 1992 Functional classes of cortical projection neurons develop dendritic distinctions by class-specific sculpting of an early common pattern. J Neurosci 12:1382–1393

Li HS, Wang D, Shen Q et al 2003 Inactivation of Numb and Numb-like in embryonic dorsal forebrain impairs neurogenesis and disrupts cortical morphogenesis. Neuron 40:1105–1118

Lopez-Bendito G, Cautinat A, Sanchez JA et al 2006 Tangential neuronal migration controls axon guidance: a role for neuregulin-1 in thalamocortical axon navigation. Cell 125:127–42

Peinado A, Katz L 1990 Development of cortical spiny stellate cells: retraction of transient apical dendrite. Abstr Soc Neurosci 16:1127

Wang Y, Thekdi N, Smallwood PM, Macke JP, Nathans J 2002 Frizzled-3 is required for the development of major fiber tracts in the rostral CNS. J Neurosci 22:8563–8573

Regulation of laminar and area patterning of mammalian neocortex and behavioural implications

Dennis D. M. O'Leary, Shen-Ju Chou, Tadashi Hamasaki[1], Setsuko Sahara, Akihide Takeuchi[2], Sandrine Thuret[3] and Axel Leingärtner

Molecular Neurobiology Laboratory, The Salk Institute, 10010 North Torrey Pines Road, La Jolla, CA 96037, USA

Abstract. We will focus on describing our recent studies on the laminar and area patterning of the mammalian neocortex. We describe a novel IgCAM, MDGA1, that is a unique laminar and area specific marker, and functional studies showing its influence on radial migration. We also describe time-lapse imaging studies showing that the preplate and its derivative, the subplate, is a cellular protomap of the cortical ventricular zone, and the implications of this finding for mechanisms of arealization and development of area-specific TCA projections. We will summarize studies of each of the four transcription factors, Emx2, Pax6, Couptf1 and Sp8, expressed by cortical progenitors and involved in specifying area patterning. Finally, we will describe studies showing that area size dictates performance at modality-specific behaviours.

2007 Cortical development: genes and genetic abnormalities. Wiley, Chichester (Novartis Foundation Symposium 288) p 141–164

The neocortex is the largest region of the cerebral cortex. In its tangential dimension, the neocortex is organized into subdivisions termed 'areas'. Areas are distinguished from one another by major differences in cytoarchitecture and chemoarchitecture, input and output connections, and patterns of gene expression. In the adult, the transition from one neocortical area to another is typically abrupt with borders defined by area differences in cytoarchitecture and chemoarchitecture,

[1] Current address: Department of Neurosurgery, Kumamoto University School of Medicine, Kumamoto 860-8556, Japan
[2] Current address: Department of Functional Genomics, Medical Research Institute, Tokyo Medical and Dental University, Bunkyo-ku, Tokyo 113-8510, Japan
[3] Current address: Centre for the Cellular Basis of Behaviour & MRC Centre for Neurodegeneration Research The James Black Centre, King's College London Institute of Psychiatry, London SE5 9NU, UK

and in some instances by the distributions of projection neurons, input projections and gene expression patterns. These attributes form a specific combination of properties that is unique for each area and determine the functional specializations that characterize and distinguish areas in the adult.

The neocortex has four 'primary' areas, as well as scores of higher order areas that are modality related to the primary areas. Three of the primary areas are sensory: the primary visual (V1), somatosensory (S1) and auditory (A1). The fourth primary area is the motor area (M1), which controls voluntary movement of body parts. These primary areas are conserved among mammals, as are their general spatial relationships: for example, V1 is positioned caudally, M1 rostrally and S1 is located between them.

Corticogenesis

Areas of the neocortex differentiate within a more or less uniform structure comprised of postmitotic neurons, termed the cortical plate (CP). Cortical layers gradually differentiate from the CP in a deep to superficial pattern. Most neocortical neurons, including all projection and glutamategic neurons, are generated in the ventricular zone (VZ) of the dorsal aspect of the lateral ventricle, or at later stages, a second germinal zone, the subventricular zone (SVZ). The first postmitotic neurons generated in the VZ accumulate on the top of it, forming the preplate (PP) positioned just beneath the pial surface. Cajal-Retzius (C-R) neurons, which are generated in germinal zones external to the neocortex, including the cortical hem and subpallial sources, are also present superficially in the PP. As CP neurons are generated, they migrate radially, accumulate in the PP, and split it into a deep subplate (SP) layer comprised of most of the PP neurons and a superficial MZ comprised largely of C-R neurons. The majority of cortical GABAergic interneurons are generated within the medial and caudal ganglionic eminences. These interneurons migrate into cortex and along tangentially aligned pathways, then eventually they turn perpendicular to their original migratory plane and migrate radially into the CP (Parnavelas 2000, Corbin et al 2001, Marin & Rubenstein 2003, Kriegstein & Noctor 2004).

Identification of novel IgCAM, MDGA1, with unique cortical expression patterns

The laminar and area pattern of the mammalian neocortex are two organizing principles that define its functional architecture. The identification of genes expressed in layer-specific and area-specific patterns is important to understanding mechanisms that regulate the fate decisions that influence cortical patterning. We have identified potential candidate genes by performing screens, including a differential display PCR screen (Liu et al 2000) and a representational difference analysis (RDA; Li et al 2006). However, we fortuitously identified a promising candidate gene, MDGA1, that appears to be involved in laminar and area patterning, from a differential display PCR screen designed to identify genes involved in the development of hindbrain nuclei and their connections (Gesemann et al 2001, Litwack et al 2004).

MDGA1 (MAM domain containing glycosylphosphatidylinositol anchor 1) is a novel glycoprotein anchored to the external surface of the neuronal membrane by a glycosylphosphatidylinositol (GPI) anchor that shares features with Ig-containing cell adhesion molecules (IgCAMs), including multiple Ig domains (six in MDGA1), a Fibronectin III domain, and uniquely for IgCAMs, a MAM domain (Litwack et al 2004). Based on its domain structure and our functional studies (see below) we conclude that MDGA1 is an IgCAM, which as a group influence neural development by regulating cell adhesion, migration, and process growth.

MDGA1 is expressed by layer 2/3 neurons throughout neocortex and transiently by Cajal-Retzius neurons

We show that MDGA1 is expressed in layers 2/3 throughout the late embryonic and postnatal neocortex to at least P21 (and in a unique laminar pattern in S1—see below), but is absent in adult neocortex (Fig. 1) (Litwack et al 2004, Takeuchi & O'Leary 2006, Takeuchi et al 2007). At earlier stages of development, MDGA1 is transiently expressed by Reelin- and Tbr1-positive C-R neurons, and marks C-R neurons recently described by others to originate from multiple sources outside of neocortex and emigrate into it, including the cortical hem, septum and a subpallial

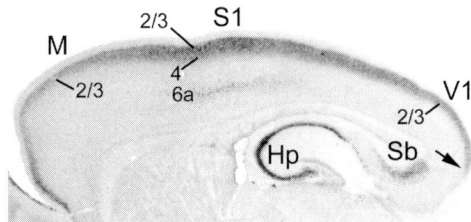

FIG. 1.   Layer and area-specific expression of MDGA1 in P7 mouse brain. MDGA1 *in situ* hybridization on a sagittal section through the primary somatosensory area (S1). MDGA1 is expressed in layer 2/3 throughout the cortex and in layer 4 and 6a limited to S1. Arrow marks posterior margin of neocotrex, marked by end of MDGA1 expression. Other abbreviations: Hp, hippocampal formation; M, motor area; Sb, subiculum; V1, visual area. See Takeuchi et al (2007).

source (Takeuchi et al 2007). At even earlier stages, MDGA1 is expressed by the earliest diencephalic and mesencephalic neurons, which appear to migrate from an MDGA1 positive domain of progenitors in the diencephalon and form a 'preplate' in these other brain vesicles (Takeuchi et al 2007). Thus, MDGA1 can be used as a unique marker for cortical lamination, and as a marker for C-R neurons originating from multiple spatially diverse sources external to the neocortex.

MDGA1 influences radial migration of superficial layer cortical neurons

CAMs play important roles modulating the interactions of migrating neurons with radial glial processes. Until recently, the integrin α3β1 was the only CAM reported to control cortical radial migration (Anton et al 1999, Dulabon et al 2000). We have recently shown, though, that MDGA1 also has a role in controlling cortical radial migration (Takeuchi & O'Leary 2006). MDGA1 is structurally similar to the IgCAMs of the L1 family and to axonin 1, which have roles in cell adhesion, migration, and process outgrowth (Walsh & Doherty 1997). In mice, MDGA1 is expressed by layer 2/3 neurons from the time they are generated and throughout their migration to the superficial aspect of the CP, consistent with a role in controlling the migration and settling of layer 2/3 neurons. To test this hypothesis, we performed loss-of-function studies using RNA interference (RNAi), with multiple siRNA (small inhibitory RNA) vectors targeting different sequences of mouse MDGA1 (Takeuchi & O'Leary 2006).

We find a direct correlation between the effectiveness of the siRNA in suppressing MDGA1 expression and its ability to disrupt migration of superficial layer neurons (Takeuchi & O'Leary 2006). A siRNA with no effect on MDGA1 levels has no effect on the migration of transfected superficial layer neurons. In contrast, each of the four siRNA's found to suppress MDGA1 expression also blocks or retards the migration of transfected superficial layer neurons. This migration defect is rescued by co-transfection of an expression construct containing rat MDGA1 cDNA that differs by only one nucleotide from the effective mouse siRNA sequence, confirming that the effect on migration is specific to diminishing MDGA1 levels in the siRNA transfected cells (Takeuchi & O'Leary 2006). siRNA transfections of deep layer neurons at E12.5, which do not express MDGA1, do not affect their migration (Takeuchi & O'Leary 2006). These findings indicate that MDGA1 acts cell autonomously to control the migration of MDGA1-expressing superficial layer cortical neurons.

We conclude that MDGA1 has an important role in controlling the radial migration of layer 2/3 neurons. MDGA1 is also expressed by a select number of other neuronal populations as they migrate, for example, by tangentially migrating populations, such as C-R neurons as well as basilar pontine neurons, L1 interneurons in the spinal cord, and a subset of neurons in dorsal root ganglia (Litwack et al 2004, Takeuchi et al 2007). We suggest that MDGA1 influences the migration of these and other neuronal populations that express it.

MDGA1 is a unique marker for S1 and a component of barrels

Within S1, MDGA1 is expressed in a unique laminar pattern that makes it useful as an area-specific marker. In addition to its expression in layers 2/3 as it is throughout cortex, within S1, MDGA1 is also expressed in layers 4 and 6a (Takeuchi et al 2007). This unique, area-specific expression of MDGA1 in S1 can be especially appreciated in tangential sections through layer 4 stained for MDGA1 or other markers, including cytochrome oxidase, Nissl, serotonin, RORβ and Couptf1 (Fig. 2). Within the main barrelfield of S1 (the posteromedial barrel subfield), MDGA1 marks a unique subset of layer 4 neurons revealed by comparison of its expression to that of the CAMs, Cadherin 6 and Cadherin 8, and other markers (Takeuchi et al 2007). These findings show that MDGA1 is useful as a selective areal marker specific for the molecular identity of S1, and suggest that it has a role in barrel formation.

The preplate/subplate is a cellular protomap of the cortical ventricular zone

The specification and differentiation of neocortical areas is presumed to be controlled by interplay between intrinsic mechanisms, i.e. genetic mechanisms that operate within the cortex and extrinsic mechanisms, e.g. thalamocortical axon (TCA) input or information relayed by it (Rakic 1988, O'Leary 1989, O'Leary & Nakagawa 2002, Sur & Rubenstein 2005). The differentiation of anatomical features that distinguish cortical areas, including architecture and distributions of output projection neurons, appears to depend upon TCA input (O'Leary & Koester 1993, Chenn et al 1997). Consistent with this role, the TCA projection exhibits area-specificity throughout its development, and the gradual differentiation of areas within CP parallels the elaboration of the TCA projection within it. In addition, a variety of manipulations, including peripheral manipulations and transplantation experiments, have demonstrated that CP exhibits considerable plasticity in the development of area-specific features, and that diverse parts of the CP initially have similar potentials to develop features unique to a specific area (Stanfield &

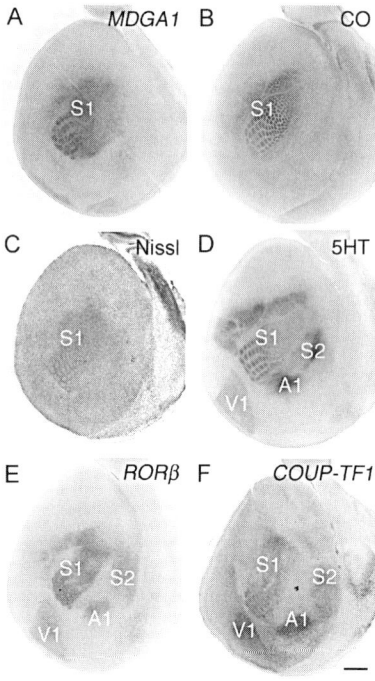

FIG. 2. Area-specific expression of MDGA1 in the layer 4 of S1. Tangential sections were cut through layer 4 of P7 flattened cortices. (A) MDGA1 *in situ* hybridization. (B) CO (cyto-chrome oxidase) staining. (C) Nissl staining. (D) Immunostaining for 5HT (serotonin). (E) *RORβ in situ* hybridization. (F) *Couptf1 in situ* hybridization. Orientation: the rostral is to the top; medial to the left. Scale bar: 1 mm. Figure reprinted by permission of Oxford University Press from Takeuchi et al (2007).

O'Leary 1985, Schlaggar & O'Leary 1991, O'Leary et al 1992). Again, TCA input has been implicated as a major influence controlling this plasticity in the differen-tiation of area-specific features.

Our previous studies indicate that the graded expression of the homeodomain protein EMX2 by progenitors in the VZ confers positional- or area-identity that is inherited by SP neurons and governs their expression of TCA guidance molecules and other properties required for cortical area-specification (Bishop et al 2000, 2002, Hamaskai et al 2004). Other transcription factors expressed by VZ progenitors and control area patterning, such as COUP-TF1, seem to operate in a similar fashion (Armentano et al 2007). However, this mechanism would be most efficient if VZ progenitors and their SP progeny maintain neighbour relationships during the generation of the PP, the precursor of the SP (Fig. 3). However, during development of the CP, studies using time lapse imaging of fluorescently labelled cells and other studies using lineage tracing retroviral tags both reveal considerable movement of cells with the VZ (Fishell et al 1993) and substantial dispersion of neuronal clones within the cortex (Walsh & Cepko 1993).

Therefore, to address the feasibility of our model, we have examined the issue of cell dispersion at earlier time points, during the period when the PP/SP is generated. We used time-lapse video microscopy to follow the movements of VZ progenitors, and the radial movement of their progeny and distribution in the PP, in wholemount or slice cortical explants from embryonic rats at stages when PP neurons are generated (O'Leary & Borngasser 2006). In control experiments, we show using the neuron-specific marker, TuJ1, and BrdU to monitor proliferation and general cell movement, that the PP develops normally under the specialized culture conditions and time-lapse imaging. We show that labelled VZ cells proliferate and retain neighbour relationships within the VZ forming clusters of progenitors (Fig. 3A–A'''). The progeny of labelled progenitors retain their labelling, and can seen to leave the VZ and move superficially along a radial column to form the overlying PP (Fig. 3B,C,C').

Thus during the period when the PP/SP is generated, VZ progenitors and their PP/SP neuronal progeny retain neighbour relationships and the PP is formed in spatial register with its VZ progenitors (Fig. 3D) (O'Leary & Borngasser 2006). This behaviour differs substantially with that reported at later stages of cortical development, when CP neurons are generated, and considerable dispersion is evident amongst both cells within the VZ and their neuronal progeny that form the CP (Fig. 3E) (Walsh & Cepko 1992, 1993). However, our findings show that at earlier stages of cortical neurogenesis, when PP/SP neurons are generated, the VZ is effectively a protomap of the PP/SP at a cellular level (O'Leary & Borngasser 2006). These findings indicate that positional information specified by the graded expression of transcription factors that control area patterning is inherited by the progeny of progenitors in the VZ and spatially mapped onto the PP/SP. This mechanism would be an efficient manner to establish within the SP a framework of genetically specified, positional information that controls the expression of axon guidance molecules that direct the development of area-specific TCA projections (Fig. 3F,G) (O'Leary 1989; also see Rakic 1988).

These findings may also have implications for the establishment and maintenance of graded distributions of transcription factors that are expressed by VZ progenitors and impart positional-identity (or area-identity) to their progeny. Although the expression of Emx2 and similar transcription factors that control area patterning are presumably under tight transcriptional control, other mechanisms might act in the VZ to equilibrate differences amongst neighbouring progenitors in their protein levels, including reciprocal activation/repression loops, as well as the potential release and local uptake of transcription factors amongst neighbouring progenitors, as described *in vitro* for the homeodomain protein engrailed (Prochiantz & Joliot 2003). Thus, our demonstration that progenitors in the VZ retain neighbour relationships would contribute to the stability and maintenance of the gradient of Emx2 and other regulatory proteins required to generate area identity (O'Leary & Borngasser 2006).

Genetic control of area identity by transcription factors expressed by cortical progenitors

The search for transcription factors that control area identity in cortical progenitors has revealed many that are differentially expressed in the VZ in a manner

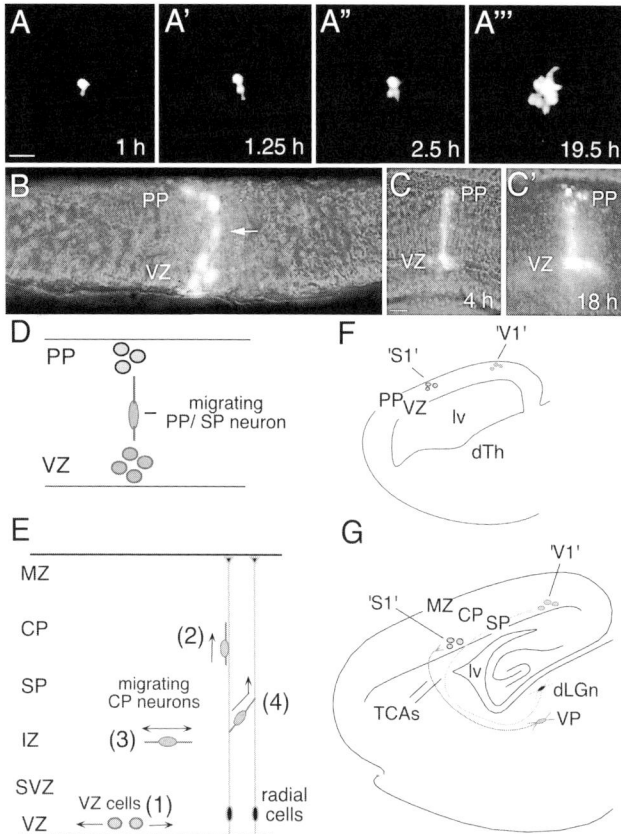

FIG. 3.   The cortical preplate/subplate is a cellular protomap of progenitors in the ventricular zone. (A–A‴) Time-lapse imaging of labelled cells in the ventricular zone (VZ) of an explant of the entire dorsal telencephalon (dTel) shows that they increase in number but remain clustered. Time-lapse images of DiI-labelled cells in the ventricular zone of an E12 dTel explant taken with low-light level rhodamine fluorescence and a SIT camera. The ventricular surface is toward the camera. One or two cells in the ventricular zone were labelled on the explant's ventricular surface at 0 h, soon after the explant was put in culture. Images were taken at 15 min intervals (0.25 h) over 24 h. The number of labelled cells increases over the period shown, presumably through cell division, and they remain clustered. Scale bar: 20 µm. (B) Labelled ventricular zone progenitors and their preplate progeny exhibit radial alignment in dTel explants. After time-lapse imaging of labelled ventricular zone cells from the ventricular surface of dTel explants, explants were sectioned to determine the radial distribution of labelled cells. The image shown is representative, and is a section through the imaged explant presented in Panel A. A small deposit of DiI was made into the ventricular surface at 0 h that focally labelled cells in the apical aspect of the ventricular zone; after 24 h of culturing, the explant was fixed and sectioned. A radial column of labelled cells is seen spanning the cross section of the explant, with accumulations deep in the VZ and superficially in the PP, and some labelled cells are found in between. Little tangential spread of labelled cells is apparent in this and adjacent sections. (C,C′) Progeny of labelled ventricular zone progenitors exhibit radial movement along a columnar path to the preplate in dorsal telencephalic slice explants. Time-lapse images of a dTel 'slice' explant. A single small deposit of DiI was made into the apical aspect of the ventricular zone (VZ) and verified. Images were taken at 15 min intervals (0.25 h). The pair of images shown were collected at 4 h and 18 h after injection. The injection labelled a small, focal group of cells deep in the VZ, which over time expanded in size and number of labelled cells, but remained focally clustered. Labelled cells migrated radially from deep to superficial, forming a columnar path with labelled cells distributed along it. The number of labelled cells beneath the pial surface in the preplate (PP) progressively increased. (D,E) Cell movements in the ventricular zone and the degree of neuronal dispersion differ between early preplate development and later cortical plate development. (D) Our findings show that early in cortical development, during the time that PP neurons are being generated, neighbouring cells within the VZ retain their neighbour relationships, and that their neuronal progeny that form the PP, maintain these neighbour relationships as they migrate superficially along a radial path and accumulate in a cluster beneath the pial surface in the PP directly above their progenitors in the VZ. This process results in a columnar arrangement of spatially related progenitors in the VZ, migrating cells, and neurons in the PP. (E) Previous studies examining cell movements using time-lapse imaging or static analyses of progressive time points in different animals, during the period of CP development, have revealed several forms of cell movements or migration. Within the VZ, time-lapse imaging has shown movements and dispersion of once neighbouring cells labelled within the VZ (Cell 1). Other studies have found that the most prominent form of migration of postmitotic neurons from the VZ and subventricular zone (SVZ) is by radially-directed migration along radial processes (Cell 2). In addition, both tangential movements within the intermediate zone (IZ) (Cell 3), and the 'jumping' of clonally-related marked neurons from one radial cell process to another as they migrate from deep to superficial (Cell 4). (F,G) Our findings show that PP neurons cluster within the PP directly above their progenitors in the VZ (F). Thus, at a cellular level, the VZ is 're-mapped' in the PP through this process. When the majority of PP neurons are 'passively' repositioned into the SP by the expanding cortical plate (CP), the neighbour relationships established in the PP are almost certainly retained (G). Thus, positional information that they inherit from their progenitors is faithfully maintained in a cellular map in the SP, providing appropriate guidance cues for the area-specific targeting of thalamocortical axons (TCAs). See text for further discussion. Abbreviations: dLGn, dorsal lateral geniculate nucleus; dTh, dorsal thalamus; lv, lateral ventricle; MZ, marginal zone; 'S1', SP neurons with positional identities appropriate for primary somatosensory area; 'V1', SP neurons with positional identities appropriate for primary visual area; VP, ventroposterior nucleus. Figures adapted from O'Leary and Borngasser (2006) with permission.

that makes them candidates to regulate area patterning, but thus far, only four have been shown to function in regulating arealization: Emx2, Pax6, Couptf1 and Sp8. Emx2 is a homeodomain transcription factor related to *Drosophila empty spiracles* (*ems*), Pax6 a paired box domain transcription factor, Couptf1 an orphan nuclear receptor, and Sp8 a zinc-finger transcription factor related to *Drosophila buttonhead* (*btd*). Here, we focus on our studies of these genes, although other groups have also contributed to understanding their roles.

These transcription factors are expressed in distinct graded patterns across the cortical VZ. Morphogens or signalling molecules secreted by patterning centres located at the perimeter of the nascent embryonic cortex establish these gradients of transcription factors (O'Leary & Nakagawa 2002, Rash & Grove 2006). The best evidence for this role is for members of the fibroblast growth factor (FGF) family, particularly FGF8, which are expressed by the Commisural Plate (CoP) located at the anterior midline of the nascent cortex. FGF8 expressed in the CoP helps establish the graded expression of Emx2, Couptf1 and Sp8 in the cortical VZ through repression (Garel et al 2003, Shimogori et al 2004, Storm et al 2006).

Emx2

Emx2 is expressed by progenitors in the cortical VZ in a low rostrolateral to high caudomedial gradient; therefore, it is expressed at highest levels by progenitors that generate caudal areas, such as V1, and at its lowest levels by those that generate more rostral areas, such as motor areas (Fig. 4). Initial studies of a role for Emx2 in area patterning were done on *Emx2* constitutive knockout mice (Bishop et al 2000, 2002, Mallamaci et al 2000). Because these mice die at birth, analyses were limited to the use of differentially expressed gene markers and the tracing of area-specific TCA projections. These findings led to the conclusion that Emx2 specifies positional identity in cortical progenitors and their progeny, preferentially imparting caudal identities. However, these conclusions were called into question because of several problems inherent to the constitutive knockouts, including their early lethality, well before cortical areas can be identified by any means, a partial loss of TCA input, and because overall cortical size is reduced by about a third. In addition, one study reported that the role of Emx2 was indirect and solely due to its repression of the morphogen Fgf8 (Fukuchi-Shimogori & Grove 2003).

Subsequent analyses though of *nestin-Emx2* transgenic (*ne-Emx2*) mice and heterozygous *Emx2* knockout (*Emx2$^{+/-}$*) mice have circumvented these problems and

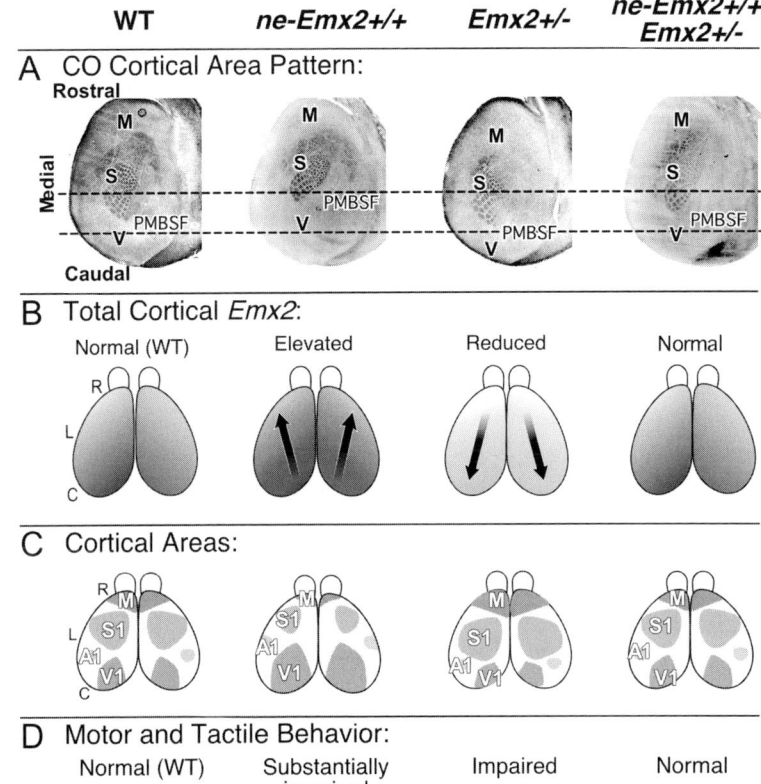

WT	ne-Emx2+/+	Emx2+/-	ne-Emx2+/+; Emx2+/-

A CO Cortical Area Pattern:

Rostral

Medial

M° M M M
S S S S
PMBSF PMBSF PMBSF
V V V V

Caudal

B Total Cortical *Emx2*:

Normal (WT) Elevated Reduced Normal

R
L
C

C Cortical Areas:

R
L
C

M M M M
S1 S1 S1 S1
A1 A1 A1 A1
V1 V1 V1 V1

D Motor and Tactile Behavior:

Normal (WT) Substantially impaired Impaired Normal

FIG. 4. Emx2 imparts area identities to cortical progenitors and their progeny: levels of Emx2 controls sizes of cortical areas. (A, from left to right) Tangential sections through layer 4 of flattened cortex of P7 wild-type (WT); homozygous *ne-Emx2* transgenics (*ne-Emx2$^{+/+}$*); *Emx2$^{+/-}$*, *ne-Emx2$^{+/+}$*;*Emx2$^{+/-}$* mice, processed for cytochrome oxidase (CO) histochemistry to reveal area patterning of neocortex by marking the primary somatosensory area (S). Other abbreviations: M, motor; PMBSF, posteromedial barrel subfield of S; V, visual areas. Dashed lines allow for alignment of area patterns and shifts across genotypes. (B) Cartoon of relative levels of graded *Emx2* expression in cortical progenitors in four genotypes shown in (A) *ne-Emx2$^{+/+}$* transgenic mice, in which an *Emx2* transgene is driven by a nestin promoter in cortical progenitors, results in a 50% increase in overall Emx2 level. *Emx2$^{+/-}$* mice have 50% of the normal (wt) level of Emx2 transcripts. *ne-Emx2$^{+/+}$*;*Emx2$^{+/-}$* mice have overall levels of *Emx2* being essentially normal due to the combination of endogenous *Emx2* and the *Emx2* transgene. The arrows show the predicted, and observed, shifts in area patterning. (C) Area patterning depicted using a cartoon of the four primary areas, motor (M), somatosensory (S1), auditory (A1), and visual (V1). The *ne-Emx2$^{+/+}$* transgenic mice exhibit decreased sizes of S1 and M, an increase in V1, and a rostral shift in all areas. The *Emx2$^{+/-}$* mice exhibit increased sizes of S1 and M and decreased size of V1. The sizes of the primary cortical areas in the *ne-Emx2;Emx2$^{+/-}$* mice revert toward their normal sizes observed in wild-type. Therefore, the size of the areas depends on the overall level of graded *Emx2* expression, and changes in area size due to increase or decreases in Emx2 can be rescued by restoring Emx2 to wild-type levels in cortical progenitors. (D) Motor and tactile behavioural performances are summarized for the four genotypes. These behaviours are substantially impaired in *ne-Emx2$^{+/+}$* mice, and are impaired in *Emx2$^{+/-}$* mice but to a lesser degree. Behavioural performances in the *ne-Emx2;Emx2$^{+/-}$* mice revert to wild-type levels, paralleling the reversion of sizes of sensorimotor areas and Emx2 levels to wild-type. Figures adapted from Hamasaki et al (2004) and Leingärtner et al (2007) with permission.

have provided conclusive evidence that although Emx2 can repress Fgf8 expression, that Emx2 directly regulates area patterning (Fig. 4) (Hamasaki et al 2004). Because these mice survive to adulthood, areas could be directly identified and analysed. In *ne-Emx2* mice, a 50% increase in Emx2 expression in neocortical progenitors by driving an *Emx2* transgene under control of a neuron-specific nestin promoter results in a significant decrease in the sizes of rostral cortical areas, such as motor and S1, and a significant increase in caudal areas, such as V1. Opposing changes in area sizes are observed in *Emx2^{+/-}* mice that have a 50% decrease in Emx2 levels in progenitors (Hamasaki et al 2004). These changes in Emx2 levels in cortical progenitors results in a complete change in the area identity of their neuronal progeny to match the contracted or expanded areas, including changes in cytoarchitecture, distributions of area-specific TCA input, output projection neurons, gene expression patterns, and functional organizations (Hamasaki et al 2004, Leingärtner et al in preparation). These findings show that Emx2 controls area patterning in a disproportionate manner and preferentially imparts caudal area identities.

The downstream events that transfer the area patterning activities of Emx2 have received little attention. As a step to address this issue, we used representational difference analysis (RDA) to screen for the genes differentially expressed in E15.5 *Emx2^{+/+}* and *Emx2^{-/-}* mouse cortices (Li et al 2006). We identified 41 unique clones. Using secondary screening by *in situ* hybridization, we selected five genes for further analysis, *Cdk4*, *Cofilin1*, *Crmp1*, *ME2* and *Odz4*, involved in neuronal proliferation, differentiation, migration and axon guidance. Each of these genes shows opposing altered changes of expression in *Emx2* null mice compared to *Pax6* (*sey*) mutant mice.

Odz4 is one of four members of a vertebrate gene family homologous to the *Drosophila* pair-rule patterning gene, *Odd Oz* (*Odz*), a transmembrane receptor (also know as *Ten-M* family in mouse). We find that the *Odz* genes are expressed in complementary areal patterns in embryonic CP and postnatal cortex. The *Odz* genes also exhibit nuclei-specific expression in thalamus that relates to their cortical expression. These findings identify potential targets of Emx2 that might account for its patterning function and other defects in *Emx2^{-/-}* cortex, and suggest that the Odz/Ten-M family of transmembrane proteins influences cortical area patterning downstream to Emx2. Further, the findings show that the Odz family can be useful and unique markers for area patterning (Li et al 2006).

Couptf1

Couptf1 is expressed in a high-caudolateral to low-rostromedial gradient in both VZ progenitors and their CP progeny. We identified *Couptf1* as a candidate areal patterning gene from a ddPCR screen for genes differentially expressed in nascent motor versus visual areas (Liu et al 2000). An analysis of a constitutive knockout for *Couptf1* provided evidence consistent with a role for *Couptf1* in area patterning (Zhou et al 2001), but the interpretations suffered from serious complications due in part to strong expression of *Couptf1* in thalamic nuclei that provide the major input to cortical areas and influence their patterning. The complications include, but are not limited to, early lethality, a near complete absence of TCA input, and a substantial loss of layer 4 neurons (Zhou et al 1999, 2001).

To circumvent these problems, Michele Studer and Maria Armentano made a floxed allele of *Couptf1*, which when crossed with the *Emx1-Cre* knock-in line made by Kevin Jones (Gorski et al 2002), results in selective deletion of *Couptf1* from cortical progenitors at E10.5 (Armentano et al 2007). Two postdoctoral fellows in my lab, Shen-ju Chou and Axel Leingärtner, and I collaborated with Michele and Maria to analyse these conditional *Couptf1* knockout mice, which live to adulthood and do not suffer from the many problems that compromised the constitutive *Couptf1* knockout previously reported by Zhou et al (2001).

We find that selective cortical deletion of *Couptf1* results in a massive expansion of anterior areas, including motor, that are normally limited to frontal cortex but expand to occupy essentially all of parietal cortex, the normal location of S1, and most of occipital cortex, which is normally occupied largely by V1 (Armentano et al 2007). This expansion of frontal/motor areas is accompanied by a substantial reduction in the sizes of the three primary sensory areas that become compressed and aligned medial-laterally along the caudal pole of the cortical hemisphere (Fig. 5). These findings are based in part on alterations in area patterning of TCA input, using serotonin (Fig. 5A,B) and axon tracing, gene markers of area identity, including RORβ (Fig. 5C,D) and MDGA1. We conclude that *Couptf1* is required to pattern the neocortex into motor and sensory areas, and acts largely through repression of rostral area identities, such as motor, within its expression domain allowing for appropriate specification of the sensory cortical areas (Armentano et al 2007). Our analyses, particularly those of the heterozygous conditional knockout, indicate that *Couptf1* also acts in arealization by participating in the specification of the primary sensory areas.

Pax6

Pax6 in expressed in the cortical VZ in opposing gradients to *Emx2*. However, roles for Pax6 in area patterning are currently vague. The initial studies that implicated Pax6 in area patterning depended upon marker analyses of small eye (*sey*)

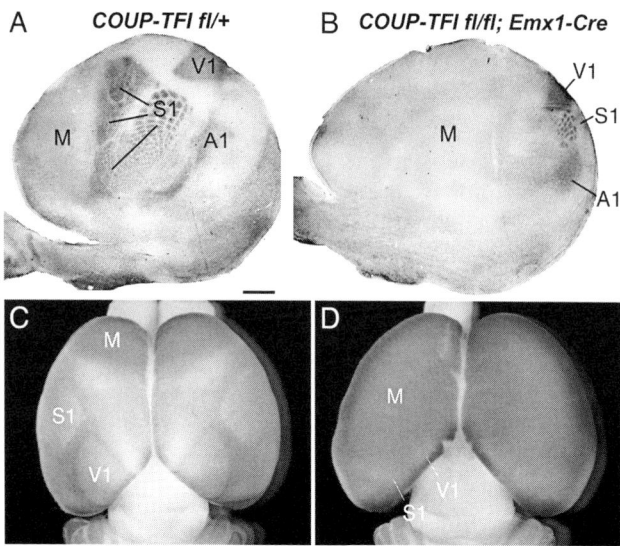

FIG. 5.   Primary sensory areas are significantly reduced in size and compressed ectopically to caudal occipital cortex in mice with selective cortical deletion of *Couptf1*. (A,B) Serotonin immunostaining on tangential sections through layer 4 of flattened cortices of P7 control (*Couptf1 fl/+*) and *Couptf1 fl/fl; Emx1-Cre* mutant cortices. Rostral is to left, medial to the top. (A) Serotonin staining reveals primary sensory areas, including primary somatosensory (S1), visual (V1) and auditory (A1) areas. V1 is caudomedial to S1 and A1 is caudolateral to S1. Frontal cortex (F) is rostral to S1. (B) In *Couptf1 fl/fl; Emx1-Cre* brains, primary sensory areas are compressed to ectopic positions at the caudal pole of occipital cortex. The S1 barrelfield retains its characteristic patterning but is substantially reduced in size and caudally shifted, while a reduced V1 is located medial and a reduced A1 lateral to the miniature S1. Scale bar: 1 mm. (C,D) *In situ* hybridization for *Cad8*, a marker of frontal/motor areas, on whole mounts of P7 control (*Couptf1 fl/+*) brains with cortex specific deletion of *Couptf1* (*Couptf1 fl/fl; Emx1-Cre*). The *Cad8* marking of the motor area is massively expanded in *Couptf1 fl/fl; Emx1-Cre* brains. In accordance with serotonin staining (B), the small domains of light *Cad8* staining is coincident with the reduced ectopic S1 and V1. Figures adapted from Armentano et al (2007).

mutant mice, which are deficient for functional Pax6 protein and die at birth, before cortical areas differentiate, have major lamination defects and a cortex reduced by a third, and in addition entirely lack TCA input. Nonetheless, the marker analyses implicated Pax6 in specifying anterior area identities associated with motor areas, consistent with its highest expression in progenitors that give rise to anterior areas (Bishop et al 2000, 2002, Muzio et al 2002). However, a recent gain-of-function study of Pax6 that used a YAK transgenic approach to overexpress Pax6 sevcralfold in cortical progenitors reports no changes in area patterning other than a small decrease in S1 size (Manuel et al 2007). This discrepancy could be explained in several ways. One appealing way is that another gene, for example, *Couptf1*, normally represses *Pax6* function in the cortical fields that would give rise to sensory areas and therefore represses the effect of *Pax6* overexpression. We are testing this possibility.

Sp8

Sp8 is expressed in a high-rostromedial to low-caudolateral gradient in the VZ and is transiently expressed in the CoP coincident with the expression domain of Fgf8 (Sahara et al 2007), a morphogen implicated in cortical area patterning (Fukuchi-Shimogori & Grove 2001, Garel et al 2003). However, in contrast to the other transcription factors described above, which are expressed throughout cortical neurogenesis, Sp8 is only expressed early, with expression being diminished by E13.5. To address roles for Sp8 in arealization and Fgf8 regulation, we employed *in utero* electroporation of expression constructs containing full length *Sp8* or *Fgf8*, or dominant active or dominant negative forms of *Sp8* (Sahara et al 2007). Using these approaches, as well as multiple *in vitro* binding assays, we show that *Sp8* and *Fgf8* exhibit reciprocal induction in the embryonic telencephalon, but that ectopic expression of *Sp8* and *Fgf8* have opposing effects on area patterning, with ectopic anterior expression of *Fgf8* inducing a posterior shift of cortical areas and ectopic anterior expression of *Sp8* inducing an anterior shift (Sahara et al 2007). We also show that Sp8 directly binds Fgf8 promoter elements and that Sp8 induction of Fgf8 can be blocked by Emx2, suggesting a mechanism to restrict Fgf8 expression to the CoP.

In a coincident study, Zembrzycki et al (2007) generated a conditional knockout of *Sp8* and by crossing with a *Bf1-Cre* line deleted the floxed alleles of *Sp8* from the telencephalon, including the CoP and progenitors in the cortical VZ, resulting in an anterior shift of cortical markers. Additional work is required to sort out the complexities of the role of Sp8 in arealization and its interactions with Fgf8 and Emx2 in this process.

Area size influences behavioural performance

Cortical areas can range in size among individuals in a species, and individuals can differ in functional efficiency and behavioural performance, but these features have not been shown to be related. For example, two- to threefold naturally occurring variations have been reported in the sizes of primary areas in human neocortex (Dougherty et al 2003, Stensaas et al 1974, White et al 1997a,b). In mice, the sizes of primary areas are very constant within an inbred strain of mice, but can vary significantly between inbred strains of mice (Airey et al 2005). These findings raise the important question of whether the size of a cortical area correlates with behavioural performance. We have addressed this question by making use of the effect of altering Emx2 levels on area size described above (Hamasaki et al 2004), and show that relatively small changes in area size can have significant effects on modality-specific behavioural performances (Leingärtner et al 2007).

Area size correlates with behavioural performance

We initially tested adult wild type, *ne-Emx2* transgenic mice and *Emx2*$^{+/-}$ knockout mice, which have different levels of cortical Emx2 and sizes of cortical areas (Fig. 4), focusing on behaviours modality-specific for sensorimotor areas (Leingärtner et al 2007). We first showed that general mobility, audition and vision, and other factors that could potentially affect performance are similar between the genetically altered and wild-type mice.

In contrast, though, we find that the *ne-Emx2* mice, which have reduced sensorimotor cortical areas compared to wild-type, have dramatically diminished performances at tests of tactile and motor behaviours, as well as balance and coordination (Leingärtner et al 2007). Surprisingly, we find that *Emx2*$^{+/-}$ mice, which have larger sensorimotor areas compared to wild-type, also have significantly reduced behavioural performance albeit not as dramatic as that seen in the *ne-Emx2* mice (Fig. 4). A straightforward interpretation of these findings is that the diminished sensorimotor behaviours are due to the changes in the sizes of sensorimotor areas relative to wild-type.

Genetic rescue experiments prove that parameters affected by area size dictate behavioural performance

To prove that area size is a critical parameter for behavioural performance, we performed genetic crosses that restore cortical Emx2 levels to wild-type levels, confirmed by real time PCR (Leingärtner et al 2007). These crosses also restore cortical areas to wild-type sizes and in parallel restore the behavioural deficiencies to wild-type performance (Fig. 4). These findings are particularly compelling because they show that the offspring of parents of distinct genotypes, but with similar behavioural deficiencies, exhibit normal behaviour. These findings demonstrate that the cortex must be the locus of the effects on behavioural performances that we describe, and importantly, that parameters affected by cortical area size are a critical determinant of behavioural proficiency (Leingärtner et al 2007). For example, changing area size alters the degree of convergence of inputs, and potentially effects intra- or inter-columnar circuitry that may result in suboptimal cortical processing.

Implications of influence of area size on behaviour

These studies indicate that areas have an optimal size, influenced by parameters of its neural system, for maximum functional efficiency and behavioural performance. Thus, these findings provide experimental support for the hypothesis that cortical areas have evolved an optimal size defined and tuned by their relationships with other components of their neural system (Krubitzer & Kaas 2005). As described above, the sizes of primary areas in humans varies two- to threefold, but the functional consequences are unknown. Our studies indicate that these differences in area size in humans likely have profound effects on the relevant behaviours. Furthermore, our findings underscore the importance of establishing during embryonic development appropriate levels of regulatory proteins that determine area sizes, thereby influencing behaviour later in life.

Acknowledgements

We thank J.C.I. Belmonte, Y. Kawakami, F. Gage and M. Studer for critical recent collaborations. Work in the O'Leary lab on the topic of this article is funded by NIH grants R37 NS31558 and R01 NS50646 (DOL).

References

Airey DC, Robbins AI, Enzinger KM, Wu F, Collins CE 2005 Variation in the cortical area map of C57BL/6J and DBA/2J inbred mice predicts strain identity. BMC Neurosci 6:18

Anton ES, Kreidberg JA, Rakic P 1999 Distinct functions of alpha3 and alpha(v) integrin receptors in neuronal migration and laminar organization of the cerebral cortex. Neuron 22:277–289

Armentano M, Chou S-J, Tomassy GS, Leingärtner A, O'Leary DDM, Studer M 2007 COUP-TFI regulates the balance of cortical patterning between frontal/motor and sensory areas. Nat Neurosci 10:1277–1286

Bishop KM, Goudreau G, O'Leary DDM 2000 Regulation of area identity in the mammalian neocortex by Emx2 and Pax6. Science 288:344–349

Bishop KM, Rubenstein JL, O'Leary DDM 2002 Distinct actions of Emx1, Emx2 and Pax6 in regulating the specification of areas in the developing neocortex. J Neurosci 22:7627–7638

Chenn A, Braisted JE, McConnell SK, O'Leary DDM 1997 Development of the cerebral cortex: mechanisms controlling cell fate, laminar and areal patterning, and axonal connectivity. In: Cowan WM, Zipursky L, Jessell T (eds) Molecular and cellular approaches to neural development. Oxford University Press p 440–473

Corbin JG, Nery S, Fishell G 2001 Telencephalic cells take a tangent: non-radial migration in the mammalian forebrain. Nat Neurosci 4 Suppl:1177–1182

Dougherty RF, Koch VM, Brewer AA, Fischer B, Modersitzki J, Wandell BA 2003 Visual field representations and locations of visual areas V1/2/3 in human visual cortex. J Vis 3:586–598

Dulabon L, Olson EC, Taglienti MG et al 2000 Reelin binds alpha3beta1 integrin and inhibits neuronal migration. Neuron 27:33–44

Fishell G, Mason CA, Hatten ME 1993 Dispersion of neural progenitors within the germinal zones of the forebrain. Nature 362:636–638

Fukuchi-Shimogori T, Grove EA 2001 Neocortex patterning by the secreted signaling molecule FGF8. Science 294:1071–1074

Fukuchi-Shimogori T, Grove EA 2003 Emx2 patterns the neocortex by regulating FGF positional signaling. Nat Neurosci 6:825–831

Garel S, Huffman KJ, Rubenstein JL 2003 Molecular regionalization of the neocortex is disrupted in Fgf8 hypomorphic mutants. Development 130:1903–1914

Gesemann M, Litwack ED, Yee KT, Christen U, O'Leary DDM 2001 Identification of candidate genes for controlling development of the basilar pons by differential display PCR. Mol Cell Neurosci 18:1–12

Gorski JA, Talley T, Qiu M, Puelles L, Rubenstein JL, Jones KR 2002 Cortical excitatory neurons and glia, but not GABAergic neurons, are produced in the Emx1-expressing lineage. J Neurosci 22:6309–6314

Hamasaki T, Leingärtner A, Ringstedt T, O'Leary DDM 2004 EMX2 regulates sizes and positioning of the primary sensory and motor areas in neocortex by direct specification of cortical progenitors. Neuron 43:359–372

Kriegstein AR, Noctor SC 2004 Patterns of neuronal migration in the embryonic cortex. Trends Neurosci 27:392–399

Krubitzer L, Kaas J 2005 The evolution of the neocortex in mammals: how is phenotypic diversity generated? Curr Opin Neurobiol 15:444–453

Leingärtner A, Thuret S, Kroll TT et al 2007 Cortical area size dictates performance at modality-specific behaviors. Proc Natl Acad Sci USA 104:4153–4158

Li H, Bishop KM, O'Leary DDM 2006 Potential target genes of EMX2 include Odz/Ten-M and other gene families with implications for cortical patterning. Mol Cell Neurosci 33:136–149

Litwack ED, Babey R, Buser R, Gesemann M, O'Leary DDM 2004 Identification and characterization of two novel brain-derived immunoglobulin superfamily members with a unique structural organization. Mol Cell Neurosci 25:263–274

Liu Q, Dwyer ND, O'Leary DDM 2000 Differential expression of COUP-TFI, CHL1, and two novel genes in developing neocortex identified by differential display PCR. J Neurosci 20:7682–7690

Mallamaci A, Muzio L, Chan CH, Parnavelas J, Boncinelli E 2000 Area identity shifts in the early cerebral cortex of Emx2−/− mutant mice. Nat Neurosci 3:679–686

Manuel M, Georgala PA, Carr CB et al 2007 Controlled overexpression of Pax6 in vivo negatively autoregulates the Pax6 locus, causing cell-autonomous defects of late cortical progenitor proliferation with little effect on cortical arealization. Development 134:545–555

Marin O, Rubenstein JL 2003 Cell migration in the forebrain. Annu Rev Neurosci 26:441–483

Muzio L, DiBenedetto B, Stoykova A, Boncinelli E, Gruss P, Mallamaci A 2002 Emx2 and Pax6 control regionalization of the pre-neurogenic cortical primordium. Cereb Cortex 12:129–139

O'Leary DDM 1989 Do cortical areas emerge from a protocortex? Trends Neurosci 12:400–406

O'Leary DDM, Schlaggar BL, Stanfield BB 1992 The specification of sensory cortex: lessons from cortical transplantation. Exp Neurol 115:121–126

O'Leary DDM, Koester SE 1993 Development of projection neuron types, axon pathways, and patterned connections of the mammalian cortex. Neuron 10:991–1006

O'Leary DDM, Nakagawa Y 2002 Patterning centers, regulatory genes and extrinsic mechanisms controlling arealization of the neocortex. Curr Opin Neurobiol 12:14–25

O'Leary DDM, Borngasser D 2006 Cortical ventricular zone progenitors and their progeny maintain spatial relationships and radial patterning during preplate development indicating an early protomap. Cereb Cortex 16 Suppl 1:i46–i56

Parnavelas JG 2000 The origin and migration of cortical neurones: new vistas. Trends Neurosci 23:126–131

Prochiantz A, Joliot A 2003 Can transcription factors function as cell-cell signalling molecules? Nat Rev Mol Cell Biol 4:814–819

Rakic P 1988 Specification of cerebral cortical areas. Science 241:170–176

Rash BG, Grove EA 2006 Area and layer patterning in the developing cerebral cortex. Curr Opin Neurobiol 16:25–34

Sahara S, Kawakami Y, Izpisua Belmonte JC, O'Leary DDM 2007 Sp8 exhibits reciprocal induction with Fgf8 but has an opposing effect on anterior-posterior cortical area patterning. Neural Develop 2:10

Schlaggar BL, O'Leary DDM 1991 Potential of visual cortex to develop an array of functional units unique to somatosensory cortex. Science 252:1556–1560

Shimogori T, Banuchi V, Ng HY, Strauss JB, Grove EA 2004 Embryonic signaling centers expressing BMP, WNT and FGF proteins interact to pattern the cerebral cortex. Development 131:5639–5647

Stanfield BB, O'Leary DDM 1985 Fetal occipital cortical neurones transplanted to the rostral cortex can extend and maintain a pyramidal tract axon. Nature 313:135–137

Stensaas SS, Eddington DK, Dobelle WH 1974 The topography and variability of the primary visual cortex in man. J Neurosurg 40:747–755

Storm EE, Garel S, Borello U et al 2006 Dose-dependent functions of Fgf8 in regulating telencephalic patterning centers. Development 133:1831–1844

Sur M, Rubenstein JL 2005 Patterning and plasticity of the cerebral cortex. Science 310:805–810

Takeuchi A, O'Leary DDM 2006 Radial migration of superficial layer cortical neurons controlled by novel Ig cell adhesion molecule MDGA1. J Neurosci 26:4460–4464

Takeuchi A, Hamasaki T, Litwack ED, O'Leary DDM 2007 Novel IgCAM, MDGA1, expressed in unique cortical area- and layer-specific patterns and transiently by distinct forebrain populations of Cajal-Retzius neurons. Cereb Cortex 17:1531–1541

Walsh C, Cepko CL 1992 Widespread dispersion of neuronal clones across functional regions of the cerebral cortex. Science 255:434–440

Walsh C, Cepko CL 1993 Clonal dispersion in proliferative layers of developing cerebral cortex. Nature 362:632–635

Walsh FS, Doherty P 1997 Neural cell adhesion molecules of the immunoglobulin superfamily: role in axon growth and guidance. Annu Rev Cell Dev Biol 13:425–456

White LE, Andrews TJ, Hulette C et al 1997a Structure of the human sensorimotor system. I: Morphology and cytoarchitecture of the central sulcus. Cereb Cortex 7:18–30

White LE, Andrews TJ, Hulette C et al 1997b Structure of the human sensorimotor system. II: Lateral symmetry. Cereb Cortex 7:31–47

Zembrzycki A, Griesel G, Stoykova A, Mansouri A 2007 Genetic interplay between the transcription factors Sp8 and Emx2 in the patterning of the forebrain. Neural Develop 2:8

Zhou C, Tsai SY, Tsai MJ 2001 COUP-TFI: an intrinsic factor for early regionalization of the neocortex. Genes Dev 15:2054–2059

Zhou C, Qiu Y, Pereira FA, Crair MC, Tsai SY, Tsai MJ 1999 The nuclear orphan receptor COUP-TFI is required for differentiation of subplate neurons and guidance of thalamocortical axons. Neuron 24:847–859

DISCUSSION

Mallamaci: I have a couple of questions about the hippocampus. What happens in *Lhx2* conditional mutants and in *Couptf1* knockouts?

O'Leary: The hippocampus is reduced in size, as you would predict from the expression patterns. Interestingly, in the *Emx* transgenics in which *Emx2* is over-expressed in progenitors that continue to produce granule neurons of the dentate gyrus of the hippocampal formation, the dentate gyrus is also smaller. But it is not significantly smaller early postnatally, whereas in an old adult, it is tiny. We are still examining the underlying mechanism, but it involves a reduction in the number of stem cells that produce additional granule cells for the dentate gyrus; they possibly also undergo apoptosis, similar to the embryonic cortical progenitor cells that overexpress *Lhx2*.

Fishell: It looks like it is time to do some epistasis analysis. You made the statement that it is the concentration of Emx2 that matters. Did you mean this just in the context of talking about Emx2, or is the implication that the way Couptf1 functions is by titrating Emx2? What would you predict from the double analysis?

O'Leary: I was remiss in not mentioning that Tsai and colleagues published a constitutive knockout of *Couptf1* in 2001 (Zhou et al 2001). This animal also died shortly after birth and has many other defects, such as defects in thalamocortical input. Their interpretation was that the *Couptf1* knockout lacked areal patterning. However, the conditional knockout of *Couptf1* that we have analysed with Michele Studer and colleagues, in which *Couptf1* was selectively deleted from cortex, does not suffer from the defects in thalamocortical input, early death, etc. that the constitutive knockout suffers. These animals can be raised to adulthood and one can delineate all of the primary areas. The markers Tsai and colleagues used didn't allow them to delineate areas and the animals were too young. To what extent are *Couptf1* and *Emx2* interacting? That's a good question. Tsai and colleagues published that eliminating *Couptf1* had no effect on *Emx2* or *Pax6* expression, and both retained their normal graded expression. On the basis of this, several groups proposed models for Couptf1 function relative to Emx2 and Pax6. One of the models was that Couptf1 was required for the interpretation of Emx2 and Pax6 to facilitate their patterning roles. We now know that this is unlikely to be the case. In the conditional *Couptf1* null we find a slight decrease in Emx2 expression and a modest increase in Pax6 expression. However, we believe that these changes in expression are not sufficient to account for the dramatic area patterning phenotypes that we see. We know this from our studies of the effect of changing levels of Emx2, or in the case of Pax6, David Price and colleagues (Manuel et al 2007) have recently published that increasing Pax6 by up to threefold has at best only a minimal effect on areal patterning. We believe that Couptf1 operates largely independently of Emx2 and activates an independent genetic pathway for areal patterning. Its main function is to repress motor/frontal area identities, whereas the main function of Emx2 seems to be to impart caudal areal identities, such as those associated with V1.

Walsh: I have a question about Lhx2. When Ed Monuki and I published on the straight *Lhx2* knockout (Monuki et al 2001) we found that in the small cortex, a large part of the cortex is converted to a roofplate identity and the rest of the entire cortex is converted to a cortical hem identity. There is almost no neocortex: only a tiny rim of the cortex is specified as neocortex way out laterally. Is this consistent with what you see in the conditional knockout?

O'Leary: Shen-ju Chou, a postdoctoral fellow in my lab, has also analysed the constitutive *Lhx2* null mice and has confirmed many of your findings. She has also shown that what we originally thought was a small cortical plate (probably the small neocortex to which you refer) is in fact filled up with Reelin-positive Cajal-Retzius neurons, probably due to the much larger cortical hem producing more Cajal-Retzius neurons, as well as the decrease in cortical volume over which they can be distributed. In addition, she finds no Tbr1-positive neurons generated from the cortical ventricular zone, and that there is effectively no proliferation ongoing in what looks like a ventricular zone.

Walsh: That is like a mediolateral or dorsoventral patterning law that may trump the rostrocaudal.

O'Leary: Possibly. Although the cortex is much smaller, it appears to be directly affecting progenitor proliferation, and potentially the initial setting up of the progenitor pool that will be neocortex. But those are issues that we are still addressing using the conditional knockout of *Lhx2* that Shen-ju has made. There is a major difference in cortical phenotypes between the conditional and constitutive knockouts. This difference is probably due in large part because we crossed it with an *Emx1-Cre* line, which deletes floxed alleles around E10 or so. This appears to be too late to affect the earlier patterning events that you are describing.

Tan: You nicely showed differences in *Emx2* expression. Have you independently confirmed this looking at different strains of mice? In humans are there quantitative trait loci (QTL) differences in single-nucleotide polymorphism (SNP) expression of *Emx2* that could regulate this?

O'Leary: There are mutations of *EMX2* in humans (Brunelli et al 1996). There probably are differences in *Emx2* expression across mouse strains, but no one has looked at this issue. There have been reports that there are differences in sizes of cortical areas in mice, not within inbred strains but by comparisons across certain inbred strains.

Price: Can you comment on the importance of the nature of the gradient of the transcription factors? The idea is that there is a gradient of transcription factor expression across the cortical sheet. My understanding is that in the *Emx* gain-of-function mice that you made (Hamasaki et al 2004), you added a similar dose of Emx2 across the whole cortex. This meant that the gradient was changed; the ratio in the levels between one end and the other would have been lower than normal.

In the *Pax6* overexpressing mice that we looked at, which involved a YAC trans-genic, the gradient was maintained and the ratio was the same as normal. Do you think the shape of the gradient of transcription factor that cells sitting next to each other see is important in regionalization?

O'Leary: I think it is important but it hasn't been shown yet. It is a potential interpretation for why we find major changes in marker expression in what is effectively a *Pax6* null that can't be explained except by true changes in positional information in the cortex. Perhaps region specific tissue loss can account for some of that change, but not for all of it. Some of the changes in differential gene expression patterns are very dramatic. All of the markers that we and Mallamaci have looked at shift anterior in the *Pax6* mutant, consistent with a loss of frontal motor identities. For example, in the wild-type the graded expression of RORβ extends back to the visual area, whereas in the *Pax6* mutant there is a major anterior shift. This can't be explained by region-specific tissue loss. This has to be due to a change in positional information that is controlling the expression of that gene. A critical role for the slope of expression is one way that we can rectify the results in the *Pax6* loss-of-function mutant with the YAC transgenic that you have made. We have also analysed the small eye mutant (Pax6) heterozygote and see no change in area patterning. If Pax6 is affecting area identity it appears not to be doing it in a straight-forward concentration-dependent manner like Emx2 is. Possibly, the slope of expression is a critical parameter, but alternatively other transcription factors, e.g. Couptf1 could repress the function of increased Pax6 in progenitors that generate sensory areas.

Stoykova: Our results on Pax6 conditional mutants are still in preparation. Indeed, the analyses performed so far indicate that after cortex-specific ablation of the function of *Pax6* in cortical progenitors at the beginning of neurogenesis (E10), the expression of caudal cortical markers is ectopically extended rostrally, which is in agreement with previously published data for homozygous embryos of the *Pax6/Small eye* mutant (Bishop et al 2000). However, the cortex of the conditional *Pax6* mutant displays apparently normal areal corticothalamic connections, suggesting preserved functional area identity. A similar cortical phenotype that includes molecular caudalization of the rostral cortex with normal thalamocortical topography was reported for the *Fgf8* hypomorphic mutant (Garel et al 2003).

O'Leary: I would say we still need to do a lot of work on Pax6 to sort out what its roles are in area patterning. There are changes in the molecular identity of the cortical plate, whereas thalamocortical input is apparently normal. I think Pax6 has a role but it appears to be more limited than that of Emx2 or Couptf1.

Crino: In the setting of transcription factor driven brain regionalization, has anyone performed a transcriptional profile to determine whether developmental alterations in these transcription factors leads to alterations in downstream genes?

O'Leary: We have just published a Representation/Difference Analysis (RDA) screen. Another group has done a micro RNA analysis. Both groups have identified a large number of genes that are both up-regulated and down-regulated when *Emx2* is removed completely. But we haven't gone beyond identifying candidate downstream targets and assessing their expression in *Emx2* and Pax6 mutants (Li et al 2006.

Crino: The reason I ask is that in human disease, many studies have looked at gene expression changes but have failed to do any kind of transcriptional mining to look upstream. Perhaps all these coordinate changes simply reflect the change in a common regional transcription factor that could have relevance for autism or epilepsy.

O'Leary: One thing that complicates this whole story is that the roles of some of these genes in area patterning have been recently questioned or greatly revised. Many of these transcription factors are involved in homeotic mutations, which is how many of them were originally identified in *Drosophila*, and they have all the hallmarks of being genes that would be predicted to impart positional or areal identity in cortex, but not all of them appear to do so. It is a conundrum.

Molnár: When you are shifting the motor and sensory identities, how is it affecting the scaling of the different specific nuclei of the thalamus? You'd expect that if you have a much smaller visual cortex you would scale down the LGN, and if you have a smaller primary somatosensory cortex you would have a smaller VB. Is there such scaling in your mutants?

O'Leary: That's exactly what we see. When thalamus is initially patterned, and before you have a retrograde effect (say at P0), it looks completely wild-type. As time goes on and there is a retrograde effect, and cell death in thalamus, in the *ne-Emx2* transgenic, which has a larger V1, you end up with a larger LGN than what you have in wild-type. There is also a smaller VP, paralleling the smaller size of S1.

Molnár: When can this be seen first?

O'Leary: By the end of the first postnatal month this form of systems matching is already evident. At the time we do the behavioural analyses the matching has taken place.

Molnár: What do these mice do with the much bigger retinotopic representation in their primary visual cortex? You showed with your elegant imaging experiments that they use their larger visual cortex and map out the visual field only once in a continuous field. Are they any better in visual acuity, object discrimination or detecting motion?

O'Leary: Maybe we should ask what a mouse normally does with its vision. They are not visually very good. I would predict that their real vision in the *ne-Emx2* transgenics with a larger V1, if it could be properly measured, is degraded compared with the wild-type. There is precedent for this from human studies. Geoff Boynton has published an fMRI study combined with analyses of visual acuity (Duncan & Boynton 2003). He has shown that humans with a bigger V1 have lower visual acuity. This is consistent with our behavioural analyses.

Goffinet: The mouse that you showed with important motor anomalies looks worse than the mouse I showed with no cortical output and no thalamocortical projections! This would suggest that impairing the balance in cortical areas is worse than having no connections at all.

O'Leary: I agree. Many cognitive scientists have suggested that if pruning fails to happen in a normal way, the retention of the exuberant connections results in aberrant behaviour or cognitive disorder. More is often not better.

Yamamoto: Couptf1 is not only areal specific but also layer specific (layer 4). How are laminar structure and laminar specificity of thalamocortical projection in the knockout mice?

O'Leary: In the conditional *Couptf1* knockout, thalamocortical projections terminate as they should in layer 4, but over a reduced tangential domain and ectopically positioned to match the reduced ectopically positioned primary sensory areas, but the cell density in layer 4 appears similar to that in wild-type. This is the appearance at P7. We have not done an appropriate analysis at later ages. In contrast, in the constitutive knockout, which has the double whammy of losing Couptf1 function in layer 4 neurons and losing thalamocortical input to layer 4, Tsai and colleagues have reported that layer 4 cell density is aberrantly low.

References

Bishop KM, Goudreau G, O'Leary DDM 2000 Regulation of area identity in the mammalian neocortex by Emx2 and Pax6. Science 288:344–349

Brunelli S, Faiella A, Capra V et al 1996 Germline mutations in the homeobox gene EMX2 in patients with severe schizencephaly. Nat Genet 12:94–96

Duncan RO, Boynton GM 2003 Cortical magnification within human primary visual cortex correlates with acuity thresholds. Neuron 38:659–671

Garel S, Hoffman KJ, Rubenstein JL 2003 Molecular regionalization of the neocortex is disrupted in FGF8 hypomorphic mutants. Development 130:1903–1914

Hamasaki T, Leingärtner A, Ringstedt T, O'Leary DDM 2004 EMX2 regulates sizes and positioning of the primary sensory and motor areas in neocortex by direct specification of cortical progenitors. Neuron 43:359–372

Li H, Bishop KM, O'Leary DDM 2006 Potential target genes of Emx2 include Odz/Ten-M and other gene families with implications for cortical patterning. Mol Cell Neurosci 33:136–149

Manuel M, Georgala PA, Carr CB et al 2007 Controlled overexpression of Pax6 in vivo negatively autoregulates the Pax6 locus, causing cell-autonomous defects of late cortical progenitor proliferation with little effect on cortical arealization. Development 134:545–555

Monuki ES, Porter FD, Walsh CA 2001 Patterning of the dorsal telencephalon and cerebral cortex by a roof plate-Lhx2 pathway. Neuron 32:591–604

Zhou C, Tsai SY, Tsai MJ 2001 COUP-TFI: an intrinsic factor for early regionalization of the neocortex. Genes Dev 15:2054–2059

Genetic regulation of prefrontal cortex development and function

Jeremy A. Cholfin*† and John L. R. Rubenstein†[1]

*Medical Scientist Training Program, Neuroscience Graduate Program, †Nina Ireland Laboratory of Developmental Neurobiology, Department of Psychiatry, University of California, San Francisco, CA 94143-2611, USA

Abstract. The prefrontal cortex (PFC) consists of multiple areas that mediate a wide range of higher-order behaviours in mammals. Despite years of intensive neuroanatomical and functional studies, little is known about the genetic mechanisms that pattern this structure during development. It has been recognized that fibroblast growth factor (FGF) signalling from the rostral patterning centre could have a central role in regulating rostral telencephalic development. A subset of FGF genes are expressed in the rostral patterning centre in the embryonic telencephalon. Recent evidence shows that FGFs, including *Fgf17*, regulate the graded expression of regulatory genes in the cortical neuroepithelium, which may specify the initial distribution of PFC regional subdivisions and ultimately mature areas.

2007 Cortical development: genes and genetic abnormalities. Wiley, Chichester (Novartis Foundation Symposium 288) p 165–177

The cerebral cortex is a highly ordered brain structure consisting of a sheet of over a billion neurons that mediates cognition and behaviour. Along its longitudinal dimension, the cortical sheet is organized into histologically discrete areas that emerge gradually during embryonic and postnatal development. The process of cortical arealization has been an area of intense investigation over the past two decades. Two models that are now largely viewed as complementary have arisen to account for this process. The protomap model explains arealization in terms of the early specification of neural precursors in the neuroepithelium that give rise to the cerebral cortex (Rakic 1988). By contrast, the protocortex model suggests that extrinsic (i.e. thalamocortical) input imparts areal characteristics onto an otherwise 'blank slate' cortex (O'Leary 1989). Evidence accumulated over the last few years has identified key genetic and non-genetic factors that regulate cortical

---

[1]This paper was presented at the symposium by John Rubenstein, to whom correspondence should be addressed.

arealization, suggesting that complex interactions between intrinsic and extrinsic mechanisms regulate this process (O'Leary & Nakagawa 2002, Grove & Fukuchi-Shimogori 2003, Sur & Rubenstein 2005).

Prior to and after the arrival of thalamocortical afferents, the main source of input to the cortex, a subset of genes are expressed in graded and coarse areal patterns within the cortical progenitor zone and emergent cortical plate (Nakagawa et al 1999, Rubenstein et al 1999). Genetic removal of thalamocortical input results in the establishment and maintenance of normal patterns of areal gene expression, providing strong evidence that early regionalization of the cortex occurs independently of thalamic input (Miyashita-Lin et al 1999, Nakagawa et al 1999). Therefore, research over the last few years has been focused on identifying genes that specify regional identity of neural progenitors and thereby contribute to cortical arealization.

FGF signalling and cortical patterning

A current model of forebrain patterning suggests that fibroblast growth factor (FGF) signalling from a rostral source (the rostral patterning centre), imparts positional information on to the adjacent neuroepithelium by regulating the expression of transcription factors and other regulatory molecules (O'Leary & Nakagawa 2002, Grove & Fukuchi-Shimogori 2003, Sur & Rubenstein 2005). The initial forebrain pattern is regulated by interactions between FGFs (in particular FGF8) from the rostral patterning centre and two other centres: the dorsal midline, which expresses members of the bone morphogenic protein (*Bmp*) and *Wnt* family of genes, and the ventral midline, which expresses *Shh* (Shimamura et al 1997, Shimamura & Rubenstein 1997, Crossley et al 2001, Shimogori et al 2004, Sur & Rubenstein 2005, Storm et al 2006). *Fgf8* exhibits dosage-dependent functions in early forebrain patterning through regulation of neural specification, progenitor proliferation and cell death (Storm et al 2003, 2006). It is unclear from these studies to what extent the neural specification function is dissociable from proliferative and apoptotic mechanisms, and whether these functions can be attributed directly to reduced *FGF* signalling, versus indirect interactions with the other patterning centres.

A nested set of FGF genes are expressed in and around the rostral patterning centre: *Fgf8*, *Fgf18*, *Fgf17* and *Fgf15* (Crossley & Martin 1995, Maruoka et al 1998, Xu et al 1999, Bachler & Neubuser 2001). The nested expression patterns may indicate that FGFs operate hierarchically by regulating the expression of other FGFs in an adjacent zone. However, the regulatory relationships among the rostral patterning centre-expressed FGFs are unknown. We have been examining the requirements of these FGFs for the expression of the others using loss-of-function mutant mice for *Fgf8*, *Fgf17* and *Fgf15*. We have found that

Fgf8 expression does not depend on *Fgf17*, while *Fgf17* expression is regulated by *Fgf8* in a dosage-dependent manner (Cholfin, Borello and Rubenstein, unpublished data).

FGF receptors (FGFRs) are receptor tyrosine kinases that are expressed in the cortical neuroepithelium and lead to the activation of the mitogen-activated protein kinase (MAPK) and phosphotidylinositol-3-kinase (PI3K) signalling pathways (Ornitz 2000, Shinya et al 2001, Hebert et al 2003). This suggests that rostral patterning centre FGFs may signal to cortical progenitors directly. FGFRs participate in patterning the telencephalon, including the olfactory bulb and ventral forebrain (Hebert et al 2003 Gutin et al 2006). Attenuated forebrain FGFRI signalling has been reported to result in loss of glutamatergic pyramidal neurons in the frontal and temporal cortex (Shin et al 2004). Microarray analysis of the *FgfR1* mutant cortical primordium suggests that FGFRI may be the key receptor that mediates Fgf8 signalling in the telencephalon *in vivo* (Sansom et al 2005). We and other groups have recently demonstrated that the MAPK pathway is a target of Fgf8 signalling in the dorsal forebrain (E. Grove laboratory and Borello, Cholfin & Rubenstein, unpublished data). Evidence in zebrafish indicates that the MAPK pathway participates in patterning the subpallial telencephalon (Shinya et al 2001), but the functional contribution of the MAPK signalling pathway to cortical patterning remains to be tested.

Fgf8 signalling induces the expression of several genes, including those that encode the ETS transcription factors Erm, Er81 and Pea3 (Fukuchi-Shimogori & Grove 2003). We have found that *Erm* and *Pea3* are highly expressed in the frontal cortex neuroepithelium and that *Fgf17* regulates their expression (Cholfin & Rubenstein, unpublished data), suggesting that these transcription factors play a role in regionalization of the frontal cortex. Fgf8 also positively regulates the expression of the general receptor tyrosine kinase signalling inhibitors Sprouty1 and 2 (Fukuchi-Shimogori & Grove 2001, 2003, Storm et al 2003), which suggests that negative feedback may regulate the extent of FGF signalling. It appears that *Fgf8* and *Fgf17* have a differential ability to regulate Sprouty expression in the forebrain (Cholfin & Rubenstein, unpublished data), consistent with previous studies of *Fgf8* and *Fgf17* in the mid-hindbrain patterning centre (Liu et al 2003).

By contrast, Fgf8 represses expression of *Emx2* and *COUP-TF1* in the cortical neuroepithelium (Crossley et al 2001, Fukuchi-Shimogori & Grove 2003, Garel et al 2003), two transcription factors with important roles in cortical patterning and arealization (Bishop et al 2000, Mallamaci et al 2000, Zhou et al 2001, Studer, O'Leary & Rubenstein, unpublished results). Unlike *Fgf8*, *Fgf17* appears not to have a major effect on gradients of these factors (Cholfin & Rubenstein, unpublished data), suggesting that *Fgf17* may act more specifically to control regional properties of the frontal cortex.

Complementary gain- and loss-of-function experiments point to a critical role for *Fgf8* in neocortical patterning and arealization (Fukuchi-Shimogori & Grove 2001, 2003, Garel et al 2003). For example, *Fgf8^{neo/neo}* mild hypomorphic mutants exhibit rostral shifts in gradients of *Emx2* and *COUP-TF1* in the cortical neuroepithelium that correlate with reduced frontal cortex size and expanded caudal cortical regions (Garel et al 2003). Ectopic expression of *Fgf8* in the caudal cortical primordium results in partial duplications of the somatosensory cortex, suggesting that *Fgf8* acts as a true neocortical patterning signal (Fukuchi-Shimogori & Grove 2001). Although the initial pattern of thalamocortical connectivity is not affected in newborn *Fgf8^{neo/neo}* mutants (Garel et al 2003), *Fgf8* can regulate neocortical cues that guide area-specific thalamic innervation postnatally (Shimogori & Grove 2005). *Fgf8* appears to regulate the early intracortical wiring pattern, another aspect of cortical arealization (Huffman et al 2004). Finally, FGF signalling is essential for differentiation of the dorsal telencephalic midline and commissural crossing (Shanmugalingam et al 2000, Huffman et al 2004, Smith et al 2006).

Emx2, which is expressed in a high-caudomedial to low-rostrolateral gradient in the cortical primordium (Simeone et al 1992, Gulisano et al 1996), regulates neocortical arealization in a direction opposite to *Fgf8*. *Emx2* mutant mice (*Emx2^{-/-}*) have reduced caudal and expanded rostral cortical areas (Bishop et al 2000, Mallamaci et al 2000), defects that were rescued by reducing FGF signalling (Fukuchi-Shimogori & Grove 2003). By contrast, overexpression of *Emx2* in neural progenitors results in expanded caudal and reduced rostral areas, despite normal *Fgf8* expression (Hamasaki et al 2004). Therefore, *Emx2* may regulate cortical arealization both by repressing *Fgf8* expression and by direct specification of neural progenitors (Fukuchi-Shimogori & Grove 2003, Hamasaki et al 2004). However, genetic interactions between endogenous FGF signalling and *Emx2* in cortical patterning have not yet been explored.

Much less is known about the roles of other FGFs that are expressed in the rostral patterning centre in cortical arealization. Although *Fgf17* over-expression was reported to have effects similar to *Fgf8* in neocortical patterning (Fukuchi-Shimogori & Grove 2003), the role of endogenous *Fgf17* in cortical development has not been studied. *Fgf17* mutant mice (*Fgf17^{-/-}*) have a small anterior cerebellar vermis and inferior colliculus, but no reported forebrain phenotype (Xu et al 2000). *Fgf8* and *Fgf17* have different effects on mid-hindbrain patterning, which may result from differences in their spatiotemporal expression patterns, ligand-receptor affinity and/or ability to regulate downstream gene expression (Xu et al 2000, Liu et al 2003, Olsen et al 2006). Therefore, it is reasonable to hypothesize that *Fgf8* and *Fgf17* may have overlapping, but distinct roles in neocortical development.

Focusing on the prefrontal cortex

Virtually all of the previous work in cortical arealization has focused on relatively large territories within the cerebral cortex (i.e. frontal, parietal, occipital), mainly due to a lack of early markers that distinguish subdivisions within a given cortical region. Do such markers exist? How are individual subdivisions of an area altered in the context of mutations in genes involved in cortical patterning? Here, we focus on subdivisions of the frontal cortex (FC), in particular the prefrontal cortex (PFC), because of its involvement in a range of important higher cognitive, motor and behavioural functions, largely unexplored development, and possible relevance to neurodevelopmental disorders.

In adult rodents, the PFC was originally described as the projection zone of the thalamic mediodorsal nucleus (Krettek & Price 1977) and can be divided into medial and orbital regions that are thought to have homologues in primate species (Zilles & Wree 1995, Heidbreder & Groenewegen 2003, Uylings et al 2003, Dalley et al 2004). The medial PFC is subdivided into dorsal (frontal association, anterior cingulate and prelimbic) and ventral (infralimbic and medial orbital) areas, while the orbital cortex is subdivided into ventral, lateral, dorsolateral and ventrolateral orbital areas.

The connectivity of the PFC follows most of the same general organizational principles as the six-layer neocortex: intracortical projections arise from layers II/III; subcortical projections to the striatum, brainstem and spinal cord layer arise from layer V; and thalamic efferents arise from layer VI. The notable exception is that the rodent PFC lacks a well-developed granular layer (layer IV), which normally receives afferent fibres from the thalamus. Instead, in the PFC thalamocortical axons terminate principally in layer III (Krettek & Price 1977, Zilles & Wree 1995).

The PFC as a whole is involved in higher order regulation of cognition and behaviour (Price 2006). However, accumulating evidence indicates that dorsal and ventral PFC each mediates distinct functions. The dorsal PFC is involved in working memory, executive function, response selection, temporal processing of information, effort-related decision making and social valuation, while ventral and orbital PFC is implicated in behavioural flexibility, emotional regulation, delay-related decision making, evaluation of rewards and autonomic control (Heidbreder & Groenewegen 2003, Uylings et al 2003, Dalley et al 2004, Price 2006). Defining how unique territories within the rodent PFC contribute to behaviour is an area of intense investigation. Currently, there is a lack of genetic mouse models that have selective dysfunction of PFC subdivisions.

The accepted anatomical subdivisions of the PFC are based on relatively subtle cytoarchitectonic characteristics (Krettek & Price 1977, Zilles & Wree 1995) that emerge gradually during postnatal development. Therefore, it is difficult to distinguish subdivisions earlier in development, which would be required to interpret regionalization in the context of patterning mutants that die perinatally. It is currently unknown whether a subset of genes, which could be used to demarcate subdivisions, is regionally expressed within the early FC.

Therefore, we initially focused on defining a novel panel of gene expression markers in the newborn (postnatal day 0 [P0]) mouse brain (Cholfin & Rubenstein 2007). None of the genes were expressed exclusively in only one PFC subdivision (note, our analysis also includes motor cortex and rostral parts of the somatosensory cortex). However, all of them were expressed in regional patterns. A subset had sharp expression borders, which were useful in mapping genetically-defined boundaries that we propose delineate PFC subdivisions. Subsequent comparison with anatomically-defined subdivisions (Zilles & Wree 1995) indicated a remarkable correlation with mature PFC areas. These results provide evidence for a genetic partitioning of the PFC that precedes overt cytoarchitectonic differentiation.

We then used this gene expression panel to determine how individual PFC subdivisions are altered in $Fgf17^{-/-}$ mice, in addition to studying effects on caudal cortical regions and connectivity (Cholfin & Rubenstein 2007). Unexpectedly, we found that $Fgf17^{-/-}$ mice have reduced size and medially shifted positions of dorsal PFC subdivisions, while ventral PFC subdivisions are normal. The reduced dorsal PFC is complemented by a rostral shift of caudal cortical areas, suggesting that the phenotype may be due to a defect in patterning. These regionalization changes persisted into adulthood. Although no qualitative changes in the pathfinding properties of PFC axons were apparent, we found evidence for a quantitative reduction in projections to the dorsolateral striatum and ventral midbrain, consistent with the reduced dorsal PFC.

What are the functional consequences of abnormal frontal cortex patterning? Recently, mutations in a G protein-coupled receptor (GPR56) were identified in human patients with bilateral fronto-parietal polymicrogyria (BFPP), a cortical malformation disorder that selectively affects the frontal lobes (Piao et al 2004). Patients with BFPP exhibit cognitive and motor dysfunction, consistent with abnormal frontal cortex function. Mouse $Gpr56$ is expressed in the cortical progenitor zones, but not in the cortical plate, suggesting abnormal specification as a potential mechanism for the regionally-selective defects (Piao et al 2004).

It has been hypothesized that a mouse with frontal cortex hypoplasia due to a weakened FGF signalling centre may exhibit 'hypofrontal' behaviours (Sur & Rubenstein 2005). Therefore we have undertaken a set of studies to examine behaviours that are associated with decreased frontal cortex function, ranging from motor function to higher order cognitive and social behaviours. We have recently identified a circumscribed set of social deficits and an associated reduction in dorsal PFC activation in $Fgf17^{-/-}$ mice (Scearce-Levie et al 2007), providing evidence for functional consequences of reduced rostral patterning centre FGF signalling during development.

Conclusions

While previous studies have yielded insight into the genetic mechanisms that govern arealization of the cerebral cortex, relatively little attention has been paid to the formation of subdivisions within a more general area. The ability to use a divide-and-conquer approach to study cortical regionalization is predicated on having a set of markers that delineate regional subdivisions. Our studies of the PFC using a new panel of gene expression markers has provided insight into the unexpectedly selective role of *Fgf17* in regionalization of the dorsal PFC. In addition, the finding of circumscribed social deficits with associated dorsal PFC hypoactivity suggests that *Fgf17* mutant mice may provide a valuable animal model for human disorders such as autism and schizophrenia that involve frontal lobe dysfunction and altered social interactions.

Acknowledgements

This work was supported by the UCSF Medical Scientist Training Program (J.A.C.), Nina Ireland (J.L.R.R.), Larry L. Hillblom Foundation (J.L.R.R.), and NIH grants: NS34661-01A1 (J.L.R.R.) and K05 MH065670 (J.L.R.R.).

References

Bachler M, Neubuser A 2001 Expression of members of the Fgf family and their receptors during midfacial development. Mech Dev 100:313–316

Bishop KM, Goudreau G, O'Leary DD 2000 Regulation of area identity in the mammalian neocortex by Emx2 and Pax6. Science 288:344–349

Cholfin JA, Rubenstein JL 2007 Patterning of frontal cortex subdivisions by Fgf17. Proc Natl Acad Sci USA 104:7652–7657

Crossley PH, Martin GR 1995 The mouse Fgf8 gene encodes a family of polypeptides and is expressed in regions that direct outgrowth and patterning in the developing embryo. Development 121:439–451

Crossley PH, Martinez S, Ohkubo Y, Rubenstein JL 2001 Coordinate expression of Fgf8, Otx2, Bmp4, and Shh in the rostral prosencephalon during development of the telencephalic and optic vesicles. Neuroscience 108:183–206

Dalley JW, Cardinal RN, Robbins TW 2004 Prefrontal executive and cognitive functions in rodents: neural and neurochemical substrates. Neurosci Biobehav Rev 28:771–784

Fukuchi-Shimogori T, Grove EA 2001 Neocortex patterning by the secreted signaling molecule FGF8. Science 294:1071–1074

Fukuchi-Shimogori T, Grove EA 2003 Emx2 patterns the neocortex by regulating FGF positional signaling. Nat Neurosci 6:825–831

Garel S, Huffman KJ, Rubenstein JL 2003 Molecular regionalization of the neocortex is disrupted in Fgf8 hypomorphic mutants. Development 130:1903–1914

Grove EA, Fukuchi-Shimogori T 2003 Generating the cerebral cortical area map. Annu Rev
    Neurosci 26:355–380
Gulisano M, Broccoli V, Pardini C, Boncinelli E 1996 Emx1 and Emx2 show different patterns
    of expression during proliferation and differentiation of the developing cerebral cortex in
    the mouse. Eur J Neurosci 8:1037–1050
Gutin G, Fernandes M, Palazzolo L et al 2006 FGF signalling generates ventral telencephalic
    cells independently of SHH. Development 133:2937–2946
Hamasaki T, Leingartner A, Ringstedt T, O'Leary DD 2004 EMX2 regulates sizes and posi-
    tioning of the primary sensory and motor areas in neocortex by direct specification of cortical
    progenitors. Neuron 43:359–372
Hebert JM, Lin M, Partanen J, Rossant J, McConnell SK 2003 FGF signaling through FGFR1
    is required for olfactory bulb morphogenesis. Development 130:1101–1111
Heidbreder CA, Groenewegen HJ 2003 The medial prefrontal cortex in the rat: evidence for a
    dorso-ventral distinction based upon functional and anatomical characteristics. Neurosci
    Biobehav Rev 27:555–579
Huffman KJ, Garel S, Rubenstein JL 2004 Fgf8 regulates the development of intra-neocortical
    projections. J Neurosci 24:8917–8923
Krettek JE, Price JL 1977 The cortical projections of the mediodorsal nucleus and adjacent
    thalamic nuclei in the rat. J Comp Neurol 171:157–191
Liu A, Li JY, Bromleigh C, Lao Z, Niswander LA, Joyner AL 2003 FGF17b and FGF18 have
    different midbrain regulatory properties from FGF8b or activated FGF receptors. Develop-
    ment 130:6175–6185
Mallamaci A, Muzio L, Chan CH, Parnavelas J, Boncinelli E 2000 Area identity shifts in the
    early cerebral cortex of Emx2−/− mutant mice. Nat Neurosci 3:679–686
Maruoka Y, Ohbayashi N, Hoshikawa M, Itoh N, Hogan BL, Furuta Y 1998 Comparison of
    the expression of three highly related genes, Fgf8, Fgf17 and Fgf18, in the mouse embryo.
    Mech Dev 74:175–177
Miyashita-Lin EM, Hevner R, Wassarman KM, Martinez S, Rubenstein JL 1999 Early neocorti-
    cal regionalization in the absence of thalamic innervation. Science 285:906–909
Nakagawa Y, Johnson JE, O'Leary DD 1999 Graded and areal expression patterns of regulatory
    genes and cadherins in embryonic neocortex independent of thalamocortical input. J Neu-
    rosci 19:10877–10885
O'Leary DD 1989 Do cortical areas emerge from a protocortex? Trends Neurosci 12:400–406
O'Leary DD, Nakagawa Y 2002 Patterning centers, regulatory genes and extrinsic mechanisms
    controlling arealization of the neocortex. Curr Opin Neurobiol 12:14–25
Olsen SK, Li JY, Bromleigh C et al 2006 Structural basis by which alternative splicing modulates
    the organizer activity of FGF8 in the brain. Genes Dev 20:185–198
Ornitz DM 2000 FGFs, heparan sulfate and FGFRs: complex interactions essential for devel-
    opment. Bioessays 22:108–112
Piao X, Hill RS, Bodell A et al 2004 G protein-coupled receptor-dependent development of
    human frontal cortex. Science 303:2033–2036
Price JL 2006 Prefrontal cortex. Boca Raton: CRC Press
Rakic P 1988 Specification of cerebral cortical areas. Science 241:170–176
Rubenstein JL, Anderson S, Shi L, Miyashita-Lin E, Bulfone A, Hevner R 1999 Genetic control
    of cortical regionalization and connectivity. Cereb Cortex 9:524–532
Sansom SN, Hebert JM, Thammongkol U et al 2005 Genomic characterisation of a Fgf-regu-
    lated gradient-based neocortical protomap. Development 132:3947–3961
Scearce-Levie K, Roberson ED, Gerstein H et al 2007 Abnormal social behaviors in mice
    lacking Fgf17. Genes Brain Behav, in press
Shanmugalingam S, Houart C, Picker A et al 2000 Ace/Fgf8 is required for forebrain commis-
    sure formation and patterning of the telencephalon. Development 127:2549–2561

Shimamura K, Rubenstein JL 1997 Inductive interactions direct early regionalization of the mouse forebrain. Development 124:2709–2718

Shimamura K, Martinez S, Puelles L, Rubenstein JL 1997 Patterns of gene expression in the neural plate and neural tube subdivide the embryonic forebrain into transverse and longitudinal domains. Dev Neurosci 19:88–96

Shimogori T, Grove EA 2005 Fibroblast growth factor 8 regulates neocortical guidance of area-specific thalamic innervation. J Neurosci 25:6550–6560

Shimogori T, Banuchi V, Ng HY, Strauss JB, Grove EA 2004 Embryonic signaling centers expressing BMP, WNT and FGF proteins interact to pattern the cerebral cortex. Development 131:5639–5647

Shin DM, Korada S, Raballo R et al 2004 Loss of glutamatergic pyramidal neurons in frontal and temporal cortex resulting from attenuation of FGFR1 signaling is associated with spontaneous hyperactivity in mice. J Neurosci 24:2247–2258

Shinya M, Koshida S, Sawada A, Kuroiwa A, Takeda H 2001 Fgf signalling through MAPK cascade is required for development of the subpallial telencephalon in zebrafish embryos. Development 128:4153–4164

Simeone A, Gulisano M, Acampora D, Stornaiuolo A, Rambaldi M, Boncinelli E 1992 Two vertebrate homeobox genes related to the Drosophila empty spiracles gene are expressed in the embryonic cerebral cortex. EMBO J 11:2541–2550

Smith KM, Ohkubo Y, Maragnoli ME et al 2006 Midline radial glia translocation and corpus callosum formation require FGF signaling. Nat Neurosci 9:787–797

Storm EE, Rubenstein JL, Martin GR 2003 Dosage of Fgf8 determines whether cell survival is positively or negatively regulated in the developing forebrain. Proc Natl Acad Sci USA 100:1757–1762

Storm EE, Garel S, Borello U et al 2006 Dose-dependent functions of Fgf8 in regulating telencephalic patterning centers. Development 133:1831–1844

Sur M, Rubenstein JL 2005 Patterning and plasticity of the cerebral cortex. Science 310:805–810

Uylings HB, Groenewegen HJ, Kolb B 2003 Do rats have a prefrontal cortex? Behav Brain Res 146:3–17

Xu J, Lawshe A, MacArthur CA, Ornitz DM 1999 Genomic structure, mapping, activity and expression of fibroblast growth factor 17. Mech Dev 83:165–178

Xu J, Liu Z, Ornitz DM 2000 Temporal and spatial gradients of Fgf8 and Fgf17 regulate proliferation and differentiation of midline cerebellar structures. Development 127:1833–1843

Zhou C, Tsai SY, Tsai MJ 2001 COUP-TFI: an intrinsic factor for early regionalization of the neocortex. Genes Dev 15:2054–2059

Zilles K, Wree A 1995 Cortex: areal and laminar structure. San Diego: Academic Press

DISCUSSION

Rakic: I am again thinking about the first of the three meetings held here at the Ciba/Novartis Foundation on cortical development, when I was arguing for the protomap hypothesis on the basis of simple experiments in which we found that depriving the part of the prospective primary visual cortex (area 17) of its input nevertheless develops some cytoarchitectural features characteristic of area 17. We did not then have such sophisticated methods of molecular genetics, but your findings supports strongly the protomap hypothesis which predicts existence of molecular gradients that are established by the cells of the cortical plate and even within the proliferative ventricular zone prior and independently of the input. You mentioned the local changes in kinetics of cell proliferation and programmed death in the ventricular zone of the most rostral area of the frontal lobe. Is this seen only in the early stages, or does it continue later on in the subventricular zone?

Rubenstein: We haven't looked carefully at older ages. The domain of Fgf expression becomes smaller and smaller as the brain grows. Therefore, the reach of Fgf into the ventricular zone away from the rostral patterning centre may become vanishingly less important over time. Unless there is an indelible vulnerability that has been induced by lower Fgf8 or Fgf17 expression at E9, I don't know if there will be a permanent effect on cell survival. Hasegawa et al (2004) have demonstrated that Fgf18 expression, unlike the other Fgfs, is maintained in post-mitotic cortical neurons. They have evidence suggesting that the Fgf18 expression in the cortical plate feeds back on the cortical ventricular zone (Hasegawa et al 2004).

Rakic: You diminished Fgf in the ventricular zone, but you did not affect ganglionic eminence.

Rubenstein: Actually, in severe *Fgf8* mutants, the medial ganglionic eminence (MGE) is reduced in size (Storm et al 2006). We are not aware of an effect of the *Fgf17* mutant on MGE development.

Rakic: Well, it would be very interesting to see whether interneurons are coming from the ganglionic eminence in the same number as in the controls.

Goffinet: Can you comment on Fgf receptors and how they fit into this system?

Rubenstein: There are believed to be four Fgf receptors, but many have a variety of splice forms that might add functional complexity. The Hebert, McConnell and Vaccarino labs have contributed papers on this subject (Gutin et al 2006, Herbert et al 2003, Shin et al 2004, Smith et al 2006). Although *in vitro* Fgfr3 seems to have the highest affinity for Fgf8, it seems that Fgfr1 may be more important. In the Fgfr1 mutant the olfactory bulb is small and the frontal cortex may also be affected. Jean has done compound Fgfr1/Fgfr2 and 1/3 double mutants which show interesting similarities to these Fgf mutants, both in the cortex and the basal ganglia.

Mallamaci: Did you find any change in rates of neuronogenesis in the early development of the Fgf loss of function mutants? And did you find any change in the lamination profile, and in particular any shrinkage of the upper layers of the cortex?

Rubenstein: We haven't looked carefully enough at either variable. At E18.5/P0 Fgf8 mild hypomorphs show relatively normal laminar information (Storm et al 2006). There was no overt effect on layers, as well as the relative size of the layers. It is still an open question.

Stoykova: In this genetic interaction you suggested for the patterning of the rostral telencephalon, where did you put Sp8?

Rubenstein: Right on the top of this list. Dennis O'Leary has been interested in Sp8 and so are we. We begin with Otx2, then Sp8 is pretty close. I'm not sure about its early expression pattern yet.

O'Leary: Sp9 is related to the sonic hedgehog expression domain in ventral telencephalon. Sp5 is related to the cortical hem and is also expressed in a gradient across the cortical ventricular zone, whereas Sp8 is expressed coincident with the Fgf8/17 domain in the commissural plate early on. Shortly thereafter Sp8 expression diminishes within the commissural plate and remains expressed in a rostrocaudal gradient in the cortical ventricular zone.

Macklis: You mentioned, Dennis, that no transcription factor gene had yet shown an areal-specific pattern: they are all graded.

O'Leary: Correct—all transcription factor genes reported to be expressed in progenitors in the ventricular zone are graded.

Macklis: Theoretically, how many axes of expression would you want? How hard have people looked to see whether there is a cross-gradient in the second dimension, for example, and would this be enough? With the temporal gradient of cellular birth, do you need three axes?

O'Leary: *Emx2* and *Couptf1* are both expressed in a low to high AP gradient, although *Couptf1* has a steeper slope within the neocortex, but they have opposing gradients of expression on the dorsoventral axis. The graded expression of *Pax6* opposes *Emx2* expression on both axes, but opposes *Couptf1* on only the rostrocaudal axis. Thus, there are many combinations of expression gradients along the cortical axes.

Macklis: So that would be enough, and you don't need a third one coming across?

O'Leary: How can we say?

Rubenstein: In *Drosophila* early embryos concentration gradients of transcription factors result in a two-dimensional progenitor map for the fly embryo.

Stoykova: To identify genes with regionalized expression in developing cortex we recently performed a large micro-array screen with samples isolated from five distinct functional areas of E16 embryonic cortex. Indeed, most of the identified genes showed a gradient-like expression along the anterior-posterior and mediolateral axis of E16.5 cortex, in ventricular zone as well as in cortical plate. However, two genes, *Hop1* (homeobox only protein, Mühlfriedel et al 2005) and *Tyrp2* (hyrosinase-related protein; *Dct*) showed a more regionalized expression confined mostly to progenitors in ventricular zone of the cingulate cortex.

Molnár: These are very general signalling pathways and cascades. You could expect them to be involved in several other developmental steps outside the brain. Interestingly many psychiatric disorders come with minor abnormalities of other parts of the body and face. Could we go back and look at these in a more systematic fashion?

Rubenstein: That is a valid point. One of the things I'm most interested in is trying to connect what we do with human birth defects.

O'Leary: That issue intrigues me also, particularly the issue that one does not require a mutation in a patterning gene itself to have an areal patterning disorder. As we were trying to show, changes in area size alone, with everything else patterned perfectly, can give major changes in behaviour. We know that there are expression level polymorphisms (ELPs) in humans. They change the expression level of a gene that still retains its normal function by roughly the same range as that which we are dealing with in these hypomorphs, heterozygous knockouts, and gain-of-function transgenic mice. There is roughly a 50% decrease or increase. One could easily imagine that ELPs in humans could affect any of these genes and have downstream effects on area patterning that then can have major effects on behaviour.

Rakic: Chris Walsh should comment here. He has done work on the cases of polymictogyria of the frontal or parietal lobes, which genetic mutation affects separately, or specifically some area but not the rest of the cortex (Piao et al 2004). This is also a strong evidence for intrinsic origin of regional cortical specification (Rakic 2004) and complements your and John's (Rubenstein) data.

Walsh: We are in the process of characterizing the genetics of a disorder that results in a small frontal cortex. There are two different cortical disorders that have these relatively specific frontal defects in humans. There are particular alleles of *Gpr56* that cause a specific disorder, but they don't have decreased social activity like your mice do. Then there is another that also has a very small frontal lobe with the central sulcus almost up at the front of the brain. This results in developmental delay, but social behaviour isn't specifically affected and isn't disproportional to activity. The mice with decreased social interaction might just have less activity generally.

Rubenstein: Their behaviour was studied by collaborators, so I can't add much more to this.

Mallamaci: Has anyone ever compared the Fgf expression levels displayed by the rostral telencephalon in rodents and primates?

Rubenstein: I don't know.

Rakic: Primates are now very expensive, but it would be interesting to look at.

References

Gutin G, Fernandes M, Palazzolo L et al 2006 FGF signalling generates ventral telencephalic cells independently of SHH. Development 133:2937–2946

Hasegawa H, Ashigaki S, Takamatsu M et al 2004 Laminar patterning in the developing neocortex by temporally coordinated fibroblast growth factor signaling. J Neurosci 24:8711

Hebert JM, Lin M, Partanen J, Rossant J, McConnell SK 2003 FGF signaling through FGFR1 is required for olfactory bulb morphogenesis. Development 130:1101–1111

Mühlfriedel S, Kirsch F, Gruss P, Stoykova A, Chowdhury K 2005 A roof plate-dependent enhancer controls the expression of Homeodomain only protein in the developing cortex. Dev Biol 283:522–534

Piao X, Hill RS, Bodell A et al 2004 G protein-coupled receptor-dependent development of human frontal cortex. Science. 303:2033–2036

Rakic P 2004 Genetic control of cortical convolutions. Science 303:1983–1984

Shin DM, Korada S, Raballo R et al 2004 Loss of glutamatergic pyramidal neurons in frontal and temporal cortex resulting from attenuation of FGFR1 signaling is associated with spontaneous hyperactivity in mice. J Neurosci 24:2247–2258

Smith KM, Ohkubo Y, Maragnoli ME et al 2006 Midline radial glia translocation and corpus callosum formation require FGF signaling. Nat Neurosci 9:787–797

Storm EE, Garel S, Borello U et al 2006 Dose-dependent functions of Fgf8 in regulating telencephalic patterning centers. Development 133:1831–1844

Self-organization and pattern formation in primate cortical networks

Henry Kennedy, Rodney Douglas*, Kenneth Knoblauch and Colette Dehay

*Inserm, U846, 18 Avenue Doyen Lepine, 69500 Bron, France; Stem Cell and Brain Research Institute, 69500 Bron, France; Université de Lyon, Université Lyon I, 69003, Lyon, France and *Institute of Neuroinformatics, University/ETH, Winterthurerstrasse 190, Zurich 8057, Switzerland*

Abstract. The primate neocortex is characterized by a highly expanded supragranular layer (SGL). The interareal connectivity of the neurons in the SLG largely determines the cortical hierarchy that constrains information flow through the cortex. Interareal connectivity is made by precise numbers of connections, raising the possibility that the physiology of a target area is dictated by the numbers of connections and hierarchical distance in each of the pathways that it receives. The developmental mechanisms ensuring the precision of these interareal networks is in part determined by (i) the numbers of SGL neurons generated by the OSVZ, a primate-specific germinal zone. Neuron generation rate in the OSVZ is determined by regulation of the G1 phase of the cell-cycle. This regulation is area-specific and is linked to thalamic projections to the OSVZ; (ii) Prolonged pre- and postnatal pruning of connections originating from the SGL when the infant monkey visually explores its environment. Remodelling serves to sharpen initial patterns of connections and establishes the adult hierarchy. These results suggest that primate cortical networks underlying high-level function undergo prolonged self-organization via regressive phenomena in the cortical plate (axon elimination) and progressive phenomena (directed growth of cortical axons).

2007 Cortical development: genes and genetic abnormalities. Wiley, Chichester (Novartis Foundation Symposium 288) p 178–198

The brain detects statistical regularities in the environment by sensing local and global correlations of neuronal activity. In this way invariant characteristics of the world can be inferred as illustrated by colour constancy and object segmentation. This notion of brain function tempers efforts to understand the brain by studying all of its components parts independently of the environment. It suggests that it is necessary to consider the statistical features of the environment that the brain is able to detect, and to directly link the structural/functional properties of the brain to its perceptual capacities (Shepard 2001). Likewise, to attempt to understand the development of the brain without taking into account the environmental factors that have moulded its phylogenetic history is to ignore the very factor that has driven its evolution. Corticogenesis can not be understood uniquely in terms of molecular pre-specification but must also take into account the environmental factors that modulate organization as cortical development unfolds.

The developing sensory apparatus produces environmental information from which the brain needs to extract behaviourally relevant patterns. By rewiring or re-weighting connections, it tunes itself to or learns about coherent (and presumably relevant) patterns in its input. This unsupervised classification procedure is used to generate self-organized maps. There is evidence that the neuronal mechanisms of ontogenetic self-organization actually persist into adulthood when they mediate adaptive changes in learning and memory.

The species as a whole is subject to environmental patterns that exert pressure through natural selection, thus promoting the development of suitable circuits and processing modules that are tuned to the exigencies that led to survival of the current generation (Geisler & Diehl 2002). This proposed process carries the prediction that corticogenesis even at very early stages of development is influenced by extrinsic factors, echoing earlier stages of phylogeny. During the early 1970s there were considerable efforts to show that corticogenesis is significantly shaped by extrinsic factors related to the sensory periphery (Van der Loos 1977). This work was largely supported by the observation that visual experience plays an important role in the elaboration of the functional architecture of the primary visual cortex (LeVay et al 1980, Thompson et al 1983). This understanding of corticogenesis was later referred to as protocortex theory but has been largely superseded by protomap theory, which postulates that corticogenesis is driven by intrinsic molecular mechanisms. While in recent years there has been overwhelming evidence in favour of a genetic specification of cortical areas, this evidence does not invalidate the numerous instances of so-called afferent specification of the cortex and points to the need for a reappraisal of self-organization (O'Leary 1989, Killackey 1990, Sur & Rubenstein 2005).

The genesis of structures that arise through co-operative mechanisms leading to optimal equilibrium of the participating forces has been referred to as self-organization (Von de Malsburg & Singer 1988). Because self-organizing systems are initially in a relatively undifferentiated state and by definition respond over time to changing signals from the environment, one might expect that they would be characterized by prolonged maturational processes. In the cortex the phenomena of self-organization has been traditionally linked to Hebbian plasticity by which competitive modification of synaptic strength underlies experience-dependent self-organization of the functional architecture of the visual cortex, so leading to postnatal plasticity in the orientation and ocular dominance domains (LeVay et al 1980, Thompson et al 1983). This form of developmental self-organization is intrinsic to the primary cortex and is largely confined to the middle layers of the cortex.

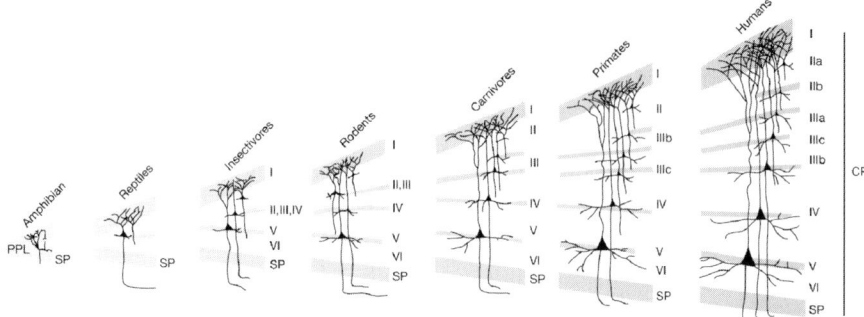

FIG. 1.   Evolution of the supragranular layers of the cortex. Supragranular cortical layers
(SGL), generated late in neurogenesis, are greatly expanded in the primate cerebral cortex,
especially in humans. In primates SGL neurons form local patchy connections and feedforward
long distance cortico-cortical connections. Taken from Hill & Walsh (2005).

   Compared to other species, the primate cortex is characterised by an over-
represented supragranular layer (SGL) compartment whose neurons are dedicated
to forming local connections as well as the transfer of information between
cortical areas (Fig. 1). In monkey cortex, the SGL is generated by a primate-
specific germinal zone (Smart et al 2002), and shows an extended maturational
period during which there is remodelling of its connectivity including during
early life when the animal is visually exploring the environment (Kennedy et al
1989, Barone et al 1995, 1996). A further argument in favour of self-organization
in the SGL is a dependency on activity for correct development. This has been
shown to be the case for cortico-cortical pathways since immature cortical
pathways are highly susceptible to manipulation of the ascending pathways
(Dehay et al 1989). In the present review we shall examine the possibility
that environmental factors might contribute to shaping the function of these
cortical layers and that this might be a characteristic feature of primate
cortical development.

Cortical hierarchy and the supragranular layers

Much of our understanding of cortical function comes from the work in the visual
cortex where stimulus response function has been most extensively studied.
Hubel and Wiesel's pioneering work showed that the receptive field structure of
neurons in the visual cortex is progressively elaborated exhibiting simple,
then complex and finally hypercomplex features. This observation leads these
authors to postulate that cortex processes afferent information through a feed-

forward (FF) hierarchy of progressive abstract detectors (Hubel & Wiesel 1968). Anatomical studies showed that FF pathways originate from SGL and terminate in layer 4, while feedback (FB) pathways originate from infragranular layers and terminate outside of layer 4 (Kennedy & Bullier 1985). In the early 1990s David Van Essen's group compiled an extensive database of the FF and FB relations of the cortical areas. Pair wise comparison of these connections revealed the dorsal and ventral streams of the visual cortices as well as a strict hierarchical organization which extended to the prefrontal cortex (Felleman & Van Essen 1991) (Fig. 2A). Malcolm Young's group performed a statistical analysis of the Van Essen database and confirmed the basic features of the Van Essen hierarchy, including the ventral and dorsal streams. However, while the Young et al organization appeared to be strictly hierarchical they found that it was highly indeterminate, in fact they found over 150 000 equally plausible solutions to the hierarchy (Fig. 2B).

Indeterminacy in the Van Essen model stems from the fact that there is no indication of hierarchical distance between nodes coupled to the fact that there are numerous parallel pathways in addition to the dorsal and ventral streams

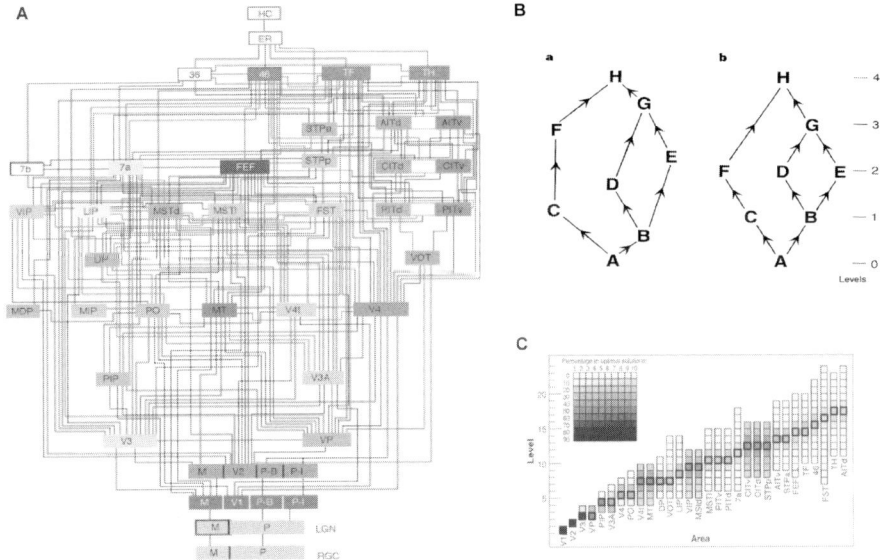

FIG. 2.   Cortical hierarchy and indeterminacy. (A) Model of visual cortex proposed by Felleman & Van Essen (1991). (B) Parallel pathways, with no distance value between stations lead to multiple possible models. (C) Area frequency distribution for 150 000 optimal hierarchical orderings (from Hilgetag et al 1996). The indeterminacy comes largely from the lack of a distance values on parallel pathways as shown in B.

(Fig. 2C). An anatomical solution to hierarchical distance is provided by the fact that long distance FF pathways arise uniquely from the SGL, and that as distance diminishes there is an increasingly important contribution to the projection from the infragranular layers (Fig. 4A). Likewise, long distance FB projections originate uniquely from infragranular layers and as distance is reduced there are increasing contributions from SGL (Kennedy & Bullier 1985, Barone et al 2000). Estimating the contributions of the SGL and infragranular layers to a given pathway involves quantitative estimations of numbers of neurons and defines the SLN% for a given pathway (SLN% = number of SGL neurons/numbers of SGL + infragranular layer neurons). Accurate SLN values makes it possible to construct a determinate model of the cortical hierarchy (Fig. 3). Furthermore, we investigated whether the number of connections in a given pathway is constant across animals (Falchier et al 2002, Vezoli et al 2004). Previous attempts to estimate whether the strength of a pathway was a characteristic feature had failed (Scannell et al 2000), largely due to the fact that they were working on a database

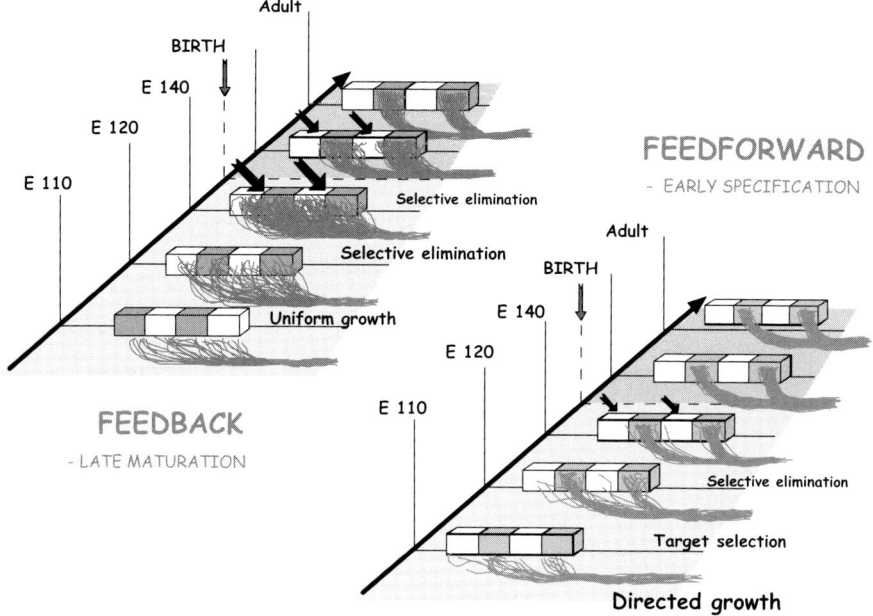

FIG. 3.   Development of feedforward and feedback connections. Feedforward connections exhibit directed growth from the earliest stages of corticogenesis, selective elimination of axonal branches is minimal and serves to refine the already segregated adult pattern of connections. By contrast, feedback projections undergo a protracted remodelling phase during which selective elimination of axonal branches plays a major role in the establishment of the adult pattern of connections.

obtained across numerous labs. While our data showed over dispersion we are able to model this over dispersion using a negative binomial distribution, which confirmed that the cortical pathways projecting to a given target area have characteristic numbers of connections.

These findings show that the connectivity signature of a cortical area is defined by the individual strengths of 10–20 cortical areas that project to it, the hierarchical distance of each of these areas as reflected by its SLN and the numerical strength of the individual pathways. This would suggest that the physiological function of, or the range of information processing performed by the target area, is constrained by the particular profile of its inputs. Projections originating from SGL terminate in layer 4 and recurrent local circuits amplify the input signals before relaying them to the output neurons in the upper and lower layers (Douglas et al 1995). The output of the cortex is modulated by the infragranular layer projections to layer 1 (Cauller 1995) and to a lesser extent to layers 5 and 6. In this way FF pathways construct the receptive field properties of the target and area while the FB pathways modulate the features of these receptive fields and the output of the target area. The numerical values and hierarchical features of the projections of a cortical area are determined during an extensive developmental period that stretches into postnatal life during which the contribution of the SGL undergoes extensive modulation (see below).

What is important here is the fact that the connectivity profile and the hierarchical organization depend on the number of neurons in the SGL and the convergence/divergence values of connectivity of these neurons. Both of these parameters are finely adjusted during development, during early corticogenesis by modulation of proliferation and at later stages by remodelling of the projections of SGL neurons.

Developmental remodelling of supragranular layer axons and formation of cortical pathways

Corticogenesis in the monkey begins at E40 with the generation of layer 1 and neuron production of the cortex is terminated by E100 (Rakic 1974). Layer 6 neurons are in place by E50 but they do not emit an axon toward distant cortical targets until E90 and when they arrive at their target area they contact preferentially the subplate, with a more modest invasion of the cortical plate. At this stage pathway length is considerably shorter given that the major phase of monkey brain expansion occurs between E120–165 (Batardiere et al 2002, Coogan & Van Essen 1996). SGL projections start to form at around E106 and reach peak levels at E120. The patchy FF projection of area V2 to V4 shows pronounced discontinuity from its onset, with only modest axon elimination at later stages which serves to sharpen the early formed pattern (Barone et al 1995), suggesting that FF pathways from

SGL are characterised by early target selection and directed growth (Barone et al 1996) (Fig. 3).

The directed growth exhibited by the FF pathways from the SGL contrasts with the growth properties of the feedback projections. Retrograde tracer injections in areas V1 and V4 in both primates and non-primates show that the characteristic SLN values in individual extrastriate areas are the consequence of a pre- and post-natal 45–90% reduction in the contribution from the SGL (Barone et al 1995, Kennedy et al 1989, Batardiere et al 1998, 2002).

In the adult cortex, cortico-cortical connections show high convergence and divergence values and therefore connect non visuotopic corresponding regions (Perkel et al 1986, Salin et al 1989, 1992). Quantification of convergence and divergence during cortical development shows that there is little modification in the topography of the pathways (Kennedy et al 1994). So it appears likely that the reduction of the projections from the SGL during development might serve to fine tune the SLN values. As discussed above the distribution of source neurons as measured by SLN values determine the cortical hierarchy. Since SLN values are changing during development it is possible that these changes reflect concomitant changes in the hierarchical organization of the cortex. However, when the SLN values in pre and postnatal cortex are analysed they show that the overall cortical hierarchy does not differ from that found in the adult, rather it would seem that the change in SLN values serves to sharpen the early-formed hierarchy (Batardiere et al 2002).

The remodelling of connections described above has been shown to result from changes in the convergence/divergence values of the SGL via pruning of axon collaterals neurons rather than neuron death (Barone et al 1995, 1998). The number

---

FIG. 4.   Laminar distribution of projection neurons determine hierarchy in macaque adult cortex. (A) Cartoon illustrating the distribution of labelled neurons in feedforward and feedback projections after injection of a retrograde tracer in the target area. Each area exhibits a specific SLN value, which determines its hierarchical distance from the target area. Distant feedforward projections have SLN values of 100 (e.g. Va). More proximal areas have lower SLN values. Distant feedback projections have SLN values of 0 (e.g. Vf). More proximal feedback areas have higher SLN values. SLN values of 40–60% correspond to lateral connections. Because of the cortical curvature, and the non-uniform distribution of labelled neurons in the projection, stable values of SLN require examining the distribution of labelled neurons at closely spaced intervals. Counts of neurons on successive sections show density profiles (B), the smoothness of which indicates appropriate sampling frequency. (C) Each area returns specific SLN values. (D) Hierarchical model of cortical areas connected to the target area according to the relationship between the SLN% and the distance rule (Barone et al 2000). (E) Total number of projections neurons is calculated for each area. This makes it possible to calculate the relative contribution of each area to the total afferent connectivity of the target area.

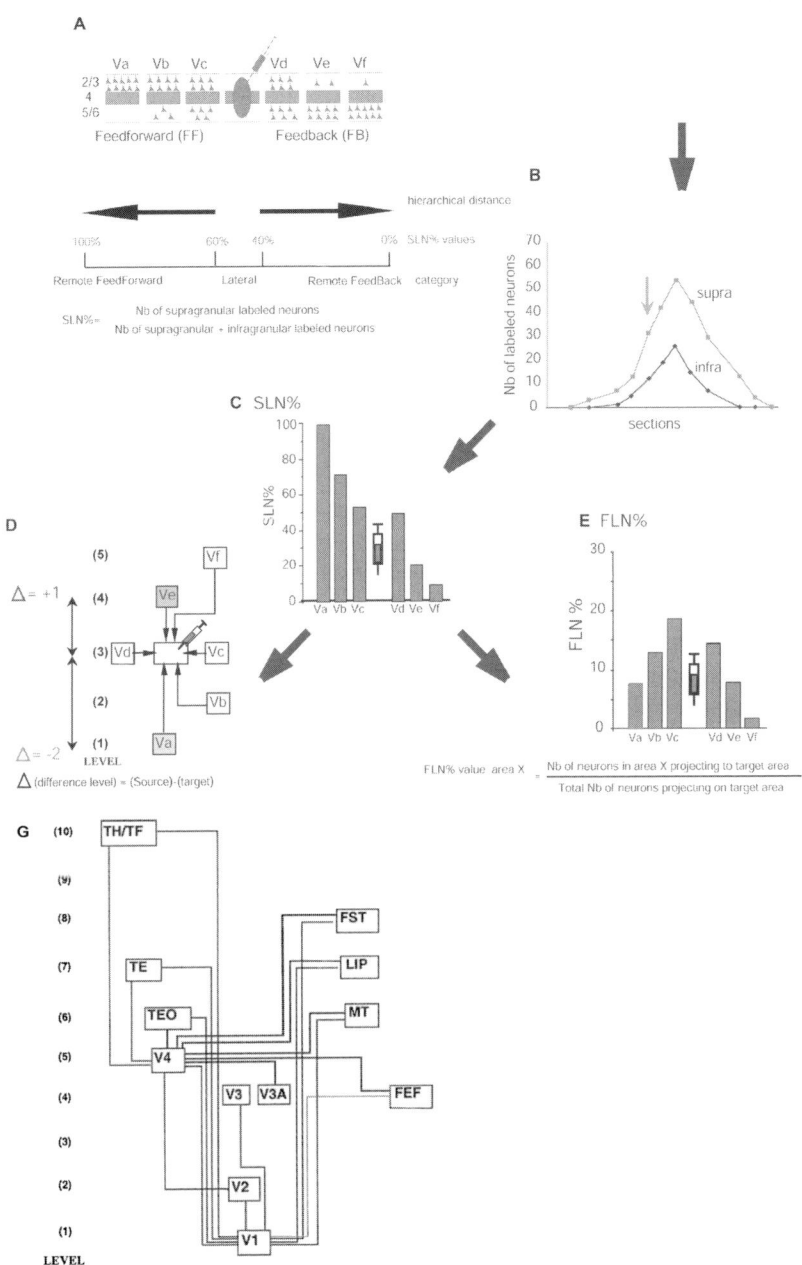

A

Va Vb Vc   Vd Ve Vf

2/3
4
5/6

Feedforward (FF)   Feedback (FB)

hierarchical distance

100%   60%   40%   0% SLN% values

Remote FeedForward   Lateral   Remote FeedBack   category

SLN%= Nb of supragranular labeled neurons
───────────────────────────────────────
Nb of supragranular + infragranular labeled neurons

B

Nb of labeled neurons

supra

infra

sections

C SLN%

SLN%

Va Vb Vc   Vd Ve Vf

D

(5)   Vf

△ = +1   (4)   Ve

(3) Vd   Vc

(2)   Vb

△ = -2   (1)   Va
        LEVEL

△ (difference level) = (Source)-(target)

E FLN%

FLN %

Va Vb Vc   Vd Ve Vf

FLN% value area X = Nb of neurons in area X projecting to target area
                       ──
                       Total Nb of neurons projecting on target area

G (10) TH/TF

(9)

(8)   FST

(7)   TE   LIP

(6)   TEO   MT

(5)   V4

(4)   V3 V3A   FEF

(3)

(2)   V2

(1)   V1

LEVEL
        (Modified from Felleman and Van Essen 1991)

of connections made by the SGL is ultimately determined by the numbers of neurons in the different sublayers of the SGL and the specification of neuron number has been shown to be area-specific (Lukaszewicz et al 2005, 2006) and to be modulated by thalamic projections to the germinal zone of the primate (Dehay et al 2001).

Generation of primate supragranular layer neurons

In the primate visual cortex, more than 75% of cortical neurons destined for the upper layers originate from subventricular zone (SVZ) precursors (Lukaszewicz et al 2005). In the primate there is an important expansion of the SVZ that includes a thick outer component (OSVZ), which is not observed in the rodent (Smart et al 2002) (Fig. 5). The OSVZ exhibits a number of unique features. Contrary to what is observed in the rodent, where the VZ is the major germinal compartment throughout corticogenesis, the primate VZ declines rapidly during the course of corticogenesis. This decline is associated with an early appearance of the SVZ followed by the OSVZ. This primate-specific organization, first described in the monkey has also been observed in the developing human cortex (Zecevic et al 2005).

Enlargement of the SVZ precursor pool in the primate might correspond to an evolutionary adaptive mechanism ensuring the increased neuronal output necessary to build a more highly developed neocortex with a pronounced cytological complexification of the SGL (Dehay et al 1993, Smart et al 2002, Lukaszewicz et al 2005, Dehay & Kennedy 2007). The rodent SVZ is only partially self-sustaining and has to receive a constant supply of precursors from the VZ (Reznikov et al 1997, Noctor et al 2004, Miyata et al 2004, Haubensak et al 2004, Wu et al 2005). In primates self-renewal (i.e. precursor division leading to an increase in numbers of precursors) would appear to be much more pronounced in the OSVZ than in rodent SVZ (Smart et al 2002, Lukaszewicz et al 2005).

One major cell type of the cortical germinal compartments is the radial glial cells (RGC) (Rakic 1972). A major research breakthrough occurred when it was shown that the RGC are not only part of a glial scaffolding, but also constitute multipotent cortical progenitors (Malatesta et al 2000, Noctor et al 2001, 2002). There is further heterogeneity of RGC in the embryonic primate, where a fraction cease dividing and function as migration scaffolding for several months before re-initiating proliferation and generating astrocytes (Schmechel & Rakic 1979, Rakic 2003).

There is a major difference in the cellular composition of the primate OSVZ with respect to the rodent SVZ. Whereas in the rodent all RGC nuclei are restricted to the VZ, RGC somata are morphologically identified in the OSVZ of the primate (Lukaszewicz et al 2005, Levitt et al 1981). We have some preliminary evidence that substantial proportions of precursors of the primate OSVZ express Pax6,

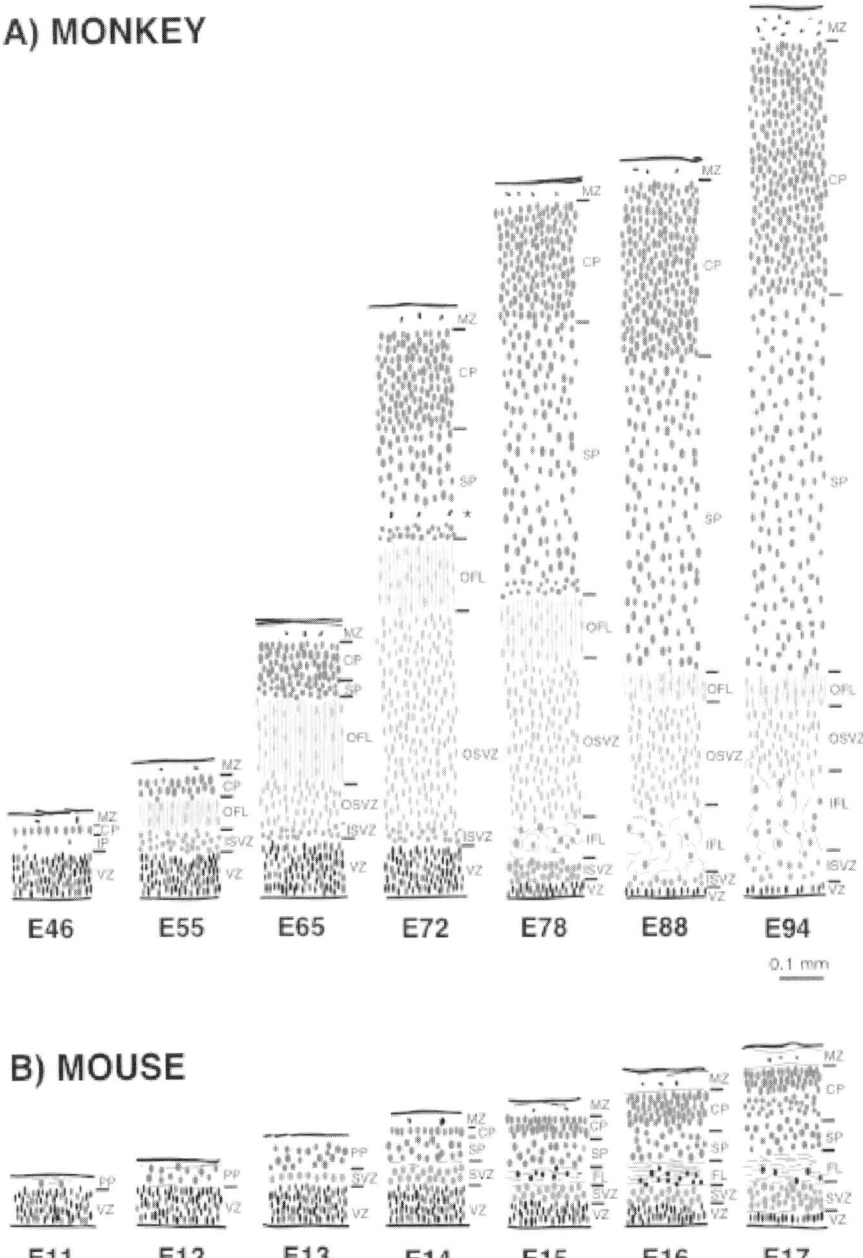

A) MONKEY

E46   E55   E65   E72   E78   E88   E94

0.1 mm

B) MOUSE

E11   E12   E13   E14   E15   E16   E17

FIG. 5.   Comparison of human and mouse germinal zones at equivalent developmental stages. These drawings are transects through presumptive Area 17 in (*A*) monkey and (*B*) mouse dorsal cortex at comparable developmental stages. The depth of each layer is drawn to a common scale. In the primate, an early appearing outer fibre layer (OFL) forms a major landmark from E55 onwards. The ventricular zone (VZ) declines progressively after E65. The subventricular zone, by contrast, increases progressively in depth and by E72 is divided into an inner subventricular zone (ISVZ) and outer subventricular zone (OSVZ) by an intruding inner fibre layer (IFL). The increase in the OSVZ is particularly important between E65 and E72 and occurs as the VZ declines.

which characterizes RGC identity in the rodent (Hartfuss et al 2001, Götz et al 1998, Malatesta et al 2000).

A number of observations link production of infragranular layers to VZ and SVZ to production of SGL. Although SVZ are derived from VZ precursors, there are clear differences in gene expression between the two precursor pools and these differences correlate with distinct neuronal progeny. For instance, Otx1 and Fezl are expressed in VZ precursors, down-regulated in SVZ and subsequently up-regulated in subsets of deep-layer neurons (Frantz et al 1994, Chen et al 2005a,b, Molyneaux et al 2005, Arlotta et al 2005). Furthemore, both Otx1 and Fezl play a crucial role in specifying the axonal projections of subsets of lower layer neurons. Recent studies in mice show that several transcription factors (Cux2, Tbr2, Satb2 and Nex) (Britanova et al 2005, Zimmer et al 2004, Nieto et al 2004, Wu et al 2005) as well as the non-coding RNA Svet1 (Tarabykin et al 2001) are selectively expressed in both the SVZ and in upper layer neurons. This congruency of expression of genes first in SVZ progenitors and subsequently in supragranular neurons together with time-lapse microscopy observations suggest that the SVZ gives rise to upper layer neurons (Tarabykin et al 2001, Zimmer et al 2004, Noctor et al 2004) and that the SVZ is not uniquely a source of glial cells as was thought until recently (Bayer & Altman 1991).

Consistent with these findings, distinct molecular mechanisms have been identified for the specification of infra- and supragranular neuronal lineages. Studies from mutant mice show that Ngn1 and Ngn2 activity is required for the specification of a subset of infragranular neurons but not for the specification of SGL neurons. In contrast, *Pax6* and *Tlx*, two genes required for the formation of the SVZ (Roy et al 2004, Nieto et al 2004, Zimmer et al 2004), are synergistically involved in the specification of SGL (Schuurmans et al 2004). It can therefore be hypothesized that the selective expansion of the SGL compartment in the primate cortex results from modifications of the *Pax6/Tlx*-related specification without modification of the Ngn specification mechanisms (Schuurmans et al 2004).

Control mechanisms of supragranular layer neuron production

The co-ordinated regulation of two cardinal cell cycle parameters of cortical precursors determines neuronal production via the regulation of the size of the precursor pool: the duration of the cell cycle and the relative frequency of cell cycle reentry compared with cell cycle exit (Dehay & Kennedy 2007). *In vivo* and *ex vivo* analysis of the cell cycle regulation of OSVZ precursors of the primate visual cortex has shed light on the molecular correlates of area-specific differences in proliferation that underlie area-specific differences in the thickness of supragranular layers.

Compared to area 18 OSVZ precursors, area 17 OSVZ precursors are characterized by both a shorter cell cycle duration—due to a reduction of the G1 phase—and an increased relative frequency of cell cycle re-entry. These areal differences on OSVZ precursor cell cycle regulation are associated with significant differences in the level of expression of molecular regulators of the G1/S transition p27^{kip1} and CyclinE. The *ex vivo* up and down modulation of their level of expression significantly affects cell cycle re-entry and the rate of cell cycle progression and stresses the role of the G1 phase regulation in corticogenesis (Lukaszewicz et al 2005, 2006). Mathematical modelling of the observed differences in both rates of cell cycle re-entry and in G1 phase duration show that the combined variation of these two parameters are sufficient to generate the enlarged SGL that distinguishes area 17 from area 18. These results show that variations of G1 phase duration and the coordinated variation in mode of division (Lukaszewicz et al 2002, Gotz & Huttner 2005) contribute directly to regulate neuron number (Dehay & Kennedy 2007).

In vitro work on mouse cortical precursors (Dehay et al 2001) indicates that thalamic afferents control corticogenesis by modulating rates of proliferation. Embryonic thalamic axons release a mitogenic factor that increases the proliferative capacity of mouse cortical precursors during generation of SGL by decreasing the G1 duration and by promoting cell cycle re-entry (Dehay et al 2001). In the monkey, the LGN axons that selectively project onto the OSVZ of area 17 could be responsible for the temporally and spatially restricted stimulation of proliferation that results in the transient upsurge of the size of SGL precursor pool in area 17 (Dehay et al 1993, Smart et al 2002, Lukaszewicz et al 2005). There is also *in vivo* suggestion in the primate that embryonic thalamic axons could impact on areal size and specification via an early influence on neuron production during cortical neurogenesis (Dehay et al 1989, 1996a,b, Rakic 1988). Because thalamic axons are precisely targeted on to distinct cortical areas (reviewed in Lopez-Bendito & Molnar 2003) they will be able to differentially affect rates of precursor proliferation and of neuron production across the germinal zones and therefore determine local areal cytoarchitectonic features.

Concluding remarks

Computation is a physical process implemented in a physical medium. The primate neocortex is the most sophisticated computational device known to humans and it has been argued that its mode of function and the local microcircuitry that subtend it might be identical across areas (Douglas et al 1989). Unlike conventional computers it does not rely on an external agency for its construction and programming. Instead, the entire circuitry is self-constructed by replication and interaction of the germinal cells and their derived neuronal types. Unlike the majority of tissues that emphasize local three-dimensional organization where cells contact their neighbours, the CNS is characterized by complex connectional topologies over very large spatial scales The underlying need for this organization is due to the fact that information processing is finally about selective communication between particular processors. Such functions can be represented as a graph-like topology composed of processing nodes (single or populations of neurons), and their connecting communication edges (axons).

One strategy for self-constructing these topologies is to assemble spatially organized cell types, constituting prototypical modules. This prototypical organization is then refined into more specific connections by both network and environmental informational data, and so provides the specific circuit functionality required for a particular information processing task. Thus, it is probable that the fundamental component of the neuronal organization arises relatively simply out of the genetic determination of cell populations, rules for migration, and cell growth. During the unfolding of this construction process, the neuronal networks in response to environmental factors begin to configure and tune themselves for effective behaviour of the organism.

An understanding of this self-construction and programming of neural computational networks would be a major contribution to the explanation of brain operation. One significant circuit regularity of the neocortex is the lattice-like patchy organization of the lateral connections between pyramidal neurons of the SGL that we refer to as Daisy architecture (Rockland & Lund 1983, Douglas & Martin 2004). With the exception of rodents (Van Hooser et al 2006), Daisy architecture is ubiquitous across areas and species and highly pronounced in primates (Douglas & Martin 2004). This lattice, plus the compexification of the SGL in carnivores and primates reaching their highest development in human, suggests that elaboration of certain common circuit principles could be associated with a huge increase in behavioural performance. These neuroanatomical clues, together with their strong implications for function, call for intensified research of the development of the SGL in species other than rodents and particularly in primates.

Here we have reviewed the evidence that thalamic afferents shape cortical cytoarchitecture via modulation of the cell cycle kinetics of the cortical precursors. There is also evidence that the pattern of activity in ascending pathways plays a key role in the elaboration of the Daisy architecture. In ferret the appropriate surgical manipulation of the fetus leads to a rewiring of the peripheral input to the thalamus such that retinal fibres innervate a deafferented auditory thalamic nuclei (Sur et al 1988). The rewired auditory thalamus acquires novel morphological features and visual responses. At the level of the rewired auditory cortex there are computational response features such as orientation selectivity and directionality. Interestingly, in the re-wired auditory cortex the Daisy architecture is expanded so that it no longer resembles auditory Daisy architecture but instead that Daisy architecture normally found in visual cortex (Gao & Pallas 1999, Sharma et al 2000). This reconfiguration of the cortex is likely the consequence of changes in the pattern of activity given that following cross-modal reorganization the auditory thalamo-cortical projections are similar to those found in the normal ferret (Pallas et al 1990). These results show that the thalamic fibres initially set up the cortical cytoarchitecture and latter patterned activity, related to the statistical properties of the visual environment, then impact on the elaboration of the computational architecture of the cortex.

Acknowledgements

This work was supported by FP6-2005 IST-1583 (HK and RD); ANR-05-NEUR-088 (HK); ANR-06-NEUR-CMMCS (CD).

References

Arlotta P, Molyneaux BJ, Chen J, Inoue J, Kominami R, Macklis JD 2005 Neuronal subtype-specific genes that control corticospinal motor neuron development in vivo. Neuron 45:207–221

Barone P, Dehay C, Berland M, Bullier J, Kennedy H 1995 Developmental remodeling of primate visual cortical pathways. Cereb Cortex 5:22–28

Barone P, Dehay C, Berland M, Kennedy H 1996 Role of directed growth and target selection in the formation of cortical pathways: prenatal development of the projection of area V2 to area V4 in the monkey. J Comp Neurol 374:1–20

Barone P, Berland M, Kennedy H 1998 Changes in convergence underly the developmental remodeling of feedback cortical connections in the monkey. Eur J Neurosci 10 Supp 10:422

Barone P, Batardiere A, Knoblauch K, Kennedy H 2000 Laminar distribution of neurons in extrastriate areas projecting to visual areas V1 and V4 correlates with the hierarchical rank and indicates the operation of a distance rule. J Neurosci 20:3263–3281

Batardiere A, Barone P, Dehay C, Kennedy H 1998 Area-specific laminar distribution of cortical feedback neurons projecting to cat area 17: quantitative analysis in the adult and during ontogeny. J Comp Neurol 396:493–510

Batardiere A, Barone P, Knoblauch K et al 2002 Early specification of the hierarchical organization of visual cortical areas in the macaque monkey. Cereb Cortex 12:453–465

Bayer SA, Altman J 1991 Neocortical development, First Edition. New York: Raven Press

Britanova O, Akopov S, Lukyanov S, Gruss P, Tarabykin V 2005 Novel transcription factor Satb2 interacts with matrix attachment region DNA elements in a tissue-specific manner and demonstrates cell-type-dependent expression in the developing mouse CNS. Eur J Neurosci 21:658–668

Cauller L 1995 Layer I of primary sensory neocortex: where top-down converges upon bottom-up. Behav Brain Res 71:163–170

Chen B, Schaevitz LR, McConnell SK 2005a Fezl regulates the differentiation and axon targeting of layer 5 subcortical projection neurons in cerebral cortex. Proc Natl Acad Sci USA 102:17184–17189

Chen JG, Rasin MR, Kwan KY, Sestan N 2005b Zfp312 is required for subcortical axonal projections and dendritic morphology of deep-layer pyramidal neurons of the cerebral cortex. Proc Natl Acad Sci USA 102:17792–17797

Coogan TA, Van Essen DC 1996 Development of connections within and between areas V1 and V2 of macaque monkeys. J Comp Neurol 372:327–342

Dehay C, Kennedy H 2007 Cell-cycle control and cortical development. Nat Rev Neurosci 8:438–450

Dehay C, Horsburgh G, Berland M, Killackey H, Kennedy H 1989 Maturation and connectivity of the visual cortex in monkey is altered by prenatal removal of retinal input. Nature 337:265–267

Dehay C, Giroud P, Berland M, Smart I, Kennedy H 1993 Modulation of the cell cycle contributes to the parcellation of the primate visual cortex. Nature 366:464–466

Dehay C, Giroud P, Berland M, Killackey H, Kennedy H 1996a Contribution of thalamic input to the specification of cytoarchitectonic cortical fields in the primate: effects of bilateral enucleation in the fetal monkey on the boundaries, dimensions, and gyrification of striate and extrastriate cortex. J Comp Neurol 367:70–89

Dehay C, Giroud P, Berland M, Killackey H, Kennedy H 1996b Phenotypic characterisation of respecified visual cortex subsequent to prenatal enucleation in the monkey: development of acetylcholinesterase and cytochrome oxidase patterns. J Comp Neurol 376:386–402

Dehay C, Savatier P, Cortay V, Kennedy H 2001 Cell-cycle kinetics of neocortical precursors are influenced by embryonic thalamic axons. J Neurosci 21:201–214

Douglas RJ, Martin KA 2004 Neuronal circuits of the neocortex. Annu Rev Neurosci 27:419–451

Douglas RJ, Martin KAC, Whitteridge D 1989 A canonical microcircuit for neocortex. Neural Comput 1:480–488

Douglas RJ, Koch C, Mahowald M, Martin KA, Suarez HH 1995 Recurrent excitation in neocortical circuits. Science 269:981–985

Falchier A, Clavagnier S, Barone P, Kennedy H 2002 Anatomical evidence of multimodal integration in primate striate cortex. J Neurosci 22:5749–5759

Felleman DJ, Van Essen DC 1991 Distributed hierarchical processing in the primate cerebral cortex. Cereb Cortex 1:1–47

Frantz GD, Weimann JM, Levin ME, McConnell SK 1994 Otx1 and Otx2 define layers and regions in developing cerebral cortex and cerebellum. J Neurosci 14:5725–5740

Gao WJ, Pallas SL 1999 Cross-modal reorganization of horizontal connectivity in auditory cortex without altering thalamocortical projections. J Neurosci 19:7940–7950

Geisler WS, Diehl RL 2002 Bayesian natural selection and the evolution of perceptual systems. Philos Trans R Soc Lond B Biol Sci 357:419–448

Gotz M, Huttner WB 2005 The cell biology of neurogenesis. Nat Rev Mol Cell Biol 6:777–788

Götz M, Stoykova A, Gruss P 1998 Pax6 controls radial glia differentiation in the cerebral cortex. Neuron 21:1031–1044

Hartfuss E, Galli R, Heins N, Gotz M 2001 Characterization of CNS precursor subtypes and radial glia. Dev Biol 229:15–30

Haubensak W, Attardo A, Denk W, Huttner WB 2004 Neurons arise in the basal neuroepithelium of the early mammalian telencephalon: a major site of neurogenesis. Proc Natl Acad Sci USA 101:3196–3201

Hilgetag CC, O'Neill MA, Young MP 1996 Indeterminate organization of the visual system. Science 271:776–777

Hill RS, Walsh CA 2005 Molecular insights into human brain evolution. Nature 437:64–67

Hubel DH, Wiesel TN 1968 Receptive fields and functional architecture of monkey striate cortex. J Physiol 195:215–243

Kennedy H, Bullier J 1985 A double-labeling investigation of the afferent connectivity to cortical areas V1 and V2 of the macaque monkey. J Neurosci 5:2815–2830

Kennedy H, Bullier J, Dehay C 1989 Transient projections from the superior temporal sulcus to area 17 in the newborn macaque monkey. Proc Natl Acad Sci USA 86:8093–8097

Kennedy H, Salin P, Bullier J, Horsburgh G 1994 Topography of developing thalamic and cortical pathways in the visual system of the cat. J Comp Neurol 348:298–319

Killackey HP 1990 Neocortical expansion: an attempt towards relating phylogeny and ontogeny. J Cogn Neurosci 2:1–17

LeVay S, Wiesel TN, Hubel DH 1980 The development of ocular dominance columns in normal and visually deprived monkeys. J Comp Neurol 191:1–51

Levitt P, Cooper ML, Rakic P 1981 Coexistence of neuronal and glial precursor cells in the cerebral ventricular zone of the fetal monkey: an ultrastructural immunoperoxidase analysis. J Neurosci 1:27–39

Lopez-Bendito G, Molnar Z 2003 Thalamocortical development: how are we going to get there? Nat Rev Neurosci 4:276–289

Lukaszewicz A, Savatier P, Cortay V, Kennedy H, Dehay C 2002 Contrasting effects of basic fibroblast growth factor and neurotrophin 3 on cell cycle kinetics of mouse cortical stem cells. J Neurosci 22:6610–6622

Lukaszewicz A, Savatier P, Cortay V et al 2005 G1 phase regulation, area-specific cell cycle control, and cytoarchitectonics in the primate cortex. Neuron 47:353–364

Lukaszewicz A, Cortay V, Giroud P et al 2006 The concerted modulation of proliferation and migration contributes to the specification of the cytoarchitecture and dimensions of cortical areas. Cereb Cortex 16 Suppl 1:i26–i34

Malatesta P, Hartfuss E, Gotz M 2000 Isolation of radial glial cells by fluorescent-activated cell sorting reveals a neuronal lineage. Development 127:5253–5263

Miyata T, Kawaguchi A, Saito K, Kawano M, Muto T, Ogawa M 2004 Asymmetric production of surface-dividing and non-surface-dividing cortical progenitor cells. Development 131:3133–3145

Molyneaux BJ, Arlotta P, Hirata T, Hibi M, Macklis JD 2005 Fezl is required for the birth and specification of corticospinal motor neurons. Neuron 47:817–831

Nieto M, Monuki ES, Tang H et al 2004 Expression of Cux-1 and Cux-2 in the subventricular zone and upper layers II-IV of the cerebral cortex. J Comp Neurol 479: 168–180

Noctor SC, Flint AC, Weissman TA, Dammerman RS, Kriegstein AR 2001 Neurons derived from radial glial cells establish radial units in neocortex. Nature 409:714–720

Noctor SC, Flint AC, Weissman TA, Wong WS, Clinton BK, Kriegstein AR 2002 Dividing precursor cells of the embryonic cortical ventricular zone have morphological and molecular characteristics of radial glia. J Neurosci 22:3161–3173

Noctor SC, Martinez-Cerdeno V, Ivic L, Kriegstein AR 2004 Cortical neurons arise in symmetric and asymmetric division zones and migrate through specific phases. Nat Neurosci 7:136–144

O'Leary DDM 1989 Do cortical areas emerge from a protocortex? Trends Neurosci 12:400–406

Pallas SL, Roe AW, Sur M 1990 Visual projections induced into the auditory pathway of ferrets. I. Novel inputs to primary auditory cortex (AI) from the LP/pulvinar complex and the topography of the MGN-AI projection. J Comp Neurol 298:50–68

Perkel DJ, Bullier J, Kennedy H 1986 Topography of the afferent connectivity of area 17 in the macaque monkey: a double-labelling study. J Comp Neurol 253:374–402

Rakic P 1972 Mode of cell migration to the superficial layers of fetal monkey neocortex. J Comp Neurol 145:61–84

Rakic P 1974 Neurons in rhesus monkey visual cortex: systematic relation between time of origin and eventual disposition. Science 183:425–427

Rakic P 1988 Specification of cerebral cortical areas. Science 241:170–176

Rakic P 2003 Developmental and evolutionary adaptations of cortical radial glia. Cereb Cortex 13:541–549

Reznikov K, Acklin SE, van der Kooy D 1997 Clonal heterogeneity in the early embryonic rodent cortical germinal zone and the separation of subventricular from ventricular zone lineages. Dev Dyn 210:328–343

Rockland KS, Lund JS 1983 Intrinsic laminar lattice connections in primate visual cortex. J Comp Neurol 216:303–318

Roy K, Kuznicki K, Wu Q et al 2004 The Tlx gene regulates the timing of neurogenesis in the cortex. J Neurosci 24:8333–8345

Salin PA, Bullier J, Kennedy H 1989 Convergence and divergence in the afferent projections to cat area 17. J Comp Neurol 283:486–512

Salin PA, Girard P, Kennedy H, Bullier J 1992 Visuotopic organization of corticocortical connections in the visual system of the cat. J Comp Neurol 320:415–434

Scannell JW, Grant S, Payne BR, Baddeley R 2000 On variability in the density of corticocortical and thalamocortical connections. Philos Trans R Soc Lond B Biol Sci 355:21–35

Schmechel DE, Rakic P 1979 Arrested proliferation of radial glial cells during midgestation in rhesus monkey. Nature 277:303–305

Schuurmans C, Armant O, Nieto M et al 2004 Sequential phases of cortical specification involve Neurogenin-dependent and -independent pathways. EMBO J 23:2892–2902

Sharma J, Angelucci A, Sur M 2000 Induction of visual orientation modules in auditory cortex. Nature 404:841–847

Shepard RN 2001 Perceptual-cognitive universals as reflections of the world. Behav Brain Sci 24:581–601; discussion 652–671

Smart IH, Dehay C, Giroud P, Berland M, Kennedy H 2002 Unique morphological features of the proliferative zones and postmitotic compartments of the neural epithelium giving rise to striate and extrastriate cortex in the monkey. Cereb Cortex 12:37–53

Sur M, Rubenstein JL 2005 Patterning and plasticity of the cerebral cortex. Science 310:805–810

Sur M, Garraghty PE, Roe AW 1988 Experimentally induced visual projections into auditory thalamus and cortex. Science 242:1437–1441

Tarabykin V, Stoykova A, Usman N, Gruss P 2001 Cortical upper layer neurons derive from the subventricular zone as indicated by Svet1 gene expression. Development 128:1983–1993

Thompson ID, Kossut M, Blakemore C 1983 Development of orientation columns in cat striate cortex revealed by 2-deoxyglucose autoradiography. Nature 301:712–715

Van der Loos H 1977 Structural changes in the cerebral cortex upon modification of the periphery: barrels in somatosensory cortex. Philos Trans R Soc Lond B Biol Sci 278:373–376

Van Hooser SD, Heimel JA, Chung S, Nelson SB 2006 Lack of patchy horizontal connectivity in primary visual cortex of a mammal without orientation maps. J Neurosci 26:7680–7692

Vezoli J, Falchier A, Jouve B, Knoblauch K, Young M, Kennedy H 2004 Quantitative analysis of connectivity in the visual cortex: extracting function from structure. Neuroscientist 10:476–482

Von de Malsburg C, Singer S 1988 Principles of cortical network organization. In: Rakic P, Singer W (eds) Neurobiology of Neocortex. John Wiley & Sons, p 69–99

Wu SX, Goebbels S, Nakamura K et al 2005 Pyramidal neurons of upper cortical layers generated by NEX-positive progenitor cells in the subventricular zone. Proc Natl Acad Sci USA 102:17172–17177

Zecevic N, Chen Y, Filipovic R 2005 Contributions of cortical subventricular zone to the development of the human cerebral cortex. J Comp Neurol 491:109–122

Zimmer C, Tiveron MC, Bodmer R, Cremer H 2004 Dynamics of Cux2 expression suggests that an early pool of SVZ precursors is fated to become upper cortical layer neurons. Cereb Cortex 14:1408–1420

DISCUSSION

Kriegstein: I am sure you have thought about what the substance or factor is that the corticothalamic fibres use to regulate progenitor cell division, and you have a good *in vitro* assay to sort this out. Has there been any progress on this?

Kennedy: We have used antibodies against FGF2, which strongly reduce the mitogenic effect in rodents. We think Sonic hedgehog might be involved. Beyond this we have no idea.

Hevner: There is some evidence that different types of progenitors may cycle at different rates, particularly neurogenic progenitors versus radial glia. In your measurements did you distinguish between VZ and SVZ, or were you pooling all the progenitors to measure the cell cycle length?

Kennedy: In the *in vitro* assay, explants of LGN are co-cultured with dissociated precursors from VZ and SVZ of area 17 and area 18. We also have made measurements both in *in vivo* and *ex vivo* on organotypic slices in which we have been able to look at the most outer part of the OSVZ. We are comparing the part of the OSVZ that is up against thalamic fibres with the part that is not. We are finding here they cycle faster with more proliferative divisions The idea that cell cycle duration is homogeneous across precursors is simply very wrong. There is strong evidence from our lab and that of Huttner that proliferative divisions are fast and the differentiation divisions are much slower (Lukaszewicz et al 2002, Calegari et al 2005).

Rakic: I agree that co-operation between specific thalamic nuclei and cortical areas is essential. But in terms of specificity, it is important to recognize that the early set of geniculo-cortical axons in monkey project only to the part of the ventricular zone that will produce neurons for area 17. This means that proliferative cells subjacent to this area have some kind of receptor or ligand that attract geniculate axons to the segment destined for area 17 but not area 18. In cat, for example, neurons of the lateral geniculate nucleus project to both areas 17 and 18. This is another proof of the early specification of these cells in the SVZ.

Kennedy: Absolutely. I showed unpublished data that dissociated cultures from area 18 that do not respond at all to the mitogenic factors released by the LGN axons.

Mallamaci: Are progenitors from other cortical areas responsive to influences coming from the LGN? In other words, if you put cells coming from, for example, the LGN and other non-visual areas in co-culture, do these progenitors respond in a similar way?

Kennedy: No.

Mallamaci: Is there a gradient of responsivity along the rostrocaudal and mediolateral axes, or is the responsivity tightly restricted to areas?

Kennedy: That's an important point. *In vivo* there is a gradient of high rates of mitogenesis in area 17 to lower values in area 18. We were initially disappointed, we had been hoping to find a boundary in the ventricular zone where there would be high rates of proliferation on one side and low rates on the other. It is not like that; there is a gradient. How we go from that gradient in the ventricular zone to the stepwise function in the cortex is a mystery. I think that the different rates of migration (fast in 17 slow in 18) coupled with differential tangential expansion must be involved. I agree with Arnold Kriegstein's idea that the thickness of the subventricular zone and tangential expansion are interconnected. Tangential expansion, differential rates of migration and proliferation all lead to determining the radial thickness and areal dimensions of the cortical areas. In passing, I'd like to add that the primate 17/18 border is a freak: it is a useful model, but there are no other borders like it.

Mallamaci: So, if you artificially combine the progenitors coming from area 18 with LGN neurons do you get a boost of proliferation?

Kennedy: We have done that, and our unpublished results are quite clear, the mitogenic effect of the LGN is specific to area 17 precursors.

Macklis: About four years ago Karl Deisseroth had a paper on mixed progenitors from forebrain that they claimed were glutamate-activated in the cell cycle increasing proliferation (Deisseroth et al 2004). This is contrary to what you were saying about the specificity. They were arguing that a proliferative zone that is activated might want to expand to receive more activity. It wasn't clear from the data whether this was a survival effect on the progenitors or proliferation.

Kriegstein: The Deisseroth paper concerned adult hippocampus. They were looking at activity-dependent regulation of cell cycle events. Some time ago we looked at cortical neuroepithelial cells and saw that they had GABA and glutamate receptors (LoTurco et al 1995). We then looked at whether these were influencing cell cycle events. Activating either GABA or glutamate receptors had a net inhibitory effect on neurogenesis. Conversely, with receptor blockade there was an increase in the number of cells entering the cell cycle.

Macklis: I remember those data, and also thinking at the time when you published yours that the activation might be saying 'let me stabilize and mature', whereas here it says 'let me divide more'. I wonder whether there could be differential effects.

Kriegstein: Pasko Rakic did a study in which he looked at the SVZ versus the VZ. He concluded that the GABA and glutamate effects were opposite in the two zones (Haydar et al 2000). The SVZ effect was the inhibitory one.

Rakic: That is correct. The study indicated that these cells are also having some kind of receptors that are differentially expressed in the progenitors in the VZ and SVZ, and made them to respond to GABA in a different way. Another interesting question that no one brings up is the length of cell cycle, which is at least four times longer in primates than in rodents (Kornack & Rakic 1998), and why does such a paradoxical thing happen? Since there are much more cells in the primate cortex, you would predict that cells should divide faster, not slower. One possible reason for this could be that it gives them time to interact. In terms of lateral expansion and vertical expansion of the cortex during evolution, I think they are separate events that are regulated by different genes. For example, dolphin has a thin cortex with more radial units and fewer cells per unit. Ungulates have a thicker cortex with more cells per unit. They are separate cellular events and are probably induced and affected by different genes. The SVZ probably plays an important role in these events, but it isn't sufficient to explain lateral tangential expansion. It is not only the superficial layers that need to expand, but also the deep layers (Rakic 1995).

Molnár: I think we can go even further back in looking at the role of thalamocortical interactions in cortical circuit formation. If we consider the whole idea of self-organization, then there are examples when just a modality-specific stimulus fed to the developing cortical circuit is enough to instruct aspects of development even without changing the thalamic relationship. This is the case for the patchy distribution of intracortical supragranular connections in the auditory cortex of the ferret. Normally, we don't see these patchy supragranular intracortical connections in auditory cortex, yet when the retina is wired into the MGN as Mriganka Sur's and Sarah Pallas' group did, then the auditory cortex starts to show the signs of patchy intracortical connectivity (Sur & Leamey 2001, Pallas 2001). With the very same thalamocortical relationship you have only changed the modality of the signals which are delivered to the forming circuitry and you will begin to see the patchy distribution of intracortical connectivity. It is the quality of the stimulus that could play on these wires. Even changing the thalamocortical relationship wouldn't be so important for these kind of self-organizations.

Kennedy: I think we need to distinguish two things here. The first is the mitogenic effect of the thalamus on the cortex that we have described in the rodent and today in my presentation. The second is the intriguing effect that Sur showed in re-wired cortex. Sur and his colleagues showed that in ferret when retinal fibres are induced to innervate the auditory thalamus, the patchy connectivity of the supragranular layers in the auditory cortex takes on the characteristics of the patchy connectivity normally found in the visual cortex and not that in the auditory cortex. Because the auditory thalamic fibres have a configuration as in the normal animal, these authors concluded that it is the firing pattern which is important.

Molnár: Sur and colleagues injected tracer into the ferret auditory cortex, which now processed visual input after their optical recording experiments. The connections showed patchiness. In normal controls in the primary auditory cortex processing normal auditory signals there are no patches.

Kennedy: Yes there are.

O'Leary: You mentioned that thalamocortical afferents in the monkey grow into the cortical primordium before neurogenesis begins. What is the relationship between the waiting period in the SVZ compartment relative to the generation of layer 4 and thalamocortical input? Do thalamocortical axons wait until layer 4 neurons are generated?

Kennedy: This is what we wanted to look at, but it wasn't that easy. It looks like nothing much is happening to the thalamocortical afferents, other than that they innervate the top part of the germinal zones. There isn't a layer 4 until about E75–E80.

O'Leary: In principal, then, the distal ends of the thalamcortical afferents are in a position where they can release a mitogenic factor.

Kennedy: Yes. Don't forget that the subplate is huge, and some of the thalamic afferents go into the subplate.

References

Calegari F, Haubensak W, Haffner C, Huttner WB 2005 Selective lengthening of the cell cycle in the neurogenic subpopulation of neural progenitor cells during mouse brain development. J Neurosci 25:6533–6538

Deisseroth K, Singla S, Toda H, Monje M, Palmer TD, Malenka RC 2004 Excitation-neurogenesis coupling in adult neural stem/progenitor cells. Neuron 42:535–552

Haydar TF, Wang F, Schwartz ML, Rakic P 2000 Differential modulation of proliferation in the neocortical ventricular and subventricular zones. J Neurosci 20:5764–5774

Kornack DR, Rakic P 1998 Changes in cell-cycle kinetics during the development and evolution of primate neocortex. Proc Natl Acad Sci USA 95:1242–1246

LoTurco JJ, Owens DF, Heath MJS, Davis MBE, Kriegstein AR 1995 GABA and glutamate depolarize cortical progenitor cells and inhibit DNA synthesis. Neuron 15:1287–1298

Lukaszewicz A, Savatier P, Cortay V, Kennedy H, Dehay C 2002 Contrasting effects of bFGF and NT3 on cell cycle kinetics of mouse cortical stem cells. J Neurosci 22:6610–6622

Pallas SL 2001 Intrinsic and extrinsic factors that shape neocortical specification. Trends Neurosci 24:417–423

Rakic P 1995 A small step for the cell—a giant leap for mankind: a hypothesis of neocortical expansion during evolution. Trends Neurosci 18:383–388

Sur M, Leamey CA 2001 Development and plasticity of cortical areas and networks. Nat Rev Neurosci 2:251–262

Molecular mechanisms of thalamocortical axon targeting

Nobuhiko Yamamoto, Takuro Maruyama, Naofumi Uesaka, Yasufumi Hayano, Makoto Takemoto and Akito Yamada

Neuroscience Laboratories, Graduate School of Frontier Biosciences, Osaka University, 1-3 Yamadaoka, Suita, Osaka 565-0871, Japan

Abstract. The thalamocortical (TC) projection in the mammalian brain is a well characterized system in terms of laminar specificity of neocortical circuits. To understand the mechanisms that underlie lamina-specific TC axon targeting, we studied the role of extracellular and cell surface molecules that are expressed in the upper layers of the developing cortex in *in vitro* culture techniques. The results demonstrated that multiple upper layer molecules co-operated to produce stop behaviour of TC axons in the target layer. Activity dependency of TC axon branching was also investigated in organotypic co-cultures of the thalamus and cortex. TC axon branches were formed dynamically by addition and elimination during the second week *in vitro*, when spontaneous firing increased in thalamic and cortical cells. Pharmacological blockade of firing or synaptic activity reduced the remodelling process, in particular branch addition, in the target layer. Together, these findings suggest that TC axon targeting mechanisms involve the regulation with multiple lamina-specific molecules and modification of the molecular mechanisms via neural activity.

2007 Cortical development: genes and genetic abnormalities. Wiley, Chichester (Novartis Foundation Symposium 288) p 199–211

The neocortex is fundamentally composed of six cell layers, which are distinguishable by the extrinsic and intrinsic circuitries they make. Such lamina-specific cortical connectivity offers a suitable model system in which to study axonal targeting mechanisms. The thalamocortical (TC) projection is well characterized in terms of the developmental time course as well as laminar specificity (López-Bendito & Molnár 2003). To this end, *in vitro* studies have demonstrated that there is a targeting mechanism by which TC axons recognize layer IV, their primary target layer (Yamamoto et al 1989, Molnár & Blakemore 1991, 1999, Bolz et al 1992, Yamamoto et al 1992, Yamamoto et al 1997). Our previous investigations have further indicated that growth-inhibitory factors in layers II/III and IV regulate termination of TC axons, whereas branch-inducing activity is enriched in layer IV (Yamamoto et al 2000a,b). On the other hand, TC axon targeting is also influenced by neural activity. One of the best examples is the effect of monocular deprivation, which induces the expansion of TC axon arbours serving the intact eye and the shrinkage of deprived eye arbours (Hubel et al 1977, Antonini et al 1999). Although such alteration is restricted to the tangential extent, a complete blockade of firing activity prevents lamina-specific TC branching (Herrmann & Shatz 1995).

Here we explored the molecular mechanisms of lamina-specific TC axon targeting and its activity dependency. First, how upper layer molecules are involved in the laminar specificity was investigated. For this purpose, the genes expressed in the upper layers (layer II/III–IV) were searched by constructing a subtraction cDNA library. Together with the molecules that had been found in the developing cortex, several molecules were raised as the candidate molecules that may be involved in TC axon growth. Subsequently, we studied the roles of these molecules in axon growth, using *in vitro* culture techniques. Second, we investigated how neural activity is involved in TC axon targeting, focusing on axon branching, in organotypic co-cultures of the thalamus and cortex by using time-lapse imaging, electrophysiological analysis and pharmacological manipulations.

The role of the molecules that are expressed in the upper layers

It has been reported that the molecules that can regulate axonal growth are expressed in the upper layers of the developing cortex. *Semaphorin 7A (Sema7A)*, a *semaphorin* family member, is expressed in layers II/III–IV of the neonatal rodent cortex (Xu et al 1998). *Cadherin 6 (Cad6)*, a classical cadherin, is also expressed in the upper layers (Suzuki et al 1997). These distributions match the laminar expression of the growth-inhibitory activity that is responsible for TC axon termination (Yamamoto et al 2000b). Moreover, *EphrinA5*, a ligand of Eph receptor tyrosine kinase, and *semaphorin 3A (Sema3A)* are expressed in the upper layers as well as the deep layers (Castellani et al 1998, Gao et al 1998, Skaliora et al 1998, Donoghue & Rakic 1999, Mackarehtschian et al 1999, Vanderhaeghen et al 2000, Yabuta et al 2000). In addition, receptors for Sema7A and EphrinA5 are expressed in the thalamic nuclei (Vanderhaeghen et al 2000, Takemoto et al 2002, Pasterkamp et al 2003). Cad6, which is known to be a homophilic binding protein, is also expressed in sensory thalamic regions (Suzuki et al 1997). Expressions of these receptor molecules imply that TC axons can respond to upper layer molecules.

To further identify the genes that are expressed in the upper layers of the developing cortex, we constructed a subtraction cDNA library in which cDNAs derived from layer IV cells were enriched (Zhong et al 2004). Layer IV and layer V strips were dissected from P7 rat somatosensory cortex. After RNA extraction and reverse transcription, the obtained layer IV and layer V cDNAs were subjected to PCR-based amplification and fragmentation with a restriction enzyme. The layer

IV DNA fragments were hybridized with an excess amount of the layer V DNA fragments. Thereafter, unhybridized DNA fragments were cloned into a plasmid vector and stored as a subtraction cDNA library. Thousands of clones from the subtraction library were subjected to differential screening in the first-round screening. Then, *in situ* hybridization was performed on P7 rat brain, in which neocortical laminar configuration is established. As a result, we have found that several genes were strongly expressed in the upper layers of the developing cortex. In particular, *Stem cell factor* (*Scf*), a ligand of c-Kit receptor, was identified as an upper layer molecule, which may affect axon growth. *Scf* expression was found in layers II/III and IV across all neocortical regions of the early postnatal rat cortex, but was much stronger in primary sensory cortices such as somatosensory and visual areas (Fig. 1). Moreover, immunohistochemistry showed that c-Kit, a receptor of SCF, was strongly expressed in TC axons. Therefore SCF, as well as Sema7A, Sema3A, Cad6 and EphrinA5, was identified as a candidate protein which may be involved in TC axon targeting.

We investigated the action of each candidate molecule for TC axon growth by culturing dissociated thalamic cells on substrate coated with each protein. After several days in culture, axons from thalamic cells were found to grow extensively on EphrinA5 substrate. SCF and Sema7A also had a similar axon growth-promoting activity. In contrast, Sema3A and Cad6 did not show prominent growth-promoting activity. Whether EphrinA5, Sema7A and SCF co-operate with each other was also examined, as these proteins are co-expressed in the upper layers when TC axons invade the cortex. Any combination of the candidate proteins produced an additive effect of axon outgrowth-promoting activity. Therefore, it is likely that EphrinA5, Sema7A and SCF can co-operate to regulate TC axon growth. To further analyse the action of EphrinA5, Sema7A and SCF in the

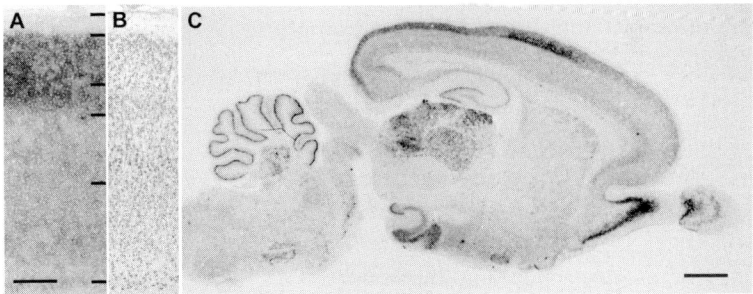

FIG. 1. *Scf* expression in the developing cortex. (A) *In situ* hybridization shows a laminar expression pattern of *Scf* mRNA in the somatosensory cortex of P7 rat. DIG-labelled probes were used for visualization. Bar represents 0.2 mm. (B) Nissl staining of the adjacent section. (C) *Scf* expression pattern on sagital sections of P7 rat brain. Bar represents 1 mm. Modified from Zhong et al 2004.

developing cortex, we tested TC axon growth on the filter membrane where lamina-specific expression of the candidate proteins was mimicked. The width of the protein-coated region was set to 200 μm, which is similar to the width of the upper layers of the early postnatal cortex. After printing the protein, the whole filter membrane was overlaid with laminin to reproduce growth-promoting activity in the deep layers. Then, thalamic explants were placed on the laminin-coated region. When Fc protein was printed as a control, TC axons grew on laminin-coated regions and further elongated into the Fc-printed region. In contrast, only a few axons entered the EphrinA5-printed region from the laminin-coated region. This inhibitory effect was dose-dependent. On the other hand, TC axon growth was promoted on either Sema7A- or SCF-printed regions. Finally, we tested whether EphrinA5, SCF and Sema7A produce co-operating activity. To examine this problem, thalamic explants were cultured on a membrane on which a mixture of the three proteins had been printed at low concentration, each of which did not influence TC axon growth. The result demonstrated that TC axons hardly invaded the printed region of the mixture.

These *in vitro* findings suggest that EphrinA5, Sema7A and SCF, which are expressed in the upper cortical layers at early postnatal stages of rodent cortex, can co-operate to regulate TC axon growth. Another interesting aspect is that a growth promoting effect of EphrinA5 was converted into growth-inhibitory activity. This may be due to alteration of growth cone responsiveness by experiencing different molecular environments (Shirasaki et al 1998, Hopker et al 1999, Nguyen-Ba-Charvet et al 2001). Indeed, dissociated cell culture demonstrated that the responsiveness of thalamic cells to EphrinA5 was altered by co-presence with laminin. Thus, the multiple molecules that are expressed with laminar specificity contribute to TC axon termination.

The role of neural activity in TC axon branching

Our previous studies have demonstrated that TC axons form branching in organotypic co-cultures of the thalamus and cortex with essentially the same laminar specificity as that *in vivo* (Yamamoto et al 1989, 1992, 1997). To investigate how TC axon branching develops during development, we examined the dynamics in the co-cultures by time-lapse imaging of single distinguishable TC axons, which were labelled with a fluorescent protein. In fact, a local electroporation method (Uesaka et al 2005) allowed us to observe a small number of TC axons in the cortical explants for more than a few weeks.

The time-lapse study demonstrated that TC axon arbours became larger and more complex mostly during the second week in culture (Fig. 2). The average number of branch points and the total axon length increased gradually as develop-

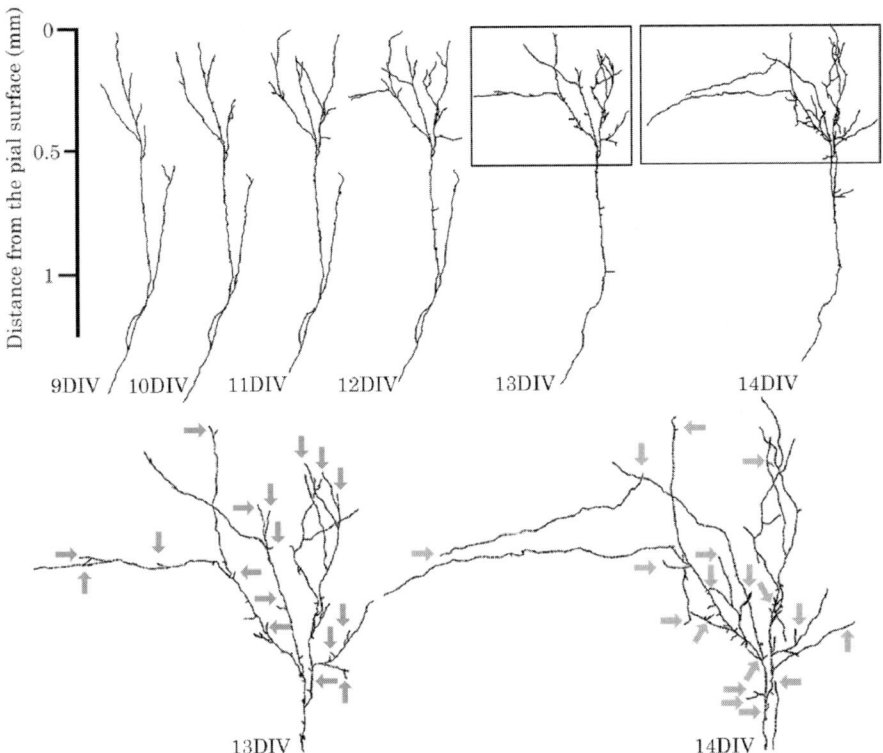

FIG. 2.   Development of TC axon arbours in an organotypic co-culture of the thalamus and cortex. TC axon was followed over 6 days by time-lapse study (top row). TC axons were labelled by introducing a plasmid containing the coding region of enhanced yellow fluorescent protein. (Lower part) Higher magnification of two boxed regions shows added (left arrows) and lost (right arrows) branches. Modified from Uesaka et al 2007.

ment progressed. Analysis of individual axons revealed that branches did not increase monotonically but remodelled actively. The majority of branches showed changes in length. Interestingly, such remodelling took place with laminar specificity. Most growing branches were primarily localized within 300–500 µm (layer IV) from the pial surface (Fig. 2). On the other hand, the distribution of eliminating branches was rather uniform. This result is consistent with our previous finding that a branch-inducing cue is enriched in layer IV while a negative regulator is uniformly distributed in all cortical layers (Yamamoto et al 2000b). As development of axon branching should be attributable to the accumulation of the difference between the number of added and eliminated branches, the daily differences were calculated in the target and non-target layers. The difference for the number of branch points in layer IV was much higher than the value in the deep layers. Thus, the number of branches in target layer gradually increased as development progressed.

We addressed whether such remodelling of TC axons is affected by electrical activity. Continuous recording with multi-electrodes implanted into the culture dish demonstrated the existence of spontaneous firing activity (Fig. 3). The activity was observed as single units and field potentials in both thalamic and cortical explants. The frequency was very low before one week *in vitro*, but was prominent at later stages. As TC axon branching was also found to develop during the second week in culture, there may be some correlation between neural activity and the remodelling process. To test this possibility, we applied antagonists of glutamate receptors or sodium channels during culturing. In drug-treated slices, the total number of branches was substantially decreased compared with the untreated group, but most branches were dynamic with both growth and retraction. The laminar distribution of growing and eliminating branches was almost constant across all cortical layers. Moreover, the difference between growing and eliminating branches in layer IV was dramatically reduced in the presence of drugs. This disruption is mediated primarily by a reduction of axonal branches in the target layer rather than an expansion of axonal branching in the non-target layers. One possibility to account for the reduction is that activity blockade merely reduces the total extent of branching by an overall decrease in growth of branches. However, this is unlikely because blockade of synaptic transmission increased the growth rates of branches. Another possible explanation is that stabilization of transient branches is promoted by neural activity, but this possibility can also be excluded because the lamina distribution of stable branches was almost constant across all cortical layers. The most plausible mechanism is that neural activity

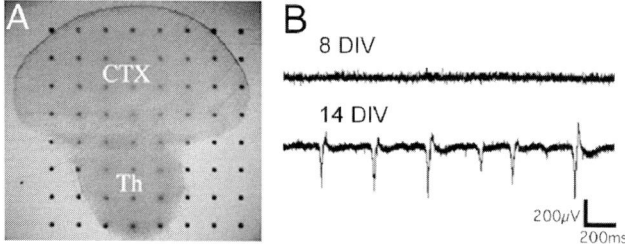

FIG. 3.   Spontaneous activity in a co-culture preparation. (A) Thalamic and cortical explants were cultured on the multi-electrode dish. Microelectrodes are arranged in an 8 × 8 array with an interpolar distance of 300 μm. (B) Spontaneous activity was recorded from both thalamic and cortical explants after 8 days and 14 days *in vitro*. Single spike and field potential activity increased as development progresses.

including synaptic transmission enhances branch dynamics including addition and elimination with a bias toward branch accumulation.

It has been demonstrated that activity blockade causes an expansion of axon arbours (Reh & Constantine-Paton 1985, Sretavan et al 1988, Cohen-Cory 1999). However, firing activity itself does not appear to be required for axonal branching, since application of glutamate receptor antagonists did not necessarily reduce the firing events. Synchronized activity between pre- and postsynaptic cells might strengthen neural connectivity.

How is laminar specificity of TC axon branching associated with neural activity? We have demonstrated that a branch-promoting molecule is expressed in the target layer while polysialic acid, a branch-inhibiting molecule, is expressed in all cortical layers (Yamamoto et al 2000b). One hypothesis is that neural activity including synaptic transmission plays a role in regulating the proper expression of these molecules (Fig. 4). Alternatively, neural activity might modify the expression and responsiveness of the receptors on growing TC axons or affect downstream signalling mechanisms. Although the molecules that underlie this process are largely unknown, an Ephrin-Eph system is one of the candidates. In the developing nervous system, the expression pattern of EphA on motoneurons has been demonstrated to be regulated by electrical activity (Hanson & Landmesser 2004).

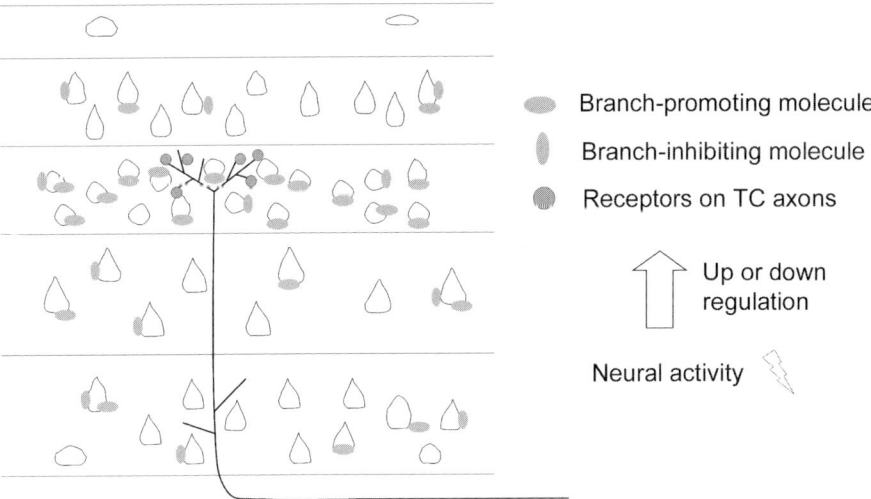

FIG. 4. Activity-dependent molecular mechanism of lamina-specific TC axon branching. Distribution of branch-promoting and branch-inhibiting factors in the cortex. Branch-promoting factor is primarily expressed in the target layer, whereas branch-inhibiting factor (PSA-NCAM) is expressed in all cortical layers. Neural activity may alter expression of these molecules or their receptor expressions.

Adhesion molecules such as N-cadherin and neural cell adhesion molecules are also candidate molecules for TC axon branching, since expressions of these molecules in the developing cortex could be affected by neural activity (Itoh et al 1997, Bozdagi et al 2000). Such interplay between guidance molecules and neural activity would be a key issue to understand neural circuit formation in the brain (Chandrasekaran et al 2005).

Conclusion

We demonstrated that TC axon termination could be regulated by co-operative activity of upper layer molecules, EphrinA5, Sema7A and SCF. The result also showed that TC axon branching involves a process of dynamic remodelling, which is regulated by branch-promoting and -inhibiting molecules. Moreover, evidence indicates that neural activity including synaptic transmission could modulate these molecular mechanisms, indicating interplay between lamina-specific targeting mechanisms and neural activity.

References

Antonini A, Fagiolini M, Stryker MP 1999 Anatomical correlates of functional plasticity in mouse visual cortex. J Neurosci 19:4388–4406

Bolz J, Novak N, Staiger V 1992 Formation of specific afferent connections in organotypic slice cultures from rat visual cortex cocultured with lateral geniculate nucleus. J Neurosci 12:3054–3070

Bozdagi O, Shan W, Tanaka H, Benson DL, Huntley GW 2000 Increasing numbers of synaptic puncta during late-phase LTP: N-cadherin is synthesized, recruited to synaptic sites, and required for potentiation. Neuron 28:245–259

Castellani V, Yue Y, Gao PP, Zhou R, Bolz J 1998 Dual action of a ligand for Eph receptor tyrosine kinases on specific populations of axons during the development of cortical circuits. J Neurosci 18:4663–4672

Chandrasekaran AR, Plas DT, Gonzalez E, Crair MC 2005 Evidence for an instructive role of retinal activity in retinotopic map refinement in the superior colliculus of the mouse. J Neurosci 25:6929–6938

Cohen-Cory S 1999 BDNF modulates, but does not mediate, activity-dependent branching and remodeling of optic axon arbors in vivo. J Neurosci 19:9996–10003

Donoghue MJ, Rakic P 1999 Molecular gradients and compartments in the embryonic primate cerebral cortex. Cereb Cortex 9:586–600

Gao PP, Yue Y, Zhang JH, Cerretti DP, Levitt P, Zhou R 1998 Regulation of thalamic neurite outgrowth by the Eph ligand ephrin-A5: implications in the development of thalamocortical projections. Proc Natl Acad Sci USA 95:5329–5334

Hanson MG, Landmesser LT 2004 Normal patterns of spontaneous activity are required for correct motor axon guidance and the expression of specific guidance molecules. Neuron 43:687–701

Herrmann K, Shatz CJ 1995 Blockade of action potential activity alters initial arborization of thalamic axons within cortical layer 4. Proc Natl Acad Sci USA 92:11244–11248

Hopker VH, Shewan D, Tessier-Lavigne M, Poo M, Holt C 1999 Growth-cone attraction to netrin-1 is converted to repulsion by laminin-1. Nature 401:69–73

Hubel DH, Wiesel TN, LeVay S 1977 Plasticity of ocular dominance columns in monkey striate cortex. Philos Trans R Soc Lond B Biol Sci 278:377–409

Itoh K, Ozaki M, Stevens B, Fields RD 1997 Activity-dependent regulation of N-cadherin in DRG neurons: differential regulation of N-cadherin, NCAM, and L1 by distinct patterns of action potentials. J Neurobiol 33:735–748

López-Bendito G, Molnár Z 2003 Thalamocortical development: how are we going to get there? Nat Rev Neurosci 4:276–289

Mackarehtschian K, Lau CK, Caras I, McConnell SK 1999 Regional differences in the developing cerebral cortex revealed by ephrin-A5 expression. Cereb Cortex 9:601–610

Molnár Z, Blakemore C 1991 Lack of regional specificity for connections formed between thalamus and cortex in coculture. Nature 351:475–477

Molnár Z, Blakemore C 1999 Development of signals influencing the growth and termination of thalamocortical axons in organotypic culture. Exp Neurol 156:363–393

Nguyen-Ba-Charvet KT, Brose K, Marillat V, Sotelo C, Tessier-Lavigne M, Chedotal A 2001 Sensory axon response to substrate-bound Slit2 is modulated by laminin and cyclic GMP. Mol Cell Neurosci 17:1048–1058

Pasterkamp RJ, Peschon JJ, Spriggs MK, Kolodkin AL 2003 Semaphorin 7A promotes axon outgrowth through integrins and MAPKs. Nature 424:398–405

Rch TA, Constantine-Paton M 1985 Eye-specific segregation requires neural activity in three-eyed Rana pipiens. J Neurosci 5:1132–1143

Shirasaki R, Katsumata R, Murakami F 1998 Change in chemoattractant responsiveness of developing axons at an intermediate target. Science 279:105–107

Skaliora I, Singer W, Betz H, Puschel AW 1998 Differential patterns of semaphorin expression in the developing rat brain. Eur J Neurosci 10:1215–1229

Sretavan DW, Shatz CJ, Stryker MP 1988 Modification of retinal ganglion cell axon morphology by prenatal infusion of tetrodotoxin. Nature 336:468–471

Suzuki SC, Inoue T, Kimura Y, Tanaka T, Takeichi M 1997 Neuronal circuits are subdivided by differential expression of type-II classic cadherins in postnatal mouse brains. Mol Cell Neurosci 9:433–447

Takemoto M, Fukuda T, Sonoda R, Murakami F, Tanaka H, Yamamoto N 2002 Ephrin-B3-EphA4 interactions regulate the growth of specific thalamocortical axon populations in vitro. Eur J Neurosci 16:1168–1172

Uesaka N, Hirai S, Maruyama T, Ruthazer ES, Yamamoto N 2005 Activity dependence of cortical axon branch formation: a morphological and electrophysiological study using organotypic slice cultures. J Neurosci 25:1–9

Uesaka N, Hayano Y, Yamada A, Yamamoto N 2007 Interplay between laminar specificity and activity-dependent mechanisms of thalamocortical axon branching. J Neurosci 27:5215–5223

Vanderhaeghen P, Lu Q, Prakash N et al 2000 A mapping label required for normal scale of body representation in the cortex. Nat Neurosci 3:358–365

Xu X, Ng S, Wu ZL et al 1998 Human semaphorin K1 is glycosylphosphatidylinositol-linked and defines a new subfamily of viral-related semaphorins. J Biol Chem 273:22428–22434

Yabuta NH, Butler AK, Callaway EM 2000 Laminar specificity of local circuits in barrel cortex of ephrin-A5 knockout mice. J Neurosci 20:RC88

Yamamoto N, Kurotani T, Toyama K 1989 Neural connections between the lateral geniculate nucleus and visual cortex in vitro. Science 245:192–194

Yamamoto N, Yamada K, Kurotani T, Toyama K 1992 Laminar specificity of extrinsic cortical connections studied in coculture preparations. Neuron 9:217–228

Yamamoto N, Higashi S, Toyama K 1997 Stop and branch behaviors of geniculocortical axons: a time-lapse study in organotypic cocultures. J Neurosci 17:3653–3663

Yamamoto N, Matsuyama Y, Harada A, Inui K, Murakami F, Hanamura K 2000a Characterization of factors regulating lamina-specific growth of thalamocortical axons. J Neurobiol
   42:56–68
Yamamoto N, Inui K, Matsuyama Y et al 2000b Inhibitory mechanism by polysialic acid for
   lamina-specific branch formation of thalamocortical axons. J Neurosci 20:9145–9151
Zhong Y, Takemoto M, Fukuda T et al 2004 Identification of the genes that are expressed in
   the upper layers of the neocortex. Cereb Cortex 14:1144–1152

DISCUSSION

Parnavelas: Have you thought about the different behaviours displayed by axons
that arise in specific and non-specific thalamic nuclei? Those from specific nuclei
go into layer IV and branch extensively, whereas the non-specific ones go straight
through layer IV giving rise only to small branches. These axons go through layer
IV, where all the branch promoting molecules are, but chose not to stop there.

Yamamoto: There are several thalamic nuclei. In our experiments we occasionally
saw thalamic axons pass through layer IV, but this was not the major population.
When we counted branching points they didn't affect branch distribution in terms
of laminar specificity. However, we don't know how to distinguish between these
two populations.

Goffinet: Is there a way of testing this idea *in vivo*?

Yamamoto: We examined thalamocortical axons in *Sema7a* knockout mice and
c-*kit* receptor knockout mice. We injected DiI (fluorescent dye) and followed
thalamocortical axons in the postnatal animals. Although this is a preliminary
experiment, the branch formation seems to be very normal. To examine whether
multiple molecules co-operate to work *in vivo*, it may be necessary for us to examine
the phenotype in double or triple knockout animals. However, it could be assessed
by examining the protein concentrations of these candidate molecules *in vivo*.

Walsh: Can you comment more on how electrical activity might be linked to receptor expression. Do you mean translation, or trafficking to the surface of the cell?

Yamamoto: For example, in axon guidance of motor neurons it has been shown
that Eph receptor expression is changed based on the activity (Hanson &
Landmesser 2004). We think a similar mechanism might be present in our case.

Rakic: It would be helpful if you could define what you mean by 'activity'. Hebb,
as well as Hubel and Wiesel, whom you mentioned, talked about the strengthening
and elimination of the already formed synapses. Whereas you are looking at something that occurs before synapse formation: it is quite a different process. What
do you mean by activity rather than just attraction of molecules?

Yamamoto: Hubel and Weisel looked at branch formation but at much later stages.
We are looking at the early stages when branch formation begins to take place.
Many targeting molecules are necessary for producing such specific patterns. I
think that such axon guidance molecules have been though to be independent of

neuronal activity. Here I mean that neuronal activity can influence molecular expression.

Rakic: What do you mean by 'neuronal activity' and 'synaptic activity'?

Yamamoto: I didn't show any electrical activity of cultured cells in the presence of blockers. Even in the presence of such glutamate receptor blockers firing frequency wasn't changed much. On the basis of this finding, the amount of firing activity itself is not important to produce such an effect. Synaptic transmission might be more important.

Molnár: I think this is extremely important work, trying to link the molecular changes elicited by neuronal activities to thalamocortical axon branching and targeting. The question on the nature of activity is extremely relevant to the work of Michael Wilson and Thomas Südhoff. These laboratories have been dissecting the different types of activity patterns by altering the function of different molecules in the synaptic release machinery (Washbourne et al 2002, Verhage et al 2000, Varoqueaux et al 2002, Molnár et al 2002). They have shown that neurons can talk to each other in various different ways. There is regulated synaptic vesicle release coupled to action potentials in the presynaptic sites. This is the classical synaptic communication. Then you can have spontaneous release that is a bit more random, but it can still elicit EPSPs. Then you can have 'oozing' of transmitters without vesicles, so called constitutive release, which can still influence the neighbouring neurons. When we talk about activity, we need to define which of these three we are talking about. With the different SNARE mutants (Snap25 knockout, Munc18 knockout and Munc13 triple knockout) it is becoming possible to dissect the contribution of the regulated, spontaneous vesicular release and the constitutive release. David Price has found that in thalamocortical co-culture that if the activity is blocked with TTX during the thalamocortical ingrowth than there is a range of interesting changes in the targeting pattern (Anderson & Price 2002). Wilkemeyer and Angelides (1995) found the same. We were interested in finding out how we could dissect these three forms of activity patterns in the Snap25, Munc18 and Munc13 knockouts. Unfortunately, all these animals die at birth. What we were trying to do with Dan Blakey in my laboratory is to coculture with E15–17 Snap25 knockout thalami and cortices from P2 wild-type mice. The fibres originating from Snap25 knockout thalamic explants invaded the wild-type cortex, and stopped and branched normally. It seems that they can stop and branch in the absence of synaptic vesicle release from their terminals, which is different from the results obtained after TTX application in the cultures by David's group.

Wilson: TTX is a strange drug that does a lot of things. It changes gene transcription enormously in cultured hippocampal neurons. I'm beginning to wonder whether the maintenance of action potentials is some part of what we call activity-dependent processes, whether or not these action potentials lead to regulated release of neurotransmitter or not. In the mice that Zoltán Molnár is talking about, the nerve terminals still undergo spontaneous vesicular release of transmitter. We call it 'humming'. But they can't evoke transmitter release in a Ca^{2+}-dependent manner, because they are missing one of the pieces of the complex. What about TTX? Do action potentials have a message encoded within them that might be mediating some of these effects? It might explain some of the experiments that Carla Shatz has done with TTX compared with the work that we have done.

Kriegstein: We are talking about signalling in two directions. There is the activity of the thalamic axon that releases transmitter or some other factor that activates target cells or cells in the vicinity, then there is the activity of the cortical cells themselves that up-regulate their growth factor secretion to promote their branching. Is that the circuit we are talking about?

Yamamoto: It is difficult to distinguish whether thalamic or cortical cells are critical. Both might be required. To discriminate between these two events, we are now trying to introduce Kir2.1, a K^+ channel, into thalamic or cortical cells differentially, because overexpression of this channel can change the firing activity.

Kriegstein: Are you talking about the firing pattern of the cortical cells?

Yamamoto: Yes. According to Hebb's idea, both activations of thalamic and cortical cells might be necessary. Many people think NMDA receptor activation is also necessary to strengthen the synaptic connections. We think a similar principle might be applicable to branch formation. We'd like to test this. For example, a plausible mechanism is that if both pre and post-synaptic cells are activated, post-synaptic cells might produce some ligand molecules for axon branching.

Macklis: When you showed the expression distributions of the Kit ligand, Sema7, and EphrinA5, what age was this?

Yamamoto: One week postnatal in rat.

Macklis: If you look much earlier, is there any transient expression, either in subplate or subventricular zone? Is there anything that would connect this biology to a transient waiting period, such as the one Dennis O'Leary was talking about?

Yamamoto: I didn't give much detail about the expression pattern. For example, c-Kit ligand is expressed in the deep layers in the early stages, and is moving to the upper layers. We should probably think about the more precise expression pattern during the first week when lamina-specific thalamocortical projections are formed.

Murakami: In your *in vivo* experiments you said that you have examined *Sema7A* knockout mice. Cortical excitatory neurons are suitable for *in utero* electroporation. So you might be able to knock down the other molecules. Do you plan to do such experiments?

Yamamoto: I agree with you in that we could perform the knockdown experiment by introducing RNAi vectors for the candidate molecules.

Molnár: You proposed that the stopping and branching signals might be different (Yamamoto et al 2000a). You also elegantly demonstrated that in fixed cortical slices the thalamic projections read the surface molecules and stop in layer IV (Yamamoto et al 2000b). Branching requires neurotrophin release from living cortical cells. Perhaps the stopping and branching signals might depend on activity in a differential manner.

Yamamoto: Yes. However, it is possible that these two phenomena are correlated. Both branching and stopping might be regulated by neural activities. David Price has shown that thalamocortical projections are also sensitive to neuronal activity. Axonal termination is probably also dependent on neuronal activity as well as branching.

References

Anderson G, Price DJ 2002 Layer-specific thalamocortical innervation in organotypic cultures is prevented by substances that alter neural activity. Eur J Neurosci 16:345–349

Hanson MG, Landmesser LT 2004 Normal patterns of spontaneous activity are required for correct motor axon guidance and the expression of specific guidance molecules. Neuron 43:687–701

Molnár Z, López-Bendito G, Small J, Partridge LD, Blakemore C, Wilson MC 2002 Normal development of embryonic thalamocortical connectivity in the absence of evoked synaptic activity. J Neurosci 22:10313–10323

Varoqueaux F, Sigler A, Rhee JS et al 2002 Total arrest of spontaneous and evoked synaptic transmission but normal synaptogenesis in the absence of Munc13-mediated vesicle priming. Proc Natl Acad Sci USA 99:9037–9042

Verhage M, Maia AS, Plomp JJ et al 2000 Synaptic assembly of the brain in the absence of neurotransmitter secretion. Science 287:864–869

Washbourne P, Thompson PM, Carta M et al 2002 Genetic ablation of the t-SNARE SNAP-25 distinguishes mechanisms of neuroexocytosis. Nat Neurosci 5:19–26

Wilkemeyer MF, Angelides KJ 1995 Adenovirus-mediated expression of a reporter gene in thalamocortical cocultures. Brain Res 703.129-138

Yamamoto N, Matsuyama Y, Harada A, Inui K, Murakami F, Hanamura K 2000a Characterization of factors regulating lamina-specific growth of thalamocortical axons. J Neurobiol 42:56–68

Yamamoto N, Inui K, Matsuyama Y et al 2000b Inhibitory mechanism by polysialic acid for lamina-specific branch formation of thalamocortical axons. J Neurosci 20:9145–9151

Genes involved in the formation of the earliest cortical circuits

Zoltán Molnár, Anna Hoerder-Suabedissen, Wei Zhi Wang, Jamin DeProto, Kay Davies*, Sheena Lee‡, Erin C. Jacobs†, Anthony T. Campagnoni†, Ole Paulsen, Maria Carmen Piñon and Amanda F. P. Cheung

Department of Physiology, Anatomy and Genetics, University of Oxford, South Parks Road, Oxford, OX1 3QX, UK, *MRC Functional Genetics Unit, Department of Physiology, Anatomy and Genetics, University of Oxford, UK, ‡The Wellcome Trust Integrative Physiology Initiative on Ion Channels (OXION), Oxford Centre for Gene Function, South Parks Road, Oxford, OX1 3QX, UK and †UCLA Semel Institute for Neuroscience, 635 Charles E Young Drive South, Los Angeles, CA 90095-7332, USA

Abstract. Building the brain is like erecting a house of cards. The early connections provide the foundation of the adult structure, and disruption of these may be the source of many developmental flaws. Cerebral cortical developmental disorders (including schizophrenia and autism) and perinatal injuries involve cortical neurons with early connectivity. The major hindrance of progress in understanding the early neural circuits during cortical development and disease has been the lack of reliable markers for specific cell populations. Due to the advance of powerful approaches in gene expression analysis and the utility of models with reporter gene expressions in specific cortical cell types, our knowledge of the early cortical circuits is rapidly increasing. With focus on the sub-plate, layer VI and layer V projection neurons, we shall illustrate the progress made in the understanding of their neurochemical properties, physiological characteristics and their integration into the early intracortical and extracortical circuitry. This field bene-fited from recent developments in mouse genetics in generating models with subtype specific gene expression patterns, powerful cell dissection and separation methods combined with microarray analysis. The emergence of cortical cell type specific biomarkers will not only help neuropathological diagnosis, but will also eventually reveal the causal relations in the pathogenesis of various cortical developmental disorders.

2007 Cortical development: genes and genetic abnormalities. Wiley, Chichester (Novartis Foundation Symposium 288) p 212–229

The billions of neurons and trillions of connections of the human brain are gener-ated from the complex interactions between our unfolding genetic programme and our environment. Development is the ultimate readout of our genome; a combina-tion of genetic susceptibility and environmental perturbations can lead to several devastating diseases. The causes of and cures for a large number of cerebral cortical developmental disorders are not known, but their prevalence in the general popula-

212

tion is high (schizophrenia [1:100]; autism [1:166]; attention deficit hyperactivity disorder [1:30]; dyslexia [1:10]; childhood epilepsy; and neural tube closure defects). In spite of recent progress, we are only beginning to understand basic neural developmental mechanisms and their involvement in the pathogenesis of several debilitating diseases. For example, several forms of childhood epilepsy have been linked to cortical neuronal migration disorders, some of which show genetic linkage (see Walsh and colleagues, this volume). The causal relationships are difficult to reveal. The single-gene determinist approach will not yield a holistic conception of neural disorders as neuronal production, migration, differentiation and development of cortical connectivity are dependent upon complex signalling cascades.

Importance of studying early cortical circuit formation

In order to gain a more comprehensive understanding of the brain as final product, we must thoroughly characterize neurodevelopmental processes contributing to its formation. In doing so, we will ascend to a platform from which we can critically analyse and perhaps treat the numerous disorders affecting the developing and mature brain. During the last decade a few pivotal studies have overturned our fundamental knowledge of the site of cortical neurogenesis, the patterns of neuronal cell migration and the genes involved in neuronal differentiation and development of connectivity (Kriegstein et al 2006, Hevner et al 2003, Price et al 2005, Rakic 2006). Recent advances have provided a wealth of information on the temporal and spatial expression of transcription and other factors and their receptors in cortical development (Guillemot et al 2006, Molyneaux et al 2007). We are beginning to understand the general mechanisms and the role of some factors that determine cortical area specialization. We have greater insight into the regulation of the process of cortical neurogenesis and the classification of laminar specific sub-classes of cells (Markram 2004, Nelson et al 2006). We also have some insight into the molecular mechanisms involved in the development of both the axonal projections and dendritic arrays of these sub-classes of neurons. The critical questions to elucidate how the expression of genes involved in these processes is regulated by discretely expressed transcription factors. Most, if not all, neurodegenerative diseases disrupt at least one of these processes. Understanding these developmental programmes is therefore of immense clinical significance, because a better understanding could aid in the diagnosis of many of the major neurological disorders (Hevner 2007). Moreover, understanding cell fate determinants and class specific biomarkers to detect, classify, and generate all classes and subclasses of neurons could help in developing cell replacement therapies.

   A century ago the advent of imaging methods allowed for sufficient contrasting of cells. Even then, based on purely morphological descriptions, it was clear that the brain is composed of a seemingly indescribable assortment of cells (Ramón y Cajal 1911, Lorente de No 1949). The view of the nature of the brain cells is increasingly more complex and context specific in contrast to numerous other organs of our body (Peters & Jones 1985). The cerebral cortex is constructed from two major neuronal types, pyramidal and non-pyramidal neurons. Yet these cells occur as numerous different sub-types, which differ from one another in laminar location of their soma, morphology, neurochemical and electrophysiological characteristics and in their connectivity (Markram 2004). The understanding of the formation of specific cortical circuits and how they effectuate specific functions is a fundamental question underpinning all aspects of complex mammalian behaviour. Developmental neurobiologists have a long standing interest in understanding cerebral cortical circuit formation and function and investigate them using anatomical, genetic, physiological, and more recently molecular approaches (Nelson et al 2006).

In this chapter we shall review recent experiments aimed at detecting and characterizing causal relationships between gene expression patterns, somato-dendritic differentiation, physiological specification and axonal target finding, using the subplate neurons and layer V projection neurons in rodents as a model system (Fig. 1).

Emerging understanding of the cellular and molecular characteristics of subplate neurons

The subplate layer, containing the earliest-generated and largely transient cortical neurons, is the foundation of the brain, and disruption of these cells may be the source of many developmental flaws, and therefore a fundamental topic to study. Cerebral cortical developmental disorders (schizophrenia, autism, attention deficit/ hyperactivity disorder [ADHD]) and perinatal hypoxic injuries (periventricular leucomalacia, PVL), dyslexia, and cerebral palsy involve cells of the subplate (McQuillen et al 2003).

Subplate cells were first described in primates (Kostovic & Rakic 1980) and carnivores (Luskin & Shatz 1985) as a transient layer of cells below layer VI. In rodents, subplate (SP) is a thin band of cells separating the white matter from layer VI. Murine SP cells, born around embryonic day (E)11, are among the earliest mature cortical neurones (Price et al 1997), and begin extending axons towards the thalamus by E13 (DeCarlos & O'Leary 1992, Molnár et al 1998a,b). The SP layer varies in size and proportions in different vertebrate species. In primates this zone is considerably larger (Kostovic & Rakic 1980), and functional equivalence of rodent subplate with that of primates or carnivores has not been established. Thus the SP layer has been variously referred to as layer VIb, layer VII, interstitial

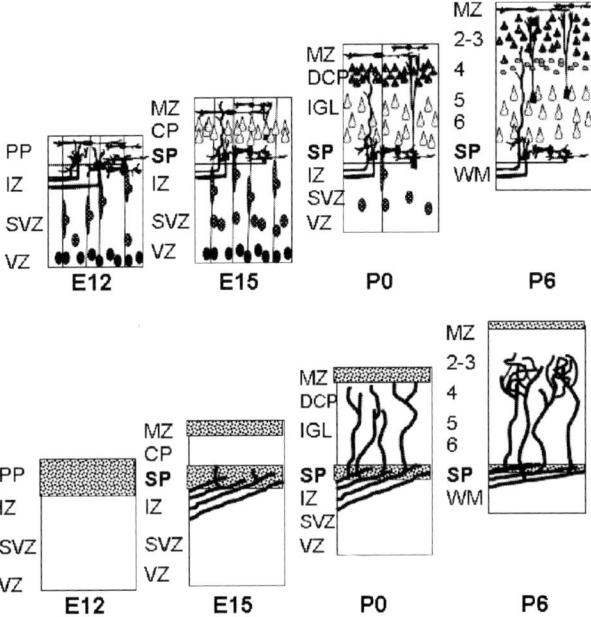

FIG. 1. Schematic summary of the early development of cortical lamination (upper row) with special attention to subplate, marginal zone and thalamocortical afferents (lower row) from E12 until P6 in the mouse. The black, round cells at the bottom of the panel represent neural progenitors in ventricular and subventricular zones (VZ, SVZ) as well as radial glia along which immature neurons migrate (black bipolar cells). At E12 the earliest generated neurons form the preplate (PP). By E15 the cortical plate (CP) splits the preplate into subplate (SP) and marginal zone (MZ). By P0 the neurogenesis is completed, the cortex is comprised of infragranular layers, (IGL, layers V and VI) and still migrating future cortical layers II–IV, the dense cortical plate (DCP). The lower, IGL cells are born at an earlier stage than the upper layers (IV–II). By postnatal day 6 most neurons complete their migration and the cortical lamination is established. Thalamocortical axons (thick black lines in lower panels) arrive to the subplate around E15. Thalamocortical projections enter the cortex by birth and begin to establish characteristic periphery related patterning in the barrel field of S1 by P3–6. Thalamocortical axons establish functional interactions in subplate as they arrive and accumulate in this layer before they enter the cortex and recognize their ultimate target cells in layer IV. Subplate neurons integrate into the overlying cortex and also develop various extracortical projections providing a stable platform for the establishment of cortical circuitry from early stages.

or even white matter cells, but the term SP is also commonly used, and will be used here. SP cells play an important role in thalamocortical axon pathfinding at the level of the initial areal targeting (Catalano & Shatz 1998, Ghosh et al 1990) as well as the eventual innervation of cortical layer IV by thalamic afferents and establishment of optical orientation columns (Kanold et al 2003, Kanold & Shatz

2006). They are also necessary for the maturation of inhibition in cortical layer IV in areas innervated by the thalamus (Kanold & Shatz 2006), and drive oscillations in the gap junction coupled early cortical syncytium (Dupont et al 2006). Little else is known about their role as part of cortical circuitry during development or in the mature cortex. During development, SP neurons are electrically active and capable of firing action potentials (Hanganu et al 2001) while incorporating (at least transiently) into the cortical and subcortical circuitry (Kanold et al 2003, Friauf & Shatz 1991, Higashi et al 2002, 2005, Hoerder et al 2006). Both electrophysiological properties and cell morphology point to a high degree of underlying diversity among SP neurons (Hanganu et al 2001, Antonini & Shatz 1990, Hoerder 2007). The diversity of SP cells is further underscored by the heterogeneity of molecular markers of glutamatergic and GABAergic cells expressed in them (Allendoerfer & Shatz 1994). Recently, two transgenic mouse models were created with green fluorescent protein (GFP) expression restricted to a subset of SP and layer VI neurons (Golli-tau-eGFP [GTE] mouse, Jacobs et al 2007; Tbr1-driven GFP, Kolk et al 2006). Both models have already helped to answer the questions surrounding the integration of the subplate neurons into the intra and extra-cortical circuitry (Table 1).

Due to the lack of sensitive purely anterograde tracers the exact timing and pattern of subcortical subplate projections were controversial. Moreover, the nature and the pattern of innervation of thalamic nuclei by various cortical projection neurons during development have not been resolved. The lack of our ability to selectively label subplate neurons and their neurites hindered our understanding of the integration of these neurons into the intra and extracortical circuitry, although numerous functional suggestions have been based on their intimate association with thalamocortical projections and their afferents on to layer 4 cortical neurons (see Table 1). The GTE mouse model proved useful to further study the remodelling associated with thalamocortical ingrowth and periphery related patterning in the barrel cortex (Jacobs et al 2007, Jetwha et al 2007). However, the nature of the cellular extension (axonal and/or dendritic) and the causal relationships, as well as the underlying molecular mechanisms are still to be investigated.

Studies suggest delayed and differential innervations of different nuclei of the dorsal thalamus by early GFP positive corticofugal projections (Jacobs et al 2007, Piñon et al 2005) and intimate association of early corticofugal projections with thalamic afferents (Piñon et al 2005). Unfortunately the selectivity, specificity (Aye et al 2006) and the exact reasons behind the reporter gene expression in a selected group of subplate and layer VI neurons in the GTE mouse are not known. Nevertheless, the GTE model is beginning to make an impact on the understanding of unresolved questions surrounding the formation of early cortical circuits sum-

TABLE 1   Questions surrounding early cortical circuit formation

---

When do the early subplate projections enter into the thalamus?

    1. Do they wait outside the LGN (Shatz & Rakic 1981)?

    2. Do they wait in TRN (Molnár & Cordery 1999)?

    3. Do they wait in white matter or internal capsule (Clasca et al 1995)?

    4. Do subplate projections enter the dorsal thalamus (Allendoerfer & Shatz 1994)?

Which cortical cells develop projections to the thalamus first?

    1. Subplate cells (McConnell et al 1989) ?

    2. Layer V cells (Clasca et al 1995)?

    3. Layer VI cells?

What is the relationship between early corticofugal and thalamic projections?

    1. Fasciculate with each other in IC and IZ (Molnár & Blakemore 1995)?

    2. Run in separate compartments (Miller et al 1993, 1995)?

    3. Interdigitate in a restricted portion (Bicknese et al 1994)?

What is the mode of integration of subplate neurites into the cortical plate prior to, during and after thalamic innervation?

    1. Ocular dominance formation (Ghosh & Shatz 1990)?

    2. Orientation column formation (Kanold et al 2003)?

    2. Barrel formation (Jethwa et al 2007)?

    3. Area-specificity (McConnell et al 1989, Catalano & Shatz 1998, Molnár & Blakemore 1995, Shimogori & Grove 2005)?

    4. Axonal and/or dendritic remodelling associated with thalamocortical ingrowth and periphery related patterning (Hoerder et al 2007, Jethwa et al 2007)?

---

marised in Table 1, and it is still one of the best models available. Unfortunately, we have relatively limited information on the neurochemical properties, somato-dendritic morphologies and physiological characteristics of subplate cells. The initial surge of publications in the early nineties (see Allendoerfer & Shatz 1994, Molnár et al 1998a) was not followed up with more detailed systematic analysis of this enigmatic cell population.

Most studies on the function of subplate cells have used ferrets or cats. These are not yet developed as genetic models. Improving our understanding of the role of subplate cells in mice is thus particularly important, as they are increasingly used as a model organism for developmental neurobiology. However, owing to a different timeline of development and the much smaller body size of mice, many of the earlier experiments in other organisms could never be extended to rodents.

Furthermore, not all subplate cells in mice disappear during cortical maturation, raising the possibility that they may have an additional role as part of the mature cortical network that is as yet completely uninvestigated. While many researchers working on cortical development acknowledge the need for further characterisation of subplate cells, most work has been severely constrained by a lack of reliable markers for subplate neurons. Only in the last few years have modern genetic tools and publication of the mouse genome provided the means for a detailed and systematic classification of cortical neurons.

We initiated a microarray project to get an insight into the gene expression patterns in subplate (Hoerder 2007). This is beginning to produce a very valuable list of genes, which includes some genes that have already been identified by others in subplate (e.g. *CTGF*, *Nurr1*; Heuer et al 2003, Arimatsu et al 2003). Our subplate microarray study highlighted other genes, some of which have known functions in the nervous system, and many whose function is thus far unknown. So far we have examined the expression levels of a handful of these markers with *in situ* hybridization and immunohistochemistry. Some of these markers are only expressed temporarily in specific stages of development, others are maintained throughout adulthood. It is likely that specification of subplate identity is achieved through a unique combination of molecules rather than a single molecule. More over, subplate is comprised of a variety of cell types with different connectivity, morphology and electrophysiological characteristics. To understand subplate gene expression, we should consider this diversity fully. The molecular data are currently being integrated with cellular and anatomical data. After this validation it will be extremely useful for functional classification, comparisons of subplate development in different species, understanding numerous mutants (Cdk5, Reeler, SRK, see Fig. 2) and clinical exploitation. There are interesting comparative neurodevelopmental questions related to subplate which might become accessible with the new markers.

Differentiation of layer V projection neurons

It is becoming evident that numerous important characteristics of projection neurons are determined very early, perhaps during the last cell division in the germinal zone. Layer V pyramidal cells form an accessible model to study the molecular determinants of target selection, dendritic and physiological differentiation because they develop their projections before the completion of dendritic remodelling and physiological specification (Koester & O'Leary 1992, Kasper et al 1994). Mature pyramidal neurons in layer V of the rodent cerebral cortex are of two main types (Larkman & Mason 1990). One has a thick apical dendrite extending into cortical layer 1 with a prominent terminal tuft, produces distinctive initial bursts of action potentials in response to current injection, and projects subcortically to the superior colliculus,

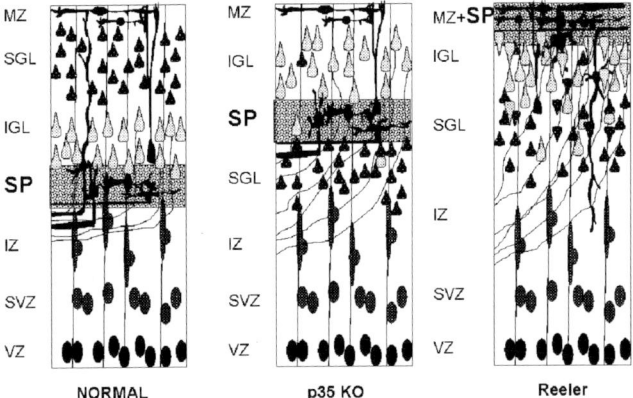

FIG. 2.   Schematic illustration of the altered position of subplate neurons in p35 knockout and Reeler cortex. Left panel: Schematic representation of the normal development of the cerebral cortex (similar labelling as seen in Fig. 1). During normal cortical development preplate splits to subplate (SP) and marginal zone (MZ) and IGL cells settle above subplate. The later generated supragranular layer (SGL, layers II–III) cells migrate through the infragranular layers to colonise the layers above them. Middle panel: in the p35 knockout cortex, the preplate is split by the cortical plate cells almost as normal, but the supragranular cells do not migrate through subplate and settle below them. Not all cortical plate cells arrive in the cortical plate, thereby also giving the appearance of a disrupted subplate, which is now scattered in between SGL and IGL. Right panel: if the Reelin signalling pathway is defective, the preplate remains un-split with the cortical plate cells settling underneath and in reversed order (IGL outside to SGL). The new markers for subplate neurons revealed by our recent experiments will be extremely valuable to understand the malpositioning of various subplate populations in Reeler and Cdk35 knockout mice and in various human pathologies.

pontine nuclei and spinal cord. The second type is characterized by a slender apical dendrite which arborizes in cortical layers II–IV without a terminal tuft, does not show burst firing, and projects intracortically. During development, all layer V neurons are generated at similar times and are initially indistinguishable. With the exception of their projections, all have tufted apical dendrites reaching layer I and no bursting firing pattern. It is only towards the end of the first postnatal week that the cortico-cortical cells selectively lose their apical tufts (Koester & O'Leary 1992) and it is not until 14 days postnatally that the first cortico-tectal cells develop the burst-firing characteristic (Kasper et al 1994; Fig. 3).

Since most cortical output is directed through layer V, understanding how the connectivity of layer V develops, and how regional variation across the cortex is programmed, are issues critical to a broader understanding of cortical function. Over the last decade, numerous laboratories have been examining the factors involved in the generation of these two distinct types of layer V neurons. Using

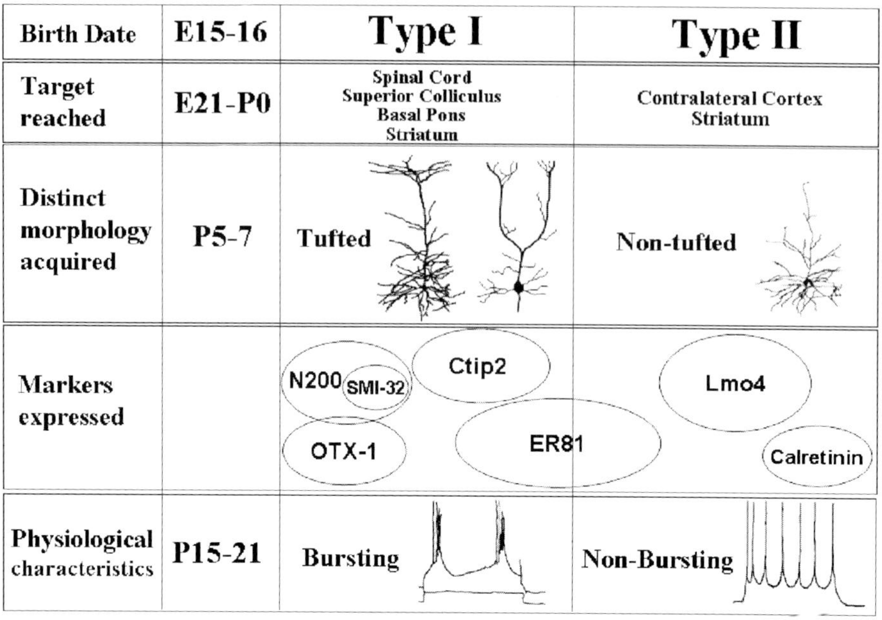

Birth Date	E15-16	Type I	Type II
Target reached	E21-P0	Spinal Cord Superior Colliculus Basal Pons Striatum	Contralateral Cortex Striatum
Distinct morphology acquired	P5-7	Tufted	Non-tufted
Markers expressed		N200 SMI-32 Ctip2 OTX-1 ER81	Lmo4 Calretinin
Physiological characteristics	P15-21	Bursting	Non-Bursting

FIG. 3.   Layer V pyramidal neurons fall into two major classes. The diagram summarizes the developmental sequence leading to target selection, dendritic differentiation and the emergence of distinct gene expression patterns and physiological properties in the rat. Based on Molnár & Cheung (2006).

antibody screening, single cell RT-PCR and library screening, reliable molecular markers have been identified to distinguish different layer V neuron subtypes. These include: neurofilament SMI-32, N200 and FNP-7 (Voelker et al 2004), transcription factors Otx1, Er81, Ctip2, Fezl, Lmo4 and many more (Weimann et al 1999, Hevner et al 2003, Yoneshima et al 2005, Arlotta et al 2005, Chen et al 2005, Mitchell et al 2006). The discovery of these marker genes has allowed us to correlate with the connectivity and physiological characteristics of layer V neurons (Christophe et al 2005), and also allowed us to look into the specific developmental steps that are involved in cortical neuron cell fate decisions (see Fig. 4 on the developmental timeline) (Guillemot et al 2006, Molnár & Cheung 2006, Nelson et al 2006, Molyneaux et al 2007).

Summary

Our challenge is now to understand the combinatorial effect of lineage- and area-specific gene expression profiles during brain development. These fundamental

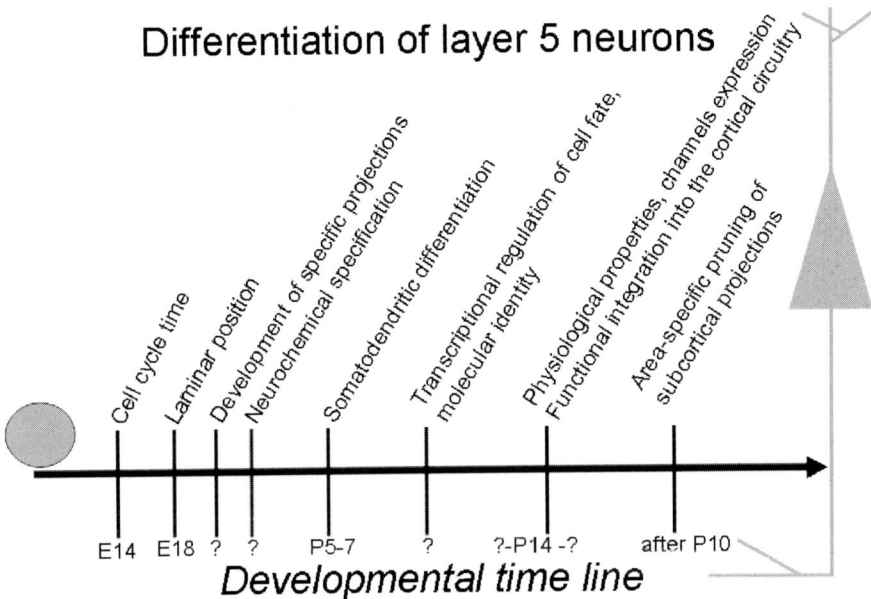

FIG. 4. Overview of the sequential steps of differentiation of layer V projection neurons. The following five areas of research are beginning to give better understanding: (1) correlation of biomarkers with specific layer V classes; (2) investigation of selected genes in the development of specific subtypes of layer V classes; (3) integration of layer V projection neuron subclasses into the functional cortical circuitry; (4) study of the factors involved in axonal and dendritic development of layer V neurons in different cortical areas; and (5) identification of components of gene cascades of transcription factors involved in layer V differentiation. Projects on early development feed into the understanding of the adult circuitry.

components drive neurogenesis, differentiation, and regional cortical connectivity. Indeed, the previous studies on morphological characterizations of the various cell types of the brain opened a window into some aspects of the development and function of the brain. Considerable advancement now requires more systematic and molecular characterizations of the various cells of the brain. The discovery of molecular markers will allow distinct subclassification of cells and functional dissection of early cortical circuits.

Acknowledgements

The laboratory of ZM was supported from grants from the MRC, Wellcome Trust, EU and Human Frontiers Science Program. We thank the Wellcome Trust Initiative in Integrative Physiology of Ion Channels (OXION) for the help with the microarray work.

References

Allendoerfer KL, Shatz CJ 1994 The subplate, a transient neocortical structure: its role in the development of connections between thalamus and cortex. Annu Rev Neurosci 17:185–218

Antonini A, Shatz CJ 1990 Relation between putative transmitter phenotypes and connectivity of subplate neurons during cerebral cortical development. Eur J Neurosci 2:744–761

Arimatsu Y, Ishida M, Kaneko T, Ichinose S, Omori A 2003 Organization and development of corticocortical associative neurons expressing the orphan nuclear receptor Nurr1. J Comp Neurol 466:180–196

Arlotta P, Molyneaux BJ, Chen J, Inoue J, Kominami R, Macklis JD 2005 Neuronal subtype-specific genes that control corticospinal motor neuron development *in vivo*. Neuron 45:207–221

Aye L, Piñon MC, Hoerder A, Jacobs E, Campagnoni A, Molnár Z 2006 Properties of layer6a and subplate neurons in the golli-tau-eGFP (GTE) mouse. Eur J Neurosci Supplement for 5[th]FENS Meeting, A196.1

Bicknese AR, Sheppard AM, O'Leary DD, Pearlman AL 1994 Thalamocortical axons extend along a chondroitin sulfate proteoglycan-enriched pathway coincident with the neocortical subplate and distinct from the efferent path. J Neurosci 14:3500–3510

Catalano S, Shatz CJ 1998 Activity dependent cortical target selection by thalamic axons. Science 281:559–562

Chen B, Schaevitz LR, McConnell SK 2005 Fezl regulates the differentiation and axon targeting of layer 5 subcortical projection neurons in cerebral cortex. Proc Natl Acad Sci USA 102:17184–17189

Christophe E, Doerflinger N, Lavery D, Molnár Z, Charpak S, Audinat E 2005 Two populations of layer V pyramidal cells of the mouse neocortex: development and sensitivity to anesthetics. J Neurophysiol 94:3357–3367

Clasca F, Angelucci A, Sur M 1995 Layer-specific programs of development in neocortical projection neurons. Proc Natl Acad Sci USA 92:11145–11149

De Carlos J, O'Leary D 1992 Growth and targeting of subplate axons and establishment of major cortical pathways. J Neurosci 12:1194–1211

Dupont E, Hanganu IL, Kilb W, Hirsch S, Luhmann HJ 2006 Rapid developmental switch in the mechanisms driving early cortical columnar networks. Nature 439:79–83

Feng G, Mellor RH, Bernstein M et al 2000 Imaging neuronal subsets in transgenic mice expressing multiple spectral variants of GFP. Neuron 28:41–51

Friauf E, Shatz CJ 1991 Changing patterns of synaptic input to subplate and cortical plate during development of visual cortex. J Neurophysiol 66:2059–2071

Ghosh A, Shatz CJ 1992 Involvement of subplate neurons in the formation of ocular dominance columns. Science 255:1441–1443

Ghosh A, Antonini A, McConnell SK, Shatz CJ 1990 Requirement for subplate neurons in the formation of thalamocortical connections. Nature 347:179–181

Guillemot F, Molnár Z, Tarabykin V, Stoykova A 2006 Molecular mechanisms of cortical differentiation. Eur J Neurosci 23:857–868

Hanganu IL, Kilb W, Luhmann HJ 2001 Spontaneous synaptic activity of subplate neurons in neonatal rat somatosensory cortex. Cereb Cortex 11:400–410

Heuer H, Christ S, Friedrichsen S et al 2003 Connective tissue growth factor: a novel marker of layer VII neurons in the rat cerebral cortex. Neuroscience 119:43–52

Hevner RF 2007 Layer-specific markers as probes for neuron type identity in human neocortex and malformations of cortical development. J Neuropathol Exp Neurol 66:101–109

Hevner RF, Daza RA, Rubenstein JL, Stunnenberg H, Olavarria JF, Englund C 2003 Beyond laminar fate: toward a molecular classification of cortical projection/pyramidal neurons. Dev Neurosci 25:139–151

Higashi S, Molnár Z, Kurotani T, Toyama K 2002 Prenatal development of neural excitation in rat thalamocortical projections studied by optical recording. Neuroscience 115:1231–46

Higashi S, Hioki K, Kurotani T, Molnár Z 2005 Functional thalamocortical synapse reorganization from subplate to layer IV during postnatal development in the reeler-like mutant rat (Shaking rat Kawasaki): an optical recording study. J Neurosci 25:1395–1406

Hoerder A 2007 Mouse cortical subplate neurones: molecular markers, connectivity and development. DPhil Thesis, University of Oxford, UK

Hoerder A, Paulsen O, Molnár Z 2006 Developmental changes in the dendritic morphology of subplate cells with known projections in the mouse cortex. Eur J Neurosci Supplement for 5th FENS Meeting, A156.10

Jacobs EC, Campagnoni C, Kampf K et al 2007 Visualization of corticofugal projections during early cortical development in a tau-GFP-transgenic mouse. Eur J Neurosci 25:17–30

Jethwa A, Piñon MC, Jacobs E, Campagnoni A, Molnár Z 2007 Dynamic integration of subplate neurons into the cortical barrel field circuitry during postnatal development in the Golli-tau-eGFP (GTE) mouse. British Neurosci Assoc Abstr 19:49

Kanold PO, Shatz CJ 2006 Subplate neurons regulate maturation of cortical inhibition and outcome of ocular dominance plasticity. Neuron 51:627–638

Kanold PO, Kara P, Reid RC, Shatz CJ 2003 Role of subplate neurons in functional maturation of visual cortical columns. Science 301:521–525

Kasper EM, Larkman AU, Lübke J, Blakemore C 1994 Pyramidal neurons in layer 5 of the rat visual cortex. III. Differential maturation of axon targeting, dendritic morphology and electrophysiological properties. J Comp Neurol 339:495–518

Koester SE, O'Leary DD 1992 Functional classes of cortical projection neurons develop dendritic distinctions by class-specific sculpting of an early common pattern. J Neurosci 12:1382–1393

Kolk SM, Whitman MC, Yun ME, Shete P, Donoghue MJ 2006 A unique subpopulation of Tbr1-expressing deep layer neurons in the developing cerebral cortex. Mol Cell Neurosci 32:200–214

Kostovic I, Rakic P 1980 Cytology and time of origin of interstitial neurons in the white matter in infant and adult human and monkey telencephalon. J Neurocytol 9:219–242

Kriegstein A, Noctor S, Martínez-Cerdeño V 2006 Patterns of neural stem and progenitor cell division may underlie evolutionary cortical expansion. Nat Rev Neurosci 7:883–890

Larkman A, Mason A 1990 Correlations between morphology and electrophysiology of pyramidal neurons in slices of rat visual cortex. I. Establishment of cell classes. J Neurosci 10:1407–1414

Lorente de No R 1949 Cerebral cortex: architecture, intracortical connections, motor projections. Oxford University Press

Luskin MB, Shatz CJ 1985 Neurogenesis of the cat's primary visual cortex. J Comp Neurol 242:611–631

Markram H 2004 Correlation maps allow neuronal electrical properties to be predicted from single-cell gene expression profiles in rat neocortex. Cereb Cortex 14:1310–1327

McConnell SK, Ghosh A, Shatz CJ 1989 Subplate neurons pioneer the first axon pathway from the cerebral cortex. Science 245:978–982

McQuillen PS, Sheldon RA, Shatz CJ, Ferriero DM 2003 Selective vulnerability of subplate neurons after early neonatal hypoxia-ischemia. J Neurosci 23:3308–3315

Miller B, Chou L, Finlay BL 1993 The early development of thalamocortical and corticothalamic projections. J Comp Neurol 335:16–41

Miller B, Sheppard AM, Bicknese AR, Pearlman AL 1995 Chondroitin sulfate proteoglycans in the developing cerebral cortex: the distribution of neurocan defines forming afferent and efferent axonal pathways. J Comp Neurol 355:615–628

Mitchell EJF, Cheung AFP, Molnár Z 2006 Expression pattern of parvalbumin, calbindin and calretinin in cortical layer V projection neurons in rat. Eur J Neurosci Supplement for 5[th]FENS Meeting, A228.10

Molnár Z, Blakemore C 1995 How do thalamic axons find their way to the cortex? Trends Neurosci 18:389–397

Molnár Z, Cordery P 1999 Connections between cells of the internal capsule, thalamus and cerebral cortex in the embryonic pallium. J Comp Neurol 413:1–25

Molnár Z, Cheung AF 2006 Towards the classification of subpopulations of layer V pyramidal projection neurons. Neurosci Res 55:105–115

Molnár Z, Adams R, Blakemore C 1998a Mechanisms underlying the establishment of topographically ordered early thalamo-cortical connections in the rat. J Neurosci 18:5723–5745

Molnár Z, Adams R, Goffinet AM, Blakemore C 1998b The role of the first postmitotic cells in the development of thalamocortical fibre ordering in the reeler mouse. J Neurosci 18:5746–5765

Molyneaux BJ, Arlotta P, Menezes JR, Macklis JD 2007 Neuronal subtype specification in the cerebral cortex. Nat Rev Neurosci 8:427–437

Nelson SB, Sugino K, Hempel CM 2006 The problem of neuronal cell types: a physiological genomics approach. Trends Neurosci 29:339–345

Peters A, Jones EG 1985 Cerebral cortex, vol. 3: Visual cortex, and vol. 4: Association and auditory cortices. Plenum Press, New York-London

Piñon MC, Jacobs E, Campagnoni A, Molnár Z 2005 Development of the cortical projections from subplate neurons to the thalamus in golli-tau-eGFP transgenic mice. British Neurosci Assoc Abstr 18:8

Price D, Aslam S, Tasker L, Gillies K 1997 Fates of the earliest generated cells in the developing murine neocortex. J Comp Neurol 37/7:414–422

Price DJ, Kennedy H, Dehay C et al 2005 The development of cortical connections. Eur J Neurosci 23:910–920

Rakic P 2006 A century of progress in corticoneurogenesis: from silver impregnation to genetic engineering. Cereb Cortex. Suppl 1:13–17

Ramón y Cajal S 1911 Histologie du Systéme Nerveaux de l'Homme et des Vertebres Vol. II. Maloine, Paris

Rolph RC, Cheung AFP, Voelker CCJ, Jessell T, Molnár Z 2005 ER81 and N200 reveal separate subpopulations of layer V pyramidal projection neurons. British Neurosci Assoc Abstr 3.04

Shatz CJ, Rakic P 1981 The genesis of efferent connections from the visual cortex of the fetal rhesus monkey. J Comp Neurol 196:287–307

Shimogori T, Grove EA 2005 Fibroblast growth factor 8 regulates neocortical guidance of area-specific thalamic innervation. J Neurosci 25:6550–6560

Voelker CCJ, Garin N, Taylor JSH, Gahwiler BH, Hornung J-P, Molnár Z 2004 Selective neurofilament (SMI-32, FNP-7 and N200) expression in subpopulations of layer 5 pyramidal neurons in vivo and in vitro. Cereb Cortex 14:1276–1286

Weimann JM, Zhang YA, Levin ME, Devine WP, Brulet P, McConnell SK 1999 Cortical neurons require Otx1 for the refinement of exuberant axonal projections to subcortical targets. Neuron 24:819–831

Yoneshima H, Yamasaki S, Voelker C et al 2005 ER81 is expressed in a subpopulation of layer 5 projection neurons in rodent cerebral cortices. Neuroscience 137:401–412

DISCUSSION

Hevner: It has clearly been a fruitful approach to use the microarrays. Did you see much temporal shift in the genes that were specifically expressed in subplate between E18 and the later time point?

Molnár: You have touched on an important question. In the search for cell type-specific markers for particular classes of neurons the most fruitful approach is a longitudinal one. This helps get the relevant patterns with more certainties. Our studies in subplate neurons have been restricted to two ages, E18 and P8, so far, but we are planning to do them at earlier and later stages as well. It is also important to perform the analysis in adult. Comparisons between different ages will not only reveal subplate markers, but will also give an insight into the functional state of subplate cells in a particular developmental period. It would be nice to see which subplate cell populations survive, and which will disappear preferentially. If we look at the gene expression patterns at early embryonic stages perhaps we will get more genes which are involved in axon elongation or synaptic remodelling, whereas later stages around P8 might show more expression of cell death genes and plasticity related genes in addition to some true subplate molecular tags. Perhaps the comparison of the embryonic and the adult markers will help us understand which subplate subtypes will disappear.

Kreigstein: I have a naïve question. My understanding of the subplate in the rodent is that postnatally, many of the cells die by apoptosis. But the suplate region in primate is much larger than rodent. Do some of these cells integrate into the lower layers of the adult cortex rather than just die or disappear?

Molnár: I think David Price is most qualified person here to answer your question on rodents. He put an end to the big dispute whether in rodent subplate neurons die or not (Price et al 1997). And the most qualified people to comment on the primate subplate are Pasko Rakic and Henry Kennedy. Henry has done an extensive birth dating on the monkey subplate (Smart et al 2002) and we have often discussed the rodent-primate differences with him while writing a review recently (Molnár et al 2006). He convinced me that the primate subplate is completely different to that in rodents because it has additional cells added throughout the time of cortical neurogenesis, whereas in rodent there is evidence that the vast majority of subplate cells is generated early with the preplate and the subplate layer derived from the preplate split. I would expect that primate subplate has some of these early generated subplate cells, but these might not constitute the major component. Our markers pulled out from the mouse screen are currently being tested in human cortex.

Kennedy: Yes, that's true. Subplate neurons in monkey are being generated throughout the whole period of neuron production from the cortical plate. Remember the neurogenic gradient is inverted in the subplate with respect to the cortical plate.

Parnavelas: Arnold has asked the question about what happens in the rodent: do these cells disappear or integrate into the other layers? There is some evidence that at least some of the cells become part of layer VIb. There is evidence that some do degenerate.

Kreigstein: I am interested in primates. There are so many more of these cells it seems unlikely that they would all involute or die. The possibility that in postnatal stages they migrate up into the cortex, or integrate into some of the other layers seems reasonable, but I have no idea that this is what happens.

Rakic: I think Arnold is right. It is different in the sense that many of these cells do not die; they stay as interstitial cells within the white matter. Once, I presented this fact at a meeting and people said it was just a small number. I then went back to New Haven and counted the number of interstitial cells in the adult primate white matter. We have more cells in the white matter subjacent to area 17 than the entire lateral geniculate body (Kostovic & Rakic 1980). It is interesting that there are many studies on the lateral geniculate nucleus (LGN) physiology, but no one studies the interstitial neurons below area 17 that are more numerous, perhaps because they are scattered in between axons of the white matter.

Kennedy: Yes, there are many more interstitial cells in the frontal cortex.

Rakic: I agree, there are probably more in the frontal cortex than in the medial geniculate body. The number of those cells is almost equal to thalamic cells, because humans have huge white matter. Between those are neurons that are scattered as individuals. No one has studied them.

Molnár: Paul Harrison's group has studied these cells. In schizophrenia there are more of these cells (Eastwood & Harrison 2005). This is very interesting.

Harrison: What is the conventional wisdom about the role of dopamine in these processes? What is the interpretation you make about the dopaminergic markers in the subplate neurons?

Molnár: Indeed, our microarray studies suggest that the enzymes are expressed in the subplate cells to produce dopamine. Anna Hoerder and Wei Zhi Wang in my laboratory are currently investigating whether some of subplate is dopaminergic. Although the possible implications for disease (schizophrenia) are obvious, the dopaminergic subplate neuron idea is uncharted territory.

Price: I want to ask about the handshake hypothesis. I am confused about what your current thinking is. Where is it taking place? I thought your hypothesis was that the handshake was taking place in the internal capsule. When you were talking I got the impression that you were now thinking about a handshake very close to the cortex.

Molnár: No, we always said that the 'handshake' takes place at the lateral part of the internal capsule, and we believe that it is essential for crossing the pallial/subpallial boundary (Molnár & Blakemore 1995, López-Bendito & Molnár 2003).

Price: Do you think it is important for guidance through the internal capsule?

Molnár: Yes, for the distal part of the internal capsule.

Macklis: I am confused by this as well. When you mentioned Reeler, the image you showed was further up. Do you conceptualise that these misplaced deep layer neurons are leading to delay in thalamocortical innervation?

Molnár: When the subplate cells are displaced to the superplate, their axons don't look back where their cell body is displaced. By the time the cortical plate has considerable size the subplate axons have descended to the pallial–subpallial boundary and started to cross it. These corticofugal pathways get laid down quite early on. In Reeler, the cortical plate develops just below the subplate (superplate) cells, whereas normally they would develop below the cortical plate. What I was emphasizing with the fibre pattern in the Reeler cortex is that when the thalamocortical projections grow through the internal capsule, they establish the handshake with the early corticofugal projections from subplate or in Reeler with superplate and then just stick to these fibres as they advance through the Reeler cortex all the way up to the superplate. This is how they can cross the pallial–subpallial boundary and continue to grow in the relatively hostile environment of embryonic cortical plate to reach the superplate in the Reeler. My slide showed the advancing growth cones associated with fibres originating from superplate within the cortical plate. I argued that without this close association thalamocortical axons could not cross the embryonic cortical plate (Molnár et al 1998). If you confront the embryonic cortical plate with thalamic fibres, there is no way they will enter. This has been shown in many co-culture systems and membrane preparations from embryonic cortical plate (Molnár & Blakemore 1999). We think this close association between the two fibre systems is essential to get through hostile territory and then accumulate in the superplate.

Macklis: So now with the GTE, do you confirm all the experiments from Pasko's lab and others which showed that these are laying down their axon as they migrate? When do they lay down their axons? When do the subplate neurons send an axon across that pallial/subpallial boundary?

Molnár: Between 13 and 13.5 in mouse. It would be interesting to breed these animals on Reeler and p35 knockout. For the last decade we have been suffering from the lack of selective tracers. Now these genetic tools will help us address these questions again. We are working on these questions in collaboration with Erin Jacobs and Tony Campagnoni (UCLA).

Macklis: Presumably your hypothesis would be that they would have their axons across that boundary just as early.

Molnár: Yes.

Rakic: It would be interesting to compare this situation in primates and mouse. Arrangement of fibres is thought to be different. For example, neostriatum is a single structure in mouse but it is separated into two distinct nuclei in humans by the large fibre tracts. These fibres penetrate the middle of the neostriatum, separating caudate nucleus from the putamen.

Molnár: Exactly. This needs to be done. With Colin Blakemore and Irina Bystron we would like to revisit this question. Henry Kennedy, Rosalind Carney and I have done quite a bit of tracing in embryonic macaque brains, although not focusing specifically on these questions (Carney et al 2002, Carney 2005). Interestingly, some aspects of primate thalamocortical development are surprisingly similar to rodents. Clearly, more work is needed in this area.

O'Leary: At the early stages of development of subplate axons and thalamocortical axon inputs, we believe that they take discrete pathways in the cortex, and that the thalamic afferents take a deep pathway beneath the subplate and subplate axons. They may intersect within the internal capsule at one point, but they don't follow one another along those trajectories intracortically. To me, this opens up the question of how important this relationship is, if it exists at all, for the earliest fibres. It seems to be an open issue.

Molnár: I agree. I think the GTE mouse (Jacobs et al 2007) is not an entirely clear model, since layer VI proper and perhaps even layer V also contribute to the labelled projections. We are currently looking at these kinds of questions with Erin and Tony, but if I showed you the digitations here, which we do get in the GTE mouse, you would say that we have contamination with other cells in addition to true subplate.

O'Leary: I am talking about stages before layer VI axons have begun to grow out.

Molnár: Then I think I have a less difficult task and I can show you interesting data on the interdigitation in the white matter from E14–15. Thalamic axons interdigitate with subplate projections (appear green) in both internal capsule and white matter. In fact, the entire depth of the cortical intermediate zone, from subplate to the germinal zone is filled with green subplate fibres in the GTE mouse (Pinon et al 2005).

Goffinet: I agree, it is still unsettled after so many years. When do the first axons project to pallidum? When is the pallidum born? At E12 if you do a DiI in pallidum there are fibres leaving pallidum: where are they going?

Molnár: Yesterday after Henry Kennedy's paper I suggested that the changing modality of input that is relayed through the thalamocortical projection changes the spatial distribution of cortico-cortical connections. Henry did not seem to be terribly convinced. Let me show some work from Mriganka Sur to convince Henry. The experiments I was referring to were those when Mriganka redirected the retinal input into the medial geniculate nucleus, and used auditory thalamus to convey visual input into the auditory cortex (Sur & Leamey 2001). Using imaging, they saw rearrangement in the cortical representation. They started to see orientation columns in these auditory cortices. If you inject a tracer into the normal visual cortex, you see a patchy distribution of back labelled cells, whereas if we do the same in the normal auditory cortex there is a much more restricted connectivity. However, if we start bombarding this putative auditory cortex with visual input through the auditory thalamus, we begin to see the patchy distribution of V1 in the putative auditory cortex. I think that these experiments are spectacular, since the only thing they changed on thalamocortical level was the modality-specific input.

References

Carney RSE 2005 Thalamocortical development and cell proliferation in fetal primate and rodent cortex. DPhil Thesis. University of Oxford, Oxford, UK

Carney RSE, Molnar Z, Giroud P et al 2002 Thalamocortical projections in the developing primate cortex. FENS Abstr 1:A007.06

Eastwood SL, Harrison PJ 2005 Interstitial white matter neuron density in the dorsolateral prefrontal cortex and parahippocampal gyrus in schizophrenia. Schizophr Res 79:181–188

Jacobs EC, Campagnoni C, Kampf K et al 2007 Visualization of corticofugal projections during early cortical development in a tau-GFP-transgenic mouse. Eur J Neurosci 25:17–30

Kostovic I, Rakic P 1980 Cytology and time of origin of interstitial neurons in the white matter in infant and adult human and monkey telencephalon. J Neurocytol 9:219–242

López-Bendito G, Molnár Z 2003 Thalamocortical development: how are we going to get there? Nat Rev Neurosci 4:276–289

Molnár Z, Blakemore C 1995 How do thalamic axons find their way to the cortex? Trends Neurosci 18:389–397

Molnár Z, Blakemore C 1999 Development of signals influencing the growth and termination of thalamocortical axons in organotypic culture. Exp Neurol 156:363–393

Molnár Z, Adams R, Goffinet AM, Blakemore C 1998 The role of the first postmitotic cells in the development of thalamocortical fibre ordering in the reeler mouse. J Neurosci 18:5746–5765

Molnár Z, Métin C, Stoykova A et al 2006 Comparative aspects of cerebral cortical development. Eur J Neurosci 23:921–934

Piñon MC, Jacobs E, Campagnoni A, Molnár Z 2005 Development of the cortical projections from subplate neurons to the thalamus in Golli-tau-eGFP transgenic mice. Br Neurosci Assoc Abstr 18:8

Price DJ, Aslam S, Tasker L, Gillies K 1997 Fates of the earliest generated cells in the developing murine neocortex. J Comp Neurol 377:414–422

Smart IH, Dehay C, Giroud P, Berland M, Kennedy H 2002 Unique morphological features of the proliferative zones and postmitotic compartments of the neural epithelium giving rise to striate and extrastriate cortex in the monkey. Cereb Cortex 12:37–53

Sur M, Leamey CA 2001 Development and plasticity of cortical areas and networks. Nat Rev Neurosci 2:251–262

Emx and *Nfi* genes regulate cortical development and axon guidance in the telencephalon

Michael Piper, Amber-Lee S. Dawson, Charlotta Lindwall, Guy Barry, Céline Plachez* and Linda J. Richards[1]

*The University of Queensland, Queensland Brain Institute and The School of Biomedical Sciences, Brisbane, 4074, Australia and * The University of Maryland School of Medicine, Baltimore, USA*

Abstract. The *Emx* and Nuclear Factor One (*Nfi*) genes encode transcription factors that regulate numerous embryonic developmental processes. The two mammalian *Emx* genes, *Emx1* and *Emx2*, are expressed in the embryonic cortex and regulate the specification of the cortex into different sensory and motor areas along the rostrocaudal axis. To date, few developmental processes have been attributed specifically to *Emx1*, with most analyses demonstrating a redundancy of function between *Emx1* and *Emx2*, with *Emx2* being most essential for development. Here we provide evidence that *Emx1* and *Emx2* regulate different developmental processes during corpus callosum formation and review how both genes function in cellular migration and the formation of cortical axon projections. The *Nfi* gene family is made up of four members, *Nfia*, *Nfib*, *Nfic* and *Nfix*. Expression analyses show that *Nfia*, *Nfib* and *Nfix* are expressed in the developing telencephalon. They play roles in patterning, glial development, cortical cell migration and axon guidance. We review the role of these genes in cortical cell migration, glial development and the formation of cortical axon projections, and examine the overlapping mutant phenotypes between the Emx and Nfi gene families.

2007 Cortical development: genes and genetic abnormalities. Wiley, Chichester (Novartis Foundation Symposium 288) p 230–245

Despite the enormous complexity inherent within the mammalian neocortex, morphogenesis of this structure involves the iteration of relatively few developmental processes. These include area specification, progenitor proliferation and differentiation, cellular migration and the formation of appropriate axonal and dendritic connections. The Emx and Nfi transcription factor families play roles in several of these developmental events.

---

[1]This paper was presented at the symposium by Linda Richards, to whom correspondence should be addressed.

The *Emx* genes are vertebrate homologues of the *Drosophila empty spiracles* gene. In the developing mouse cortex, *Emx2* is expressed in a high caudomedial/low rostrolateral gradient (Fig. 1A–C), and current data suggest that this expression gradient is involved in the specification of neocortical areas, with areas of high *Emx2* expression destined to become visual cortex (Bishop et al 2000).

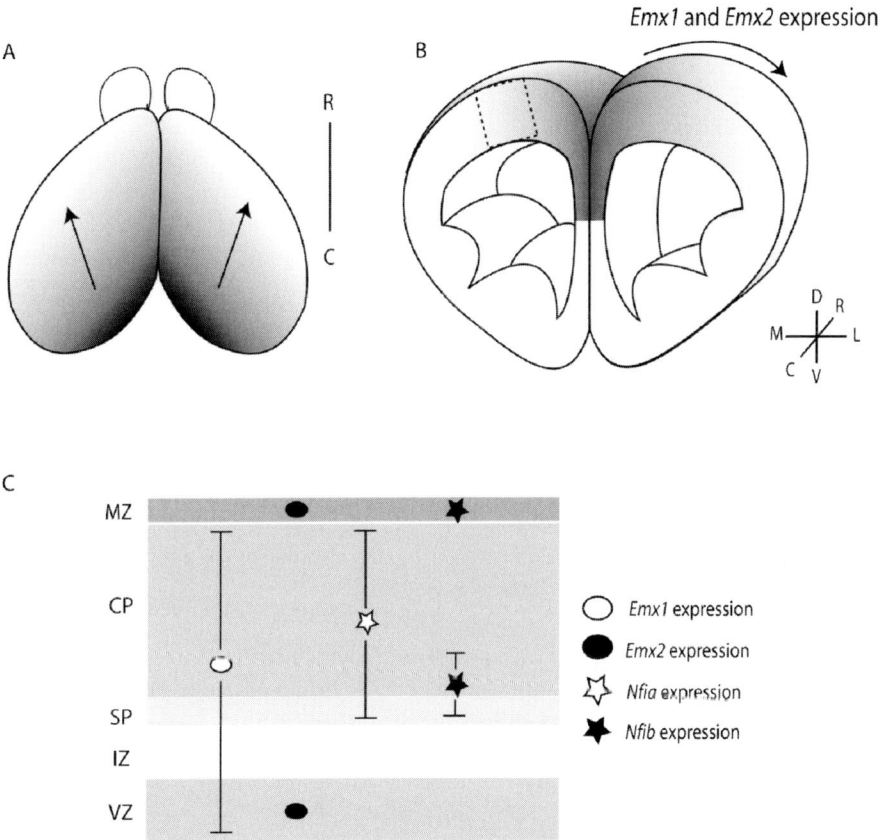

FIG. 1.   Expression patterns of *Emx1*, *Emx2*, *Nfia* and *Nfib* genes in the cortex of embryonic mouse embryos at E15.5. (A) Dorsal view of the high caudomedial/low rostrolateral gradient displayed by the *Emx1* and *Emx2* genes (grey). It is not known whether *Nfia* or *Nfib* are expressed in a graded manner across the cortex. (B) Coronal view of an E15.5 mouse brain, showing the medial to lateral gradient of *Emx* genes. (C) Schematic of the boxed area in B, showing layer specific expression of the *Emx* and *Nfi* genes within the cortex. Note that the expression of the *Nfi* genes is dynamic and changes depending on the stage of development. This schematic represents their expression at E15.5 in mice. VZ, ventricular zone; IZ, intermediate zone; SP, subplate; CP, cortical plate; MZ, marginal zone; R, rostral; C, caudal; D, dorsal; V, ventral; M, medial; L, lateral.

Experiments which alter the *Emx2* gradient by either knockout or over-expression in more rostral cortical areas demonstrate that *Emx2* has the capacity to specify visual cortical areas recognised by thalamic afferents from the lateral geniculate nucleus (Leingartner et al 2003, Hamasaki et al 2004). *Emx1* is also expressed in a gradient similar to *Emx2* across the cerebral cortex (Fig. 1A–C) but *Emx1* mutants do not exhibit defects in arealization, suggesting that *Emx1* does not play a role in this process (Bishop et al 2002). The role of *Emx2* in regulating cortical arealization is reviewed elsewhere (O'Leary & Nakagawa 2002, Rash & Grove 2006). Here we focus on the role that *Emx1* and *Emx2* play in later aspects of corticogenesis, as knockout studies have implicated these genes in lamination and in the targeting of cortical projections.

The *Nfi* genes are a phylogenetically conserved family of transcription factors that act as homo- or heterodimers to activate or repress gene expression during the development of multiple organ systems (Gronostajski 2000). Nfi family members are expressed throughout cortical development in a spatially and temporally dynamic fashion (Fig. 1C; Plachez et al submitted) and have been identified as potential downstream targets of the cortical arealization genes *Emx2* and *Pax6*. Mice lacking either *Nfia* or *Nfib* display a spectrum of abnormalities, including defects in the formation of forebrain commissures and the loss of specific midline glial populations, which we review here.

Emx and *Nfi* genes regulate cortical cell migration

Cortical neuron migration is determined by a combination of complex developmental events. These involve intrinsic neuronal mechanisms, such as dynamic cytoskeletal regulation leading to edge extension and nucleokinesis, and extrinsic signals that modulate the environment through which neurons migrate. Neurons destined to populate the mammalian neocortex originate from progenitor cells within the telencephalic proliferative zones, and their final position within the cortex is reached by different modes of directional migration (Fig. 2A). Glutamatergic projection neurons migrate radially from the ventricular zone (VZ) and subventricular zone (SVZ) along a radial glial scaffold to form the layers of the cortical plate (CP) in a typical inside-out fashion. In contrast, GABAergic interneurons take a tangential route from the ventrally located ganglionic eminences. These neurons reach their cortical destination by migrating through either the intermediate zone (IZ) or via the marginal zone (MZ) after which they turn and migrate radially into the CP (Metin et al 2006).

Emx1 and *Emx2* are involved in regulating cortical neuron migration (Fig. 2B, C) (Mallamaci et al 2000, Shinozaki et al 2002, Bishop et al 2003). *Emx2* is expressed in Cajal-Retzius (CR) cells of the developing mouse cortex (Mallamaci et al 1998). CR cells secrete Reelin, a key molecule that regulates neuronal migra-

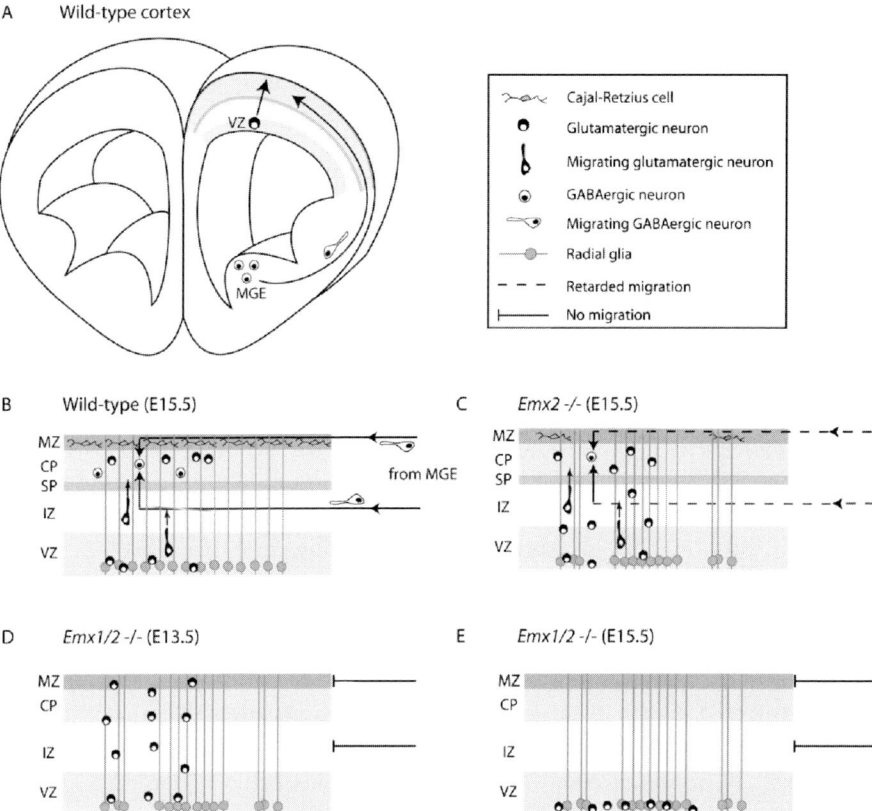

FIG. 2. *Emx2* in neuronal migration in the developing cortex. (A) Schematic overview of a coronal section of the mouse brain (E15.5) illustrating the origin and migration of glutamatergic projection neurons and GABAergic interneurons. (B) In wild-type mice, projection neurons migrate from the VZ along radial glial fibres to reach their final destinations in the superficial layers of the CP. Interneurons are born in the medial ganglionic eminences (MGE) and migrate tangentially to the CP via the MZ or the IZ. Cajal-Retzius (CR) cells, which express Reelin, reside in the MZ. (C) In the cortex of *Emx2*-deficient mice the glial scaffold is disrupted with large diameter fascicles separated by wide spaces. The number of CR cells, and consequently Reelin expression, is severely reduced. Cortical layering of projection neurons migrating from the VZ is disrupted with cells found scattered throughout the CP. Some neurons do however reach the superficial layers of the CP. Migration of GABAergic interneurons is retarded. (D, E) *Emx1/2* double mutant mice display a more severe phenotype than the *Emx2* single mutant with a disrupted laminar architecture and absence of a subplate. These mutants are also devoid of CR cells and completely lack tangential migration of GABAergic neurons. Neurons born at E13.5 (D) can be found scattered throughout the cortex, whereas those born at E15.5 (E) completely fail to migrate and remain in the periventricular zone. Panels B–E are adapted from Mallamaci et al (2000). VZ, ventricular zone; IZ, intermediate zone; SP, subplate; CP, cortical plate; MZ, marginal zone.

tion (D'Arcangelo et al 1995). In wild-type mice, CR cells express Reelin from embryonic (E) day 11.5 onward. In $Emx2^{-/-}$ mutant mice however, Reelin mRNA is expressed normally until E13.5 after which it is virtually absent from the MZ (Mallamaci et al 2000). Cortical architecture is thus disrupted in $Emx2$ mutant cortices, as sequential layering that occurs after E13.5 is misregulated. Furthermore, the $Emx2$ mutant cortex displays subtle radial glial scaffold malformations (Fig. 2C compared to 2B). This radial glial defect, in addition to the deficiency of late gestational CR cells and subsequent lack of Reelin expression, could underlie the specific migration deficits displayed in the $Emx2$ mutant mice (Mallamaci et al 2000).

In contrast to the developmental defects found in the $Emx2$ mutant mice, malformations of the $Emx1^{-/-}$ neocortex are subtle (Yoshida et al 1997). However, two separate studies recently demonstrated a co-operative role for $Emx1$ and $Emx2$, implying that these transcription factors may act in concert to control cortical development (Shinozaki et al 2002, Bishop et al 2003). $Emx1/2$ double mutant mice have a more severe phenotype than either of the single mutants, including abnormalities of cortical laminar architecture. In $Emx1/2$ double mutants, CR cells are initially formed (E11.5) in the MZ, although at a reduced number compared to wild-type cortex, but they completely disappear at E13.5 (Shinozaki et al 2002). Also, neurons born at E13.5 were observed in a scattered pattern throughout the cortex, while birth-dating indicated those neurons born at E15.5 remained within the periventricular zone, indicating that they failed to migrate (Fig. 2D, E). Defects in subplate formation were also observed (Shinozaki et al 2002, Bishop et al 2003). $Emx2$ single mutants exhibit retarded migration of GABAergic interneurons (Fig. 2C), but in $Emx1/2$ double mutants interneuron migration is completely absent (Fig. 2D and E), demonstrating that both radial and tangential neuronal migration in the cortex is regulated by the Emx genes (Shinozaki et al 2002).

Intrinsic neuronal properties, such as impaired cytoskeletal function, could account for the cellular migration defects in Emx mutants. In this context, a recent representational difference analysis revealed a number of potential $Emx2$ target genes, including $Cdk4$, $ME2$, $Odz4$ and $Cofilin1$ (Li et al 2006). The actin-depolymerizing protein Cofilin1 regulates actin cytoskeletal dynamics and misregulation of this process could cause migration defects in $Emx2$ mutants. However, recent transplantation experiments imply that extrinsic factors may also regulate Emx-dependent cortical migration. Wild-type medial ganglionic eminence (MGE) neurons transplanted into the MGE of $Emx1/2$ double mutant mice did not migrate to the cortex, whereas $Emx1/2$ double mutant MGE neurons were able to move into the cortex when transplanted into wild-type MGE (Shinozaki et al 2002). The failure of wild-type neurons to migrate in the mutant cortex may arise from defects in the cortical environment similar to the malformed radial glial scaffold observed in the *Emx2* mutant cortex. More work is required to delineate the contribution of intrinsic versus extrinsic factors with respect to cortical migration in these mutant mice.

Little is known about the role of *Nfi* genes in cortical neuron migration. However, one report has implicated *Nfib* as a downstream effector of Neurogenin2, a basic-helix-loop-helix transcription factor that regulates neuronal lineage determination and the dorsoventral specification of neuronal identity during embryogenesis (Fode et al 2000). Using a subtractive hybridisation approach, *Nfib* was shown to be down-regulated in post-mitotic neurons in the CP of *Neurogenin2*-deficient mice, indicating that *Neurogenin2* may regulate *Nfib* expression, thus implicating *Nfib* in the molecular cascade driving corticogenesis (Mattar et al 2004).

Emx and *Nfi* genes in glial development and their role in axon guidance at the midline

Glia are essential for multiple aspects of corticogenesis. Radial glia, whose processes span the width of the cortex, act as scaffolds for neuronal progenitors to migrate into the CP, and are themselves multipotent progenitor cells that can differentiate into other cell types, including neurons (Gotz & Huttner 2005). Regulation of axon guidance is another process mediated by glia. For instance, glia localized to specific locations in the developing forebrain mediate axon pathfinding extrinsically by providing molecular signals to extending axons (Richards et al 2004, Plachez & Richards 2005). This is illustrated during development of the corpus callosum (CC), the fibre tract joining neurons in the left and right cerebral hemispheres. The glial wedge (GW) and the indusium griseum glia (IGG), positioned ventral and dorsal to the CC respectively (Shu & Richards 2001), mediate formation of this interhemispheric fibre tract via expression of guidance cues such as *Slit2* (Shu et al 2003a, b) and *Wnt5A* (Keeble et al 2006). An indication that the *Nfi* genes may directly regulate glial development has come from a recent study investigating their function in the embryonic spinal cord (Deneen et al 2006). Both *Nfia* and *Nfib* are expressed in the VZ of the spinal cord at the onset of gliogenesis. Using both *in vitro* and *in vivo* approaches designed to abrogate or over-express these genes, Anderson and colleagues demonstrated that *Nfia* and *Nfib* promote glial cell fate specification. *Nfia* was also shown to promote the maintenance of the progenitor pool in the VZ, thereby indirectly inhibiting neurogenesis. At later stages of development, *Nfia* and *Nfib* promoted astrocyte differentiation, and the activity of *Nfia* in the oligodendrocyte lineage was repressed by *Olig2*, possibly via direct interaction between the two proteins (Deneen et al 2006).

These data correlate with findings from our laboratory on the role of the *Nfi* genes in cortical development, as midline glial development is aberrant in *Nfi* mutant mice (Fig. 3). During embryonic development in wild-type mice, *Nfia* is

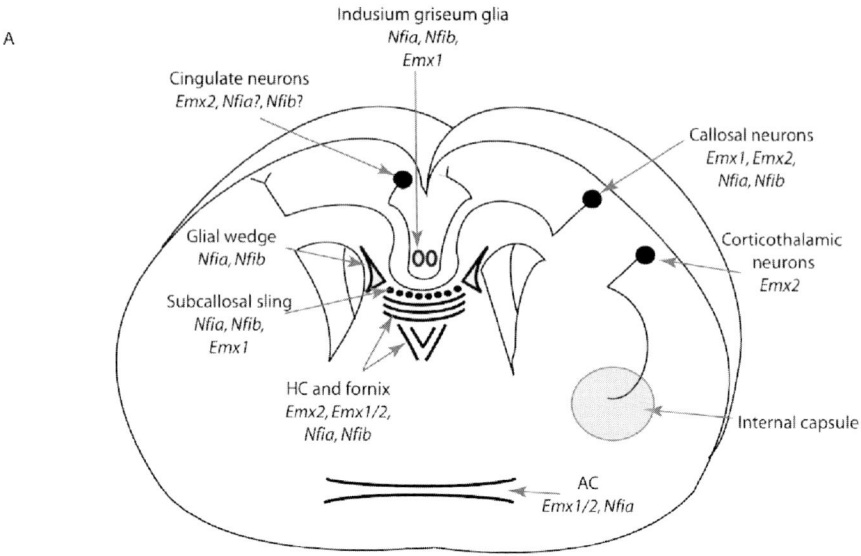

FIG. 3. Axonal projection and glial defects in *Emx* and *Nfi* mutants. (A) Schematic of a coronal section of an E17 mouse cortex detailing glial populations and telencephalic projections that develop aberrantly in knockout mutants. (B) Comparison of glial and axonal projection defects observed in *Emx1*, *Emx2*, *Nfia* and *Nfib* knockout mice. CC, corpus callosum; HC, hippocampal commissure; AC, anterior commissure; TC, thalamocortical projection; CT, corticothalamic projection; IGG, indusium griseum glia; GW, glial wedge; SS, subcallosal sling.

expressed in the GW, IGG and in a population of migrating neurons, the subcallosal sling. In *Nfia$^{-/-}$* mice the GW and the IGG are absent or greatly reduced and midline *Slit2* expression is reduced (Shu et al 2003c). *Nfia* mutant mice exhibit ACC with callosal axons reaching the midline but entering the septum instead of crossing the midline, which may arise from changes in the extrinsic environmental substrate at the midline. It remains to be determined if ACC in *Nfia*-deficient mice results from reduced glial numbers *per se*, or if the defect arises from reduced expression of *Slit2* in the GW and IGG. Rescue experiments where *Slit2* is over-expressed in the GW region of *Nfia* mutants may clarify this question. *Nfib* mutants also display ACC and have defects in midline glial development, with a reduction in the GW, IGG and in the midline zipper glia (MZG), a third midline glial population (Steele-Perkins et al 2005). Furthermore, *Emx1* mutants, which have abnormal CC formation, also have defects in glial formation at the midline, with the IGG and taenia tecta absent (Yoshida et al 1997). Thus, the Emx and Nfi families of transcription factors play an important role in midline glial formation and so may extrinsically regulate formation of the CC.

What lies downstream of these transcription factors to direct midline glial development and differentiation? One possible candidate is the FGF receptor, *fibroblast growth factor receptor 1 (Fgfr1)*, as mutations to this gene cause septal and midline glial defects (Smith et al 2006, Tole et al 2006). The IGG develop from radial glia within the GW that retract their ventricular endfeet and translocate towards the pia during embryogenesis. Using *Fgfr1* mutants, together with knockdown and over-expression studies in cortical explants, Smith and colleagues elegantly demonstrated that *Fgfr1*-dependent translocation of radial glial cells to form the IGG is required for CC development. As the IGG is reduced or absent in *Nfi* and *Emx1* mutants, it will be important to ascertain if these transcription factors drive the expression of *Fgfr1 in vivo*.

Emx and *Nfi* genes regulate the formation of cortical axon projections

The role played by *Emx1* in regulating cortical axon guidance has been investigated previously by three groups that independently generated *Emx1*-deficient mice. Two groups reported a primary defect in CC development that manifested as complete or partial agenesis, with callosal axons forming pathological longitudinal fascicles, known as Probst bundles, lateral to the midline, instead of crossing (Qiu et al 1996, Yoshida et al 1997). In contrast, Li and colleagues reported that the CC formed normally in their *Emx1* mutants (Guo et al 2000). While apparently contradictory, these data may be reconciled by considering the potential presence of genetic modifiers due to differences in the strain backgrounds of the mice used; mixed 129sv/C57BL/6 (Qiu et al 1996) versus inbred C57BL/6 (Guo et al 2000). Embryonic stem cells from the 129sv strain are commonly used to generate knockout mice but this strain can display spontaneous ACC whereas the C57BL/6 strain does not display this defect. Of note is that Guo et al (2000) primarily analysed the phenotype of their mice using sagittal bisection of the brain at the midline, which does not enable the visualisation of more laterally located Probst bundles. While the authors did investigate the morphology of the cortex using coronal sections of the brain, the cortical ultrastructure was characterised using a stain that labelled cell bodies (Nissl). Preliminary data from our laboratory using *Emx1*$^{-/-}$ mice (Qiu et al 1996) backcrossed onto a C57BL/6 background for over ten generations demonstrates that, while much of the CC forms normally, small Probst bundles do form lateral to the midline. This suggests that *Emx1* does play a role in the development of a subpopulation of callosal axons (Fig. 3; Dawson & Richards, in preparation), although the majority of the corpus callosum forms normally. As *Emx1* is expressed in both cortical progenitors and in their post-mitotic neuronal progeny (Simeone et al 1992), the contribution of *Emx1* to extrinsic guidance versus intrinsic development of callosal neurons in CC formation is yet to be resolved.

Emx2-deficient mice display a more severe phenotype than *Emx1* knockouts, with the CC being thinner as a result of fewer axons being present in this tract (Yoshida et al 1997). *Emx2*-deficient mice have a severely reduced cortical size, which suggests that the CC phenotype may be a secondary defect arising from reduced cortical proliferation, migration and differentiation. Evidence suggests that *Emx2* may regulate cortical size by promoting symmetric precursor cell division (Heins et al 2001), as *Emx2*-deficient mice exhibit a reduction in progenitor proliferation accompanied by early exit from the cell cycle (Muzio et al 2005). The thinner CC displayed by *Emx2* mutants may also arise from extrinsic effects, as the size of the cingulate cortex is reduced in these mice (Pellegrini et al 1996). As the axons that pioneer the CC arise from neurons in the cingulate cortex (Koester & O'Leary 1994, Rash & Richards 2001), abnormal development of this area may significantly affect the formation of this fibre tract. Unlike *Emx1* knockout mice, *Emx2* mutants also display defects in other fibre tracts, such as the hippocampal commissure, fimbria and fornix and the thalamocortical projection (Fig. 3) (Yoshida et al 1997, Lopez-Bendito et al 2002). Aberrant thalamocortical projections in *Emx2* mutants arise from early guidance defects at the boundary between the diencephalon and telencephalon, where thalamic axons misroute. Compensatory mechanisms do enable many thalamocortical axons to eventually ascend to the cortex, but they are shifted in their targeting (Lopez-Bendito et al 2002). As previously described, this suggests that *Emx2* expression imparts positional targeting information in the cortex that is used by thalamocortical axons.

Functional redundancy between *Emx1* and *Emx2* with respect to the formation of long-range axonal projections is also evident, as double mutants have a more severe phenotype than single null mutants, including hypoplasia of all the telencephalic commissures. Thus, these genes are likely to co-operate in regulating multiple aspects of cortical development including formation of the CC and the thalamocortical pathway (Bishop et al 2003). Such recent findings have increased our knowledge of the role of the Emx genes in cortical development. The extent to which the defects in these mutants arise from cell-autonomous effects within the projection neurons and those that arise from extrinsic changes to the environmental milieu through which their axons must navigate are currently under investigation.

Finally, knockout studies have demonstrated that the *Nfi* genes also regulate cortical axonogenesis (Fig. 3). *Nfia*$^{-/-}$ mice display aberrant forebrain commissure formation. Defects include ACC, with callosal axons entering the septum instead of crossing the midline, and a reduction in, or absence of, the hippocampal commissure. The ipsilaterally projecting perforating pathway, however, appears normal (Shu et al 2003c). Axonal projections in *Nfib*-deficient mice are also malformed, with ACC and a reduction in the fimbria seen in null mutants (Steele-Perkins et al 2005). The ACC observed in *Nfia* and *Nfib* mutants may be due to either the failure of midline glial structures to develop appropriately or intrinsic defects inherent to the mutant cortical axons themselves. Alternatively, these genes may regulate the development of the cingulate pioneering axons. Both *Nfia* and *Nfib* mutants have an enlarged cingulate cortex (unpublished observation). Efferent projections from the cingulate cortex pioneer the CC (Koester & O'Leary 1994, Rash & Richards 2001), and so play an integral role in its development. Studies are currently underway to ascertain how the Nfi genes regulate the formation of cortical projections such as the CC, and to identify some of the downstream targets of these transcription factors to gain a clearer view of how they may regulate axonogenesis.

Conclusion

The *Emx* and *Nfi* genes have been shown to participate in many different aspects of cortical development, including arealization, neuronal migration and cortical lamination, glial development and the formation of long-range axonal projections. One interesting feature of these studies has been the observation that many of the same structures are disrupted in *Nfi* and *Emx* mutants, such as the malformation of the CC, dentate gyrus, hippocampal commissure and anterior commissure (Fig. 3B). This raises the question of whether these genes act in a genetic cascade to control the development of these structures. Evidence in support of this has come from the study of *Emx2*-deficient neurospheres, in which *Nfia* expression is altered (Gangemi et al 2006). Both *in vitro* and *in vivo* approaches will be needed to determine if there are regulatory interactions between these gene families. Furthermore, our understanding of key issues, such as how these genes are themselves regulated, which genes they in turn activate/repress and how these genetic programmes are translated into developmental outcomes need to be addressed. As expression of these genes is dynamic and widespread, it is also likely that their regulation and function in different populations of cells will vary considerably. Investigating these issues, as well as whether the cortical defects observed in the knockouts represent cell intrinsic or extrinsic effects of these genes, will clarify the role of these genes in cortical development.

Acknowledgements

We thank the following funding agencies for supporting this work: NH&MRC grant 401616, the March of Dimes Foundation for Birth Defects grant FY05-119, the National Institutes of Heath grant NS44054. LJR is an NH&MRC Senior Research Fellow, MP is an NH&MRC Howard Florey Centenary Research Fellow. ALSD is supported by an NH&MRC postgraduate award.

References

Bishop KM, Goudreau G, O'Leary DD 2000 Regulation of area identity in the mammalian neocortex by Emx2 and Pax6. Science 288:344–349

Bishop KM, Rubenstein JL, O'Leary DD 2002 Distinct actions of Emx1, Emx2, and Pax6 in regulating the specification of areas in the developing neocortex. J Neurosci 22:7627–7638

Bishop KM, Garel S, Nakagawa Y, Rubenstein JL, O'Leary DD 2003 Emx1 and Emx2 cooperate to regulate cortical size, lamination, neuronal differentiation, development of cortical efferents, and thalamocortical pathfinding. J Comp Neurol 457:345–360

D'Arcangelo G, Miao GG, Chen SC, Soares HD, Morgan JI, Curran T 1995 A protein related to extracellular matrix proteins deleted in the mouse mutant reeler. Nature 374:719–723

Deneen B, Ho R, Lukaszewicz A, Hochstim CJ, Gronostajski RM, Anderson DJ 2006 The transcription factor NFIA controls the onset of gliogenesis in the developing spinal cord. Neuron 52:953–968

Fode C, Ma Q, Casarosa S, Ang SL, Anderson DJ, Guillemot F 2000 A role for neural determination genes in specifying the dorsoventral identity of telencephalic neurons. Genes Dev 14:67–80

Gangemi RM, Daga A, Muzio L et al 2006 Effects of Emx2 inactivation on the gene expression profile of neural precursors. Eur J Neurosci 23:325–334

Gotz M, Huttner WB 2005 The cell biology of neurogenesis. Nat Rev Mol Cell Biol 6:777–788

Gronostajski RM 2000 Roles of the NFI/CTF gene family in transcription and development. Gene 249:31–45

Guo H, Christoff JM, Campos VE, Jin XL, Li Y 2000 Normal corpus callosum in Emx1 mutant mice with C57BL/6 background. Biochem Biophys Res Commun 276:649–653

Hamasaki T, Leingartner A, Ringstedt T, O'Leary DD 2004 EMX2 regulates sizes and positioning of the primary sensory and motor areas in neocortex by direct specification of cortical progenitors. Neuron 43:359–372

Heins N, Cremisi F, Malatesta P et al 2001 Emx2 promotes symmetric cell divisions and a multipotential fate in precursors from the cerebral cortex. Mol Cell Neurosci 18:485–502

Keeble TR, Halford MM, Seaman C et al 2006 The Wnt receptor Ryk is required for Wnt5a-mediated axon guidance on the contralateral side of the corpus callosum. J Neurosci 26:5840–5848

Koester SE, O'Leary DD 1994 Axons of early generated neurons in cingulate cortex pioneer the corpus callosum. J Neurosci 14:6608–6620

Leingartner A, Richards LJ, Dyck RH, Akazawa C, O'Leary DD 2003 Cloning and cortical expression of rat Emx2 and adenovirus-mediated overexpression to assess its regulation of area-specific targeting of thalamocortical axons. Cereb Cortex 13:648–660

Li H, Bishop KM, O'Leary DD 2006 Potential target genes of EMX2 include Odz/Ten-M and other gene families with implications for cortical patterning. Mol Cell Neurosci 33:136–149

Lopez-Bendito G, Chan CH, Mallamaci A, Parnavelas J, Molnar Z 2002 Role of Emx2 in the development of the reciprocal connectivity between cortex and thalamus. J Comp Neurol 451:153–169

Mallamaci A, Iannone R, Briata P et al 1998 EMX2 protein in the developing mouse brain and olfactory area. Mech Dev 77:165–172

Mallamaci A, Mercurio S, Muzio L et al 2000 The lack of Emx2 causes impairment of Reelin signaling and defects of neuronal migration in the developing cerebral cortex. J Neurosci 20:1109–1118

Mattar P, Britz O, Johannes C et al 2004 A screen for downstream effectors of Neurogenin2 in the embryonic neocortex. Dev Biol 273:373–389

Metin C, Baudoin JP, Rakic S, Parnavelas JG 2006 Cell and molecular mechanisms involved in the migration of cortical interneurons. Eur J Neurosci 23:894–900

Muzio L, Soria JM, Pannese M, Piccolo S, Mallamaci A 2005 A mutually stimulating loop involving emx2 and canonical wnt signalling specifically promotes expansion of occipital cortex and hippocampus. Cereb Cortex 15:2021–2028

O'Leary DD, Nakagawa Y 2002 Patterning centers, regulatory genes and extrinsic mechanisms controlling arealization of the neocortex. Curr Opin Neurobiol 12:14–25

Pellegrini M, Mansouri A, Simeone A, Boncinelli E, Gruss P 1996 Dentate gyrus formation requires Emx2. Development 122:3893–3898

Plachez C, Richards LJ 2005 Mechanisms of axon guidance in the developing nervous system. Curr Top Dev Biol 69:267–346

Qiu M, Anderson S, Chen S et al 1996 Mutation of the Emx-1 homeobox gene disrupts the corpus callosum. Dev Biol 178:174–178

Rash BG, Richards LJ 2001 A role for cingulate pioneering axons in the development of the corpus callosum. J Comp Neurol 434:147–157

Rash BG, Grove EA 2006 Area and layer patterning in the developing cerebral cortex. Curr Opin Neurobiol 16:25–34

Richards LJ, Plachez C, Ren T 2004 Mechanisms regulating the development of the corpus callosum and its agenesis in mouse and human. Clin Genet 66:276–289

Shinozaki K, Miyagi T, Yoshida M et al 2002 Absence of Cajal-Retzius cells and subplate neurons associated with defects of tangential cell migration from ganglionic eminence in Emx1/2 double mutant cerebral cortex. Development 129:3479–3492

Shu T, Richards LJ 2001 Cortical axon guidance by the glial wedge during the development of the corpus callosum. J Neurosci 21:2749–2758

Shu T, Puche AC, Richards LJ 2003a Development of midline glial populations at the corticoseptal boundary. J Neurobiol 57:81–94

Shu T, Sundaresan V, McCarthy MM, Richards LJ 2003b Slit2 guides both precrossing and postcrossing callosal axons at the midline in vivo. J Neurosci 23:8176–8184

Shu T, Butz KG, Plachez C, Gronostajski RM, Richards LJ 2003c Abnormal development of forebrain midline glia and commissural projections in Nfia knock-out mice. J Neurosci 23:203–212

Simeone A, Gulisano M, Acampora D, Stornaiuolo A, Rambaldi M, Boncinelli E 1992 Two vertebrate homeobox genes related to the Drosophila empty spiracles gene are expressed in the embryonic cerebral cortex. EMBO J 11:2541–2550

Smith KM, Ohkubo Y, Maragnoli ME et al 2006 Midline radial glia translocation and corpus callosum formation require FGF signaling. Nat Neurosci 9:787–797

Steele-Perkins G, Plachez C, Butz KG et al 2005 The transcription factor gene Nfib is essential for both lung maturation and brain development. Mol Cell Biol 25:685–698

Tole S, Gutin G, Bhatnagar L, Remedios R, Hebert JM 2006 Development of midline cell types and commissural axon tracts requires Fgfr1 in the cerebrum. Dev Biol 289:141–151

Yoshida M, Suda Y, Matsuo I et al 1997 Emx1 and Emx2 functions in development of dorsal telencephalon. Development 124:101–111

DISCUSSION

Parnavelas: I would like to ask Antonello Mallamaci to describe his observations. The differences in the reported presence or absence of corpus callosum in *Emx1* mutants is rather disturbing.

Mallamaci: I did not check carefully the development of corpus callosum in both *Emx1* and *Emx2* mutants. What impressed me in Linda Richard's talk is the *Emx1* areal phenotype, i.e. the fact that, in the absence of *Emx1*, there is a slight but reproducible reduction of caudal areas. And, even more remarkable, the intermediate phenotype displayed by heterozygous mice.

Richards: That's right. We see an intermediate phenotype in the heterozygotes.

Mallamaci: Exactly like in *Emx2* mutants.

Walsh: Aren't a lot of the Reelin neurons generated from the midline hem? This suggests that *Nfib* is expressed in the glial cells that are giving rise to a lot of the CR neurons.

Richards: I have checked whether *Nfib* is expressed in the hem. They have a hippocampal phenotype that is very similar to the *Emx2* mouse. *Nfia* is expressed all the way down and into the hem, but *Nfib* doesn't seem to be expressed in the hem. They must turn it on after they migrate.

Rakic: It is very interesting that you can rescue some fibres but not all. Do you differentially affect rostral and posterior neurons? There are two fundamentally different axonal classes in the corpus callosum: those that are myelinated and those that are non-myelinated, which are considered as fast-conducting and slow-conducting, respectively. They are differentially distributed in the caudal and rostral portion of the corpus callosum (LaMatia & Rakic 1990). Have you looked to see whether ones that are rescued are myelinated or not?

Richards: I haven't looked at this, but it would be interesting to study.

Molnár: It would also be interesting to look at the cell populations. Lots of different cells contribute to the callosum including layers II–III, IV, V and VI. It would be interesting to look at these questions in the context of the gene expression of these subpopulations in the cortex and the signals they read when their axons cross. This is a crucial question to address: which cell populations can make it?

Richards: Absolutely.

O'Leary: You described that the neuropilin 1 staining and a bit of residual cingulate cortex were present caudally in *Emx2* null. This is where the *Emx2* expression is highest, even in cingulate cortex. What do you think is occurring?

Richards: I don't know. It seems to me there is no gradient in the rostro-caudal axis at the midline. It is high across the entire rostro-caudal extent.

O'Leary: It may be overlapping in expression with another gene that has redundant function within that select domain.

Hevner: Was it the *Nfib* mutant that also had the enhanced subplate?

Richards: I wouldn't say it had more cells, but it was more diffuse.

Hevner: I am wondering whether that and Reelin might be related. There could be an increased early neurogenesis, but this would imply that the *Nfib* is in the proliferative compartments.

Richards: *Nfib* is expressed at low levels in the proliferative zones of the brain. It increases when the cells are differentiating. It is potentially regulated by Pax6, though, so there may be a link there.

Walsh: Getting back to the question of Reelin, we used a GF7-Cre with a LacZ tag in a paper in *Neuron* some years ago (Monuki et al 2001). We also saw labelled marginal zone cells. I think there is also a source of Reelin-producing cells that is medial to the hem. This might still be consistent with *Nfib* having a direct effect in repressing the formation of marginal zone cells. I have another comment. There are five patients who have been described by Richard Maas's lab who have *Nfia* abnormalities resulting in partial or complete agenesis of the corpus callosum. The weird thing is that it is not a consistent abnormality. Sometimes it is thin, other times the fibres are missing.

Richards: In children with ACC there is a wide variety of phenotypes. You can find complete absence, but often there is retention of the rostral part of corpus callosum and also a thinning. This is an interesting phenotype and we do not know why it is that we are losing the more caudal region of the corpus callosum in these patients.

Macklis: Everyone has been splitting the callosal neurons into different sets across axes. I know you said that it is hard from the corpus bundles to do retrograde labelling, but do you have any insight from failed or partial attempts about the layer II/III versus V? This might be either in the *Emx1* mutants or also the other mutants where there are partial thinnings or crossings. Or, in the purely to cortical callosals, or those with striatal or largely striatal projections, when you look later do you see something pruning back from one population to another? Can you split these populations?

Richards: This is one of very few mice that have some axons crossing and some that don't. Most often we see complete agenesis of the corpus callosum, with the fibres failing to cross very early on, prior to when the projection from layer II/III is formed. The projections from layer V are important in making sure the callosal axons cross the midline.

Macklis: I guess I am asking if your experiments will help to dissociate whether it is purely pioneering and following along, or whether those might be distinct populations: even if you had a pioneer population someone without the right gene and pathfinding might say that they don't care whether layer V crosses or not, I'm not crossing.

Richards: If it turns out that in the *Emx1* mutant the fibres that don't cross come from the cingulate cortex (the origin of the pioneering axons), this is the best piece of evidence against the hypothesis that the pioneering axons are required for the corpus callosum to form, since the rest of the corpus callosum forms normally.

Macklis: With your anterograde labelling, do you ever see misrouting, especially of the deep layer ones?

Richards: What we really need are callosal-specific markers to address these questions.

Harrison: Your *Emx* experiments show the importance of genetic background on the phenotype. Presuming that backcrossing is a kind of asymptotic process, how does one know when genetic background stops being an explanation for a residual phenotype?

Richards: The gold standard is 10 backcrosses. We could never rule out the possibility that a modifying gene that is very close to the gene of interest could still be present.

Harrison: So is the default interpretation that after five generations a phenotype is due to the gene itself? And what is the probability that it is instead due to the background?

Richards: What I can say is that the phenotype is consistent. It was influenced by the background. I'd now like to cross back to 129 mice to see whether we get the phenotype back.

Rakic: Good luck!

O'Leary: A while back Pasko mentioned the cost of doing primate work, where one monkey costs US$10 000. Especially in these times where funding is tight, the number of generations estimated by some to be the gold standard for backcrossing, at the Salk and many institutions, costs more than I can afford in money and time, if one needs to stretch the money to study multiple mutants, not even considering the high costs of making the knockout and then the personnel and reagent costs to do the analyses. In the end perhaps it evens out.

Richards: There is a new way of speeding up backcrossing, called speed congenics, which uses genomic markers to determine how much of the BL/6 background is present. This approach will be helpful in the future for speeding up the time it takes to backcross lines onto an inbred background

O'Leary: As you mentioned, when you have a line that doesn't breed very well, it becomes frustrating and expensive.

Murakami: Are there any other studies showing that NF1 is involved in cell migration. Is it possible that midline crossing by migrating cells is affected by NF1s, and that what you see is a result of that?

Richards: I was referring to the subcallosal sling cells. In the NF1A mice there is a migration defect of subcallosal sling cells that enter the septum. In the NF1B mice there is also a migration defect. It is clear that there is a correlation between the formation of the corpus callosum and the correct formation of the subcallosal sling. We know that the first axons to cross the midline cross before the sling actually forms. The cells are closely associated with the axons, and potentially might use the axons to form the sling. In acallosal mutants the sling cells get caught up in Probst bundles or fail to migrate.

References

LaMantia AS, Rakic P 1990 Cytological and quantitative characteristics of four cerebral commissures in the rhesus monkey. J Comp Neurol 291:520–537
Monuki ES, Porter FD, Walsh CA 2001 Patterning of the dorsal telencephalon and cerebral cortex by a roof plate-Lhx2 pathway. Neuron 32:591–604

Schizophrenia susceptibility genes and their neurodevelopmental implications: focus on neuregulin 1

Paul J. Harrison

University of Oxford, Department of Psychiatry, Warneford Hospital, Oxford OX3 7JX, UK

Abstract. On the basis of epidemiological as well as neurobiological evidence, schizophrenia has been conceptualized as a neurodevelopmental disorder. It is also known to have a large heritable component and a complex genetic architecture. Many putative susceptibility genes have recently been identified, arising both from positional cloning and candidate gene approaches. The evidence is strong for neuregulin 1, dysbindin and DISC1, and moderate for several others. However, there are key unanswered questions. For example, concerning the molecular basis of genetic association, multiple, mostly non-coding, variants have been found within the genes, complicating discussion as to the strength and interpretation of the data. Second, there is speculation whether the genes converge on common pathways, notably glutamatergic synaptic transmission. Additional questions concern the emerging evidence for epistasis, the clinico-genetic correlates, and the extent to which the genes confer schizophrenia risk via their roles in neurodevelopment. Here, the genetic advances and their neurodevelopmental implications are summarised, with a particular focus on neuregulin 1.

2007 Cortical development: genes and genetic abnormalities. Wiley, Chichester (Novartis Foundation Symposium 288) p 246–259

Schizophrenia: a genetic disorder of neurodevelopment

A meta-analysis of population-based twin studies estimated the heritability of schizophrenia at 81% (95% confidence interval: 73–91%; Sullivan et al 2003). This strongly suggests (though does not prove) that there is a predominant genetic aetiology to the disorder, consistent with the results of family and adoption studies. The latter also indicate, in agreement with epidemiological findings, that a significant proportion of the environmental risk factors for the disorder act pre- or perinatally. For example, maternal influenza, obstetric complications, urban birth and winter birth all confer small excess risks (McGrath & Murray 2003). The nature of these risk factors provided one of the main pieces of support for the neurodevelopmental hypothesis of schizophrenia (Table 1), which has become

TABLE 1 Evidence adduced to support a neurodevelopmental origin of schizophrenia

- Usual age of onset in late adolescence and early adulthood
- Most known environmental risk factors are pre- and perinatal
- Neuromotor, behavioural and social differences in children many years prior to onset of schizophrenia
- Structural brain changes and cognitive impairment present at, and prior to, onset of illness—and limited progression thereafter
- Neuropathological findings—cytoarchitectural disturbances and lack of gliosis imply prenatal origin
- Morphological features suggestive of early developmental anomaly (e.g. abnormal dermatoglyphics, craniofacial measures)
- Animal models—early lesions and other interventions lead to delayed 'schizophrenia-like phenotypes'

For review, see Weinberger (1995), Harrison (1997) and Lewis & Levitt (2002).

the prevailing pathogenic model of the disorder since its contemporary exposition by Weinberger (1987) and Murray & Lewis (1987). It has remained disappointingly vague in terms of timing and molecular mechanisms, in part because of the lack of identified genes or unequivocal neuropathological or biochemical correlates of the disorder (Harrison 1999); a second trimester origin, perhaps with a second 'hit' in adolescence, together impacting adversely on the fine details of microcircuitry and functioning of the dorsolateral prefrontal cortex and hippocampus is usually emphasized. As schizophrenia genes have begun to be identified, the notion that they operate, at least partly, via their influence on neurodevelopment (Jones & Murray 1991) has been reactivated, and has begun to be investigated empirically, as described below.

Returning to the genetic basis of schizophrenia, various findings imply strongly that this is complex (i.e. non-Mendelian), and that there are multiple susceptibility genes which are all of small effect (viz., no locus likely to confer a relative risk to siblings of more than 3 [λs<3]; see Kirov et al 2005). Given this genetic architecture, it is not surprising that linkage studies have had limited success. Nevertheless, two meta-analyses of the genome-wide linkage scans of schizophrenia identified several loci which met, or came close to, accepted criteria for significance (Badner & Gershon 2002, Lewis et al 2003), encouraging renewed efforts to locate the gene(s) responsible for the linkage signal. Since that time, a rapidly expanding number (dozens) of 'schizophrenia genes' have been reported based on genetic association (from fine mapping of the locus, and from candidate gene approaches) in both case-control and family studies. Whilst the evidence for many of the genes is wholly inconclusive, the evidence for several genes is strong, especially when considered in tandem with the linkage evidence and the biological plausibility of the gene. Table 2 summarizes the leading susceptibility genes for schizophrenia. It is a matter of active debate as to how many (if any) can be considered schizophrenia genes beyond reasonable doubt, and how many (more) should be included in a Table such as this. For reviews illustrating the range of opinions regarding schizophrenia genes and genetics, see Harrison & Owen (2003), Kirov et al (2005), Gogos & Gerber (2006), and Crow (2007). If one takes the criterion of a positive meta-analysis, then neuregulin 1 (*NRG1*), D-amino acid oxidase activator (*DAOA/ G72*) and, equivocally, regulator of G protein signalling 4 (*RGS4*) and catechol-*O*-methyltransferase (*COMT*) come into that category. The data for dysbindin (*DTNBP1*) and disrupted-in-schizophrenia-1 (*DISC1*) have yet to be subjected to meta-analysis, but appear at least as strong as that for *DAOA*, and considerably stronger than for other putative genes.

Speculation has arisen as to how genetic variation impacts on schizophrenia risk, and whether the known functions of the genes suggest a convergence upon core biochemical pathways (Harrison & Weinberger 2005). In particular, many of the genes influence NMDA receptor-mediated synaptic transmission (Harrison & Owen 2003, Harrison & West 2006), which is of note given that hypofunction of NMDA receptor signalling was already a prominent biochemical hypothesis for schizophrenia pathogenesis (see Coyle 2006). Genetic convergence on other related, functional domains has also been proposed, such as synaptic plasticity and

TABLE 2   Schizophrenia susceptibility genes

Gene name	Symbol	Locus	Linkage[a]	Association[b]	Neurodevelopment[c]
Neuregulin 1	*NRG1*	8p12-21	Strong	Strong	Strong
Dysbindin 1	*DTNBP1*	6p22	Moderate	Strong	Weak
D-amino acid oxidase activator	*DAOA/G72*	13q33	Moderate	Strong	None
Disrupted-in-schizophrenia-1	*DISC1*	1q42	Moderate	Strong	Strong
Regulator of G protein signalling 4	*RGS4*	1q22	Weak	Moderate	None
mGluR3	*GRM3*	7q	Weak	Moderate	None
Catechol-*O*-methyl transferase	*COMT*	22q11	Strong	Weak	Weak
ErbB4	*ERBB4*	2q34	Weak	Moderate	Strong
Proline dehydrogenase	*PRODH*	22q11	Strong	Moderate	None

[a] Strength of evidence of linkage of gene locus to schizophrenia.
[b] Strength of evidence of genetic associations with polymorphisms at the gene locus.
[c] Strength of evidence that the gene has a role in neurodevelopment.

myelination (Moises et al 2002, Carter 2006). All such speculation is however hindered by the uncertain candidacy of many of the genes and by another critical issue: within each gene, multiple risk alleles have been reported in different populations, with no clear causal (or coding) variant found. Whether this reflects true allelic heterogeneity, false positives, or linkage disequilibrium with as yet unidentified mutations is a moot point (Harrison 2007). At present, it is assumed that much of the genetic association *is* due to non-coding variation and, as such, the mechanism of association is via altered expression of the gene, in one way or another. Evidence in support of this proposal is beginning to emerge for several of the genes, with demonstrable effects of schizophrenia and/or the risk polymorphisms on the abundance, splicing, or location of the encoded gene products. A good example is *NRG1*, which is discussed in detail below. Indeed, pending definitive statistical genetic data (that may never be forthcoming; Paige et al 2003), a better understanding of the expression and function of each gene may be an important component to determining if, as well as how, a gene is a risk gene for schizophrenia.

A final general issue to note is that of clinico-genetic correlations. Schizophrenia is a syndrome still defined by symptoms and behaviours, not by aetiology, pathology or biological markers. The syndrome has good reliability, but unknown and probably limited validity. It is unlikely that there are any 'schizophrenia genes' *per se*; rather, there are genes which contribute either to specific elements or endophenotypes (e.g. to delusions, or to working memory deficits, or to age of onset), or which operate across diagnostic boundaries to confer vulnerability to a broader spectrum of psychosis (e.g. bipolar disorder). For example, *NRG1* appears to act primarily on subjects with schizophrenia who also have affective symptoms (i.e. towards the bipolar disorder end of the psychosis spectrum), whereas *DTNBP1* is associated with cognitive impairment and negative symptoms. Future work needs to pay as much attention to the clinical profile as to the molecular complexities of schizophrenia genetics, particularly in order to maximize the potential of this work to inform diagnosis, prognosis and treatment.

Neuregulin 1 and schizophrenia

NRG1 was first associated with schizophrenia by Stefansson et al (2002) after fine mapping of the 8p locus which showed linkage to the disorder in their Icelandic population. They found association to a risk haplotype in the 5′ region of the gene, and replicated this in a Scottish population. Subsequently, association of the *NRG1* locus with schizophrenia has been reported in different populations in most but not all studies (Harrison & Law 2006). The majority of the positive studies find association to the 5′ region, but others are located further downstream, and there is no consensus as to the critical allele(s). Nevertheless, two meta-analyses confirm association, albeit with very different significance levels, in part reflecting the uncertain methods of handling the diverse risk alleles and haplotypes (Li et al 2006, Munafo et al 2006). Figure 1 shows a schematic of the *NRG1* gene and the various polymorphisms and haplotypes associated with schizophrenia.

Although *NRG1* had never been investigated in, nor considered relevant to, schizophrenia prior to the report of Stefansson et al (2002), a lot was known about the gene and its functions, both in the nervous system and elsewhere (Falls 2003). *NRG1* is a growth factor that, in its canonical signalling mode, binds via its epidermal growth factor (EGF) domain to ErbB4 tyrosine kinase receptors on adjacent cells and thereby affects a range of downstream pathways. *NRG1* has key roles in many aspects of neurodevelopment and neural plasticity that with hindsight made it an attractive candidate for involvement in schizophrenia pathogenesis (Corfas et al 2004, Scolnick et al 2006). Equally, its multiple roles make this proposal difficult to specify more precisely, or to refute. The functional diversity of *NRG1* is reflected in its structural complexity. A large number of isoforms are

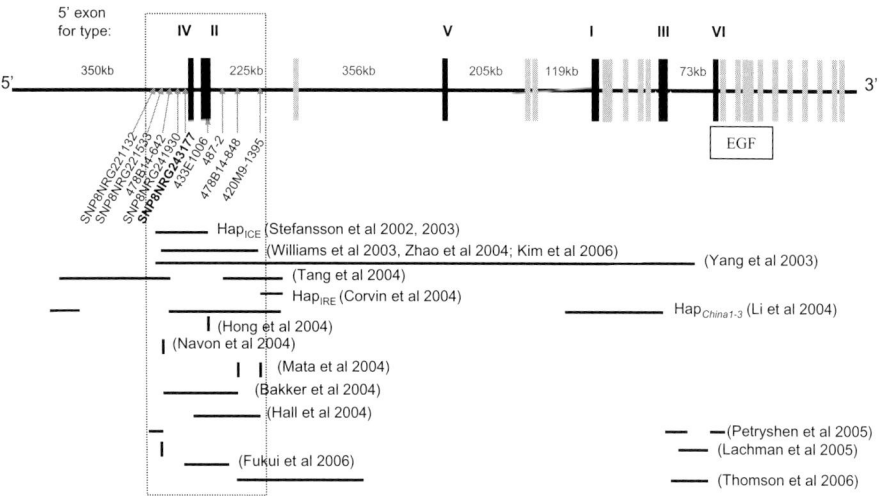

FIG. 1.   The human *NRG1* gene: exon–intron structure, and genetic associations with schizophrenia. Exons are shown as vertical lines. 5′ exons that define the six isoforms (I–VI) of *NRG1* are in black; others are in grey. The large dashed rectangle encompasses the 5′ region of the gene wherein association to schizophrenia has been most commonly found, including the core Icelandic risk haplotype (Hap$_{ICE}$). The SNPs and microsatellites studied within this region are also shown, with SNP8NRG243177 highlighted in bold. EGF: approximate position of the encoded EGF domain. The location of the SNPs and haplotypes associated with schizophrenia in different studies are shown below the gene with the study name. Citations of these studies can be found in Harrison & Law (2006), from which the figure has been adapted and updated.

generated from the gene, both by alternative promoter usage, and alternative splicing of several domains. The former give rise to the nomenclature of *NRG1* isoforms based on 5′ exon inclusion, with the 'types' of *NRG1* shown in Fig. 1. Functional differences between some isoforms are increasingly recognised; for example, type I and III *NRG1* contribute differentially to myelination (Michailov et al 2004) and interneuron migration (Flames et al 2004).

Studies have begun to investigate the pathobiology of *NRG1* in schizophrenia, and the functional effects of allelic variation within the gene. First, *NRG1* mRNA and protein were shown to be present in neuronal populations in multiple regions of the adult human brain (Law et al 2004). This indicates that the involvement of *NRG1* in schizophrenia may be neurodevelopmental, but may also occur later in life. Quantitative gene expression studies show that *NRG1* mRNA (Hashimoto et al 2004, Law et al 2006) and protein (Hahn et al 2006) are not grossly altered in the brain in schizophrenia. However, there are differences in *NRG1* isoform expression. Hashimoto et al (2004) and Law et al (2006) found, in independent cohorts and in different brain regions (prefrontal cortex and hippocampus, respectively), that type I *NRG1* mRNA is increased in subjects with the disorder compared to matched controls. The other isoforms studied (types II–IV) did not differ. Perhaps of greater importance, Law et al (2006) also found that type IV *NRG1* mRNA is increased in subjects carrying the core risk *NRG1* haplotype originally identified by Stefansson et al (2002), most significantly for the risk allele at SNP8NRG243177. This SNP is located just upstream of the type IV 5′ exon (Fig. 1), making it a plausible candidate to regulate its transcription. The suggestion that SNP8NRG243177 may be functional (and thus relevant to why variation at that allele impacts schizophrenia risk), is bolstered by two findings: the SNP is predicted bioinformatically to determine three transcription factor binding sites (Law et al 2006), and it is associated with cognitive performance, cortical activity, and risk of progression to psychosis in a high risk population (Hall et al 2006). These data illustrate how the functionality and pathogenic potential of non-coding genetic variants in schizophrenia genes are being investigated. Ongoing studies are underway to further test the importance of this and other *NRG1* SNPs.

Investigations are also moving downstream from *NRG1* to consider ErbB receptors and the functional status of *NRG1*/ErbB signalling. Hahn et al (2006) presented evidence that *NRG1* activation of ErbB4 receptors is increased in schizophrenia, and that this leads to an impairment of NMDA receptor function, consistent with other data showing a *NRG1*/NMDA receptor link (via PSD-95) and suggesting a contribution of *NRG1* to the NMDA receptor hypofunction of schizophrenia mentioned earlier. The ErbB4 receptor has become a focus of interest in its own right: it may be a schizophrenia risk gene, and its expression is affected in an isoform-specific fashion in the disease (Silberberg et al 2006, Law et al 2007), and in relation to the ErbB4 risk SNPs (Law et al 2007). These findings emphasise that schizophrenia genes do not work, nor should be studied, in isolation, but as part of biochemical pathways or networks, with both epistatic and functional interactions between genes being actively sought. Indeed, an increasing number of gene–gene (as well as gene–environment) interactions are being identified (see Harrison 2007).

Do schizophrenia susceptibility genes work neurodevelopmentally?

At present, whilst these findings begin to shed light upon how *NRG1* is affected in subjects with schizophrenia and in those who carry risk alleles, it is difficult to relate them back to the broader functional roles of *NRG1* and in particular the neurodevelopmental implications. Interpretation of the type IV isoform changes is hindered by the fact that the isoform was only discovered in 2004, and nothing is known of its specific functions. Also, the schizophrenia findings have inevitably been made in subjects with established disease, and it is not yet clear at what point in life the critical events occur; however, the data of Hall et al (2006) suggest that at least some of the impact occurs prior to the onset of illness. Clarifying the roles of *NRG1* in disease pathogenesis and its neurodevelopmental component will be challenging. One approach is to use transgenic mice; some relevant phenotypes have already been reported in *NRG1* mutants (Falls 2003, O'Tuathaigh et al 2006), but it will also be important to study conditional transgenics, and isoform-selective and overexpressing mice, that better reproduce the alterations in *NRG1* expression seen in the disease.

The difficulty in establishing whether *NRG1* influences schizophrenia risk as a consequence of its neurodevelopmental functions, rather than because of its ongoing roles later in life (or elsewhere in the body, for that matter) also pertains to the other susceptibility genes. For some, such as *DISC1*, there is direct evidence that it is involved in brain development (Ishizuka et al 2006); for others, such as *COMT*, there is circumstantial evidence, in that its expression (Tunbridge et al 2007) and functional significance (Gothelf et al 2005) change through adolescence and early adulthood. For the remainder of the genes, there is at present no substantive evidence that they impact on brain maturation. As with *NRG1*, a range of experimental approaches will be needed to show whether and how a neurodevelopmental pathogenesis explains the disease susceptibility conferred by each gene.

A final aspect of the interface between schizophrenia genes and neurodevelopment concerns velocardiofacial syndrome (VCFS; also called DiGeorge syndrome). VCFS is caused by hemideletion of chromosome 22q11 and produces a complex range of congenital abnormalities including a characteristic facial dysmorphology, cleft palate, congenital heart defects and mild learning disability (Murphy 2002). Notably, schizophrenia-like psychoses are increased more than 20-fold in VCFS, whilst the rate of 22q11 microdeletions is increased in patients with schizophrenia (Bassett & Chow 1999, Murphy 2002). Moreover, 22q is strongly implicated by the linkage scans of schizophrenia, and at least three genes within the 22q11 region have been associated with schizophrenia, including *COMT* and *PRODH* (Table 1) and *ZDHHC8* (see Gogos & Gerber 2006). These findings provide convergent evidence in support of 22q11 as a locus harbouring schizophrenia genes. Of greater relevance here, the link with VCFS indicates that some cases of schizophrenia are unequivocally both genetic and developmental in origin. What is not known is the proportion of schizophrenia that will eventually prove to come into this category.

Conclusions

Schizophrenia remains an enigmatic disorder. There is little doubt that it has a predominant genetic and neurodevelopmental basis, but it continues to be difficult to pin down the specific details. Nevertheless, the evidence is now strong for several susceptibility genes, with *NRG1* the leading candidate and the strongest neurodevelopmental pedigree. Equally, the complexities of the genetic association, and the pleiomorphic nature of *NRG1*, preclude firm understanding of how and why the gene contributes to schizophrenia risk, and the extent to which this occurs via *NRG1*'s influences on neurodevelopment. An iterative and multifaceted research approach is now required, whereby epidemiology, phenotypic character-ization, statistical genetics, and studies of gene expression and gene function play crucial and complementary roles in clarifying the genetic basis of schizophrenia and the neurodevelopmental and other mechanisms by which the genes exert their pathogenic effects.

Acknowledgements

Work in the author's laboratory is supported by the Medical Research Council, Stanley Medical Research Institute, Wellcome Trust, and the National Alliance for Research into Schizophrenia and Depression. Thanks to all group members for their contributions, especially Amanda Law and Inga Deakin for their work on *NRG1*.

References

Badner JA, Gershon ES 2002 Meta-analysis of whole-genome linkage scans of bipolar disorder and schizophrenia. Mol Psychiatry 7:405–411
Bassett AS, Chow EW 1999 22q11 deletion syndrome: a genetic subtype of schizophrenia. Biol Psychiatry 46:882–891
Carter CJ 2006 Schizophrenia susceptibility genes converge on interlinked pathways related to glutamatergic transmission and long-term potentiation, oxidative stress and oligodendrocyte viability. Schizophr Res 86:1–14

Corfas G, Roy K, Buxbaum JD 2004 Neuregulin 1–erbB signaling and the molecular/cellular basis of schizophrenia. Nat Neurosci 7:575–580

Coyle JT 2006 Glutamate and schizophrenia: beyond the dopamine hypothesis. Cell Mol Neurobiol 26:363–382

Crow TJ 2007 How and why genetic linkage has not solved the problem of psychosis: review and hypothesis. Am J Psychiatry 164:13–21

Falls DL 2003 Neuregulins and the neuromuscular system: 10 years of answers and questions. J Neurocytol 32:619–647

Flames N, Long JE, Garratt AN et al 2004 Short- and long-range attraction of cortical GABAergic interneurons by Neuregulin-1. Neuron 44:251–261

Gogos JA, Gerber DJ 2006 Schizophrenia susceptibility genes: emergence of positional candidates and future directions. Trends Pharmacol Sci 27:226–233

Gothelf D, Eliez S, Thompson T et al 2005 COMT genotype predicts longitudinal cognitive decline and psychosis in 22q11.2 deletion syndrome. Nat Neurosci 8:1500–1502

Hahn CG, Wang HY, Cho DS et al 2006 Altered neuregulin 1–erbB4 signaling contributes to NMDA receptor hypofunction in schizophrenia. Nat Med 12:824–828

Hall J, Whalley HC, Job DE et al 2006 A neuregulin 1 variant associated with abnormal cortical function and psychotic symptoms. Nat Neurosci 9:1477–1478

Harrison PJ 1997 Schizophrenia: a disorder of neurodevelopment? Curr Opin Neurobiol 7:285–289

Harrison PJ 1999 The neuropathology of schizophrenia—a critical review of the data and their interpretation. Brain 122:593–624

Harrison PJ 2007 Schizophrenia genes: searching for common features, functions, and mechanisms. In: O'Donnell P (ed) Cortical deficits in schizophrenia: from genes to function. Springer, Norwell MA (in press)

Harrison PJ, Owen MJ 2003 Genes for schizophrenia? Recent findings and their pathophysiological implications. Lancet 361:417–419

Harrison PJ, Weinberger DR 2005 Schizophrenia genes, gene expression, and neuropathology: on the matter of their convergence. Mol Psychiatry 10:40–68

Harrison PJ, Law AJ 2006 Neuregulin 1 and schizophrenia: genetics, gene expression, and neurobiology. Biol Psychiatry 60:132–140

Harrison PJ, West VA 2006 Six degrees of separation: on the prior probability that schizophrenia susceptibility genes converge on synapses, glutamate and NMDA receptors. Mol Psychiatry 11:981–983

Hashimoto R, Straub RE, Weickert CS, Hyde TM, Kleinman JE, Weinberger DR 2004 Expression analysis of neuregulin-1 in the dorsolateral prefrontal cortex in schizophrenia. Mol Psychiatry 9:299–307

Ishizuka K, Paek M, Kamiya A, Sawa A 2006 A review of Disrupted-In-Schizophrenia-1 (DISC1): neurodevelopment, cognition, and mental conditions. Biol Psychiatry 59:1189–1197

Jones P, Murray RM 1991 The genetics of schizophrenia is the genetics of neurodevelopment. Br J Psychiatry 158:615–623

Kirov G, O'Donovan MC, Owen MJ 2005 Finding schizophrenia genes. J Clin Invest 115:1440–1448

Law AJ, Weickert CS, Hyde TM, Kleinman JE, Harrison PJ 2004 Neuregulin-1 (NRG-1) mRNA and protein in the adult human brain. Neuroscience 127:125–136

Law AJ, Lipska BK, Weickert CS et al 2006 Neuregulin 1 transcripts are differentially expressed in schizophrenia and regulated by 5′ SNPs associated with the disease. Proc Natl Acad Sci USA 103:6747–6752

Law AJ, Kleinman JE, Weinberger DR, Weickert CS 2007 Disease associated intronic variants in the ErbB4 gene are related to altered ErbB4 splice variant expression in the brain in schizophrenia. Hum Mol Genet 16:129–141

Lewis DA, Levitt P 2002 Schizophrenia as a disorder of neurodevelopment. Annu Rev Neurosci 25:409–432

Lewis CM, Levinson DF, Wise LH et al 2003 Genome scan meta-analysis of schizophrenia and bipolar disorder, part II: Schizophrenia. Am J Hum Genet 73:34–48

Li D, Collier DA, He L 2006 Meta-analysis shows strong positive association of the neuregulin 1 (NRG1) gene with schizophrenia. Hum Mol Genet 15:1995–2002

McGrath JJ, Murray RM 2003 Risk factors for schizophrenia: from conception to birth. In: Hirsh SR, Weinberger D (eds) Schizophrenia, 2nd ed. Blackwell Publishing, p 232–250

Michailov GV, Sereda MW, Brinkmann BG et al 2004 Axonal neuregulin-1 regulates myelin sheath thickness. Science 304:700–703

Moises HW, Zoetga T, Gottesman II 2002 The glial growth factors deficiency and synaptic destabilization hypothesis of schizophrenia. BMC Psychiatry 2:8

Munafo MR, Thiselton DL, Clark TG, Flint J 2006 Association of the NRG1 gene and schizophrenia: a meta-analysis. Mol Psychiatry 11:539–546

Murphy KC 2002 Schizophrenia and velo-cardio-facial syndrome. Lancet 359:426–430

Murray RM, Lewis SW 1987 Is schizophrenia a neurodevelopmental disorder? BMJ 295:681–682

O'Tuathaigh CM, O'Sullivan GJ, Kinsella A et al 2006 Sexually dimorphic changes in the exploratory and habituation profiles of heterozygous neuregulin-1 knockout mice. Neuroreport 17:79–82

Paige GP, George V, Go RC, Page PZ, Allison DB 2003 'Are we there yet?': deciding when one has demonstrated specific genetic causation in complex diseases and quantitative traits. Am J Hum Genet 73:711–719

Scolnick EM, Petryshen T, Sklar P 2006 Schizophrenia: do the genetics and neurobiology of neuregulin provide a pathogenesis model? Harv Rev Psychiatry 14:64–77

Silberberg G, Darvasi A, Pinkas-Kramarski R, Navon R 2006 The involvement of *ErbB4* with schizophrenia: association and expression studies. Am J Med Genet B Neuropsychiatr Genet 141B:142–148

Stefansson H, Sigurdsson E, Steinthorsdottir V et al 2002 Neuregulin 1 and susceptibility to schizophrenia. Am J Hum Genet 71:877–892

Sullivan PF, Kendler KS, Neale MC 2003 Schizophrenia as a complex trait—evidence from a meta-analysis of twin studies. Arch Gen Psychiatry 60:1187–1192

Tunbridge EM, Weickert CS, Kleinman JE et al 2007 Catechol-O-methyltransferase enzyme activity and protein expression in human prefrontal cortex across the postnatal lifespan. Cereb Cortex 17:1206–1212

Weinberger DR 1987 Implications of normal brain development for the pathogenesis of schizophrenia. Arch Gen Psychiatry 44:660–669

Weinberger DR 1995 From neuropathology to neurodevelopment. Lancet 346:552–557

DISCUSSION

Crino: Among the likely candidate loci, within those regions are there other genes which fall within the ErbB4/neuregulin pathway? Is it similar to tuberous sclerosis where we had two genes on two different chromosomes which turned out to be protein binding partners?

Harrison: Certainly, under the linkage peaks, which are frequently broad, there can be association of two or three genes. It may just be a searchlight effect: you are looking in more detail there so you are seeing more associations. In terms of pathways, neuregulin/ErbB4 is probably the best example of linked genes. It is unclear whether there is any connection genetically between dysbindin and neuregulin pathways. At the moment the biology of these two genes, and their clinical correlates, appear to be rather different. I don't know whether this is telling us at the end there will be different aetiological subtypes of schizophrenia.

Crino: When you were doing the SNP locus analysis, were these all-comer studies for anyone with a clinical diagnosis of schizophrenia, or has there been any attempt to subgroup clinically?

Harrison: Most of these studies are all-comers. Some groups divide patients by whether the family history is positive or negative, but most are working with anyone who has a DSM IV diagnosis. Classic clinical subtypes (e.g. paranoid and hebephrenic) are usually ignored. When researchers start to go back and look within their samples to see where the genetic signal is coming from, some interesting genotype–phenotype relationships are observed. For example, the dysbindin genetic signal in schizophrenia seems to be strong in those with negative symptoms and severe disease, whereas neuregulin is associated more with affective symptoms, and spills over into bipolar disorder. I think we will end up fragmenting the syndrome, but not along the classic clinical subtypes.

Walsh: When I think of developmental disorders in humans or mice, most of the genes act in a recessive fashion and are heterogeneous. There are a lot of epidemiological studies showing that mental retardation is more common in parents that are related. What is the outcome of similar kinds of epidemiological studies in schizophrenia? Is it more common in parents who are related?

Harrison: I can't think of good studies addressing that question, although anecdotally there does appear to be high rates (and early onset) of psychosis in populations with high levels of consanginuity.

Walsh: People from the Middle East have told me this casually but I don't have data.

Wilkins: I was struck by the fact that there are no postnatal risk factors. There were news stories a while back indicating that rates of schizophrenia were sixfold higher in highly disadvantaged groups of society.

Harrison: There is an association between schizophrenia and social deprivation, but the question is whether this is causal (as opposed to being consequential, or epiphenomenal).

Richards: Can you say more about the expression of neuregulin in human pathological specimens? What do you think it is doing? Is it involved in GABAergic interneuron development?

Harrison: In terms of cellular expression profile in human brain there is one study, only in adults, and it doesn't distinguish among the isoforms (Law et al

2004). It shows that neuregulin (mRNA and protein) overall looks to be localized mainly in pyramidal neurons in the cortex, in Purkinje and Golgi cells of the cerebellum, and in various brainstem nuclei. We don't know whether the isoforms have different expression patterns, nor the situation during development. The abundance of each isoform is quite low, so it is not a trivial question to answer in post mortem material.

Richards: What is the connection with the Reelin story?

Harrison: There are many molecules whose expression is clearly changed in schizophrenia. Reelin is one. They are nearly always down-regulated. My guess is that most of these are simply downstream of whatever the key genetic (and epigenetic) mechanisms are. If signalling has been different throughout development, the brain is just wired up a bit differently. The changes in Reelin may well come into this category, and be downstream to some of the risk genes.

Rakic: To link your findings on arrest of the cells in their migration in the prospective white matter to neuregulins is interesting because there is some evidence that this class of molecules is involved in neuronal migration (Anton et al 1997). Neuregulin could also be related to Notch signalling.

Harrison: One approach is to go back to the material on which the classic morphometric and neuropathological studies were done, genotype them, and see if genotype predicts the abnormalities better than (or in addition to) diagnosis.

Macklis: I recently heard Gabriel Corfas speak on neuregulin and schizophrenia. Both of you have pointed out that the developmental aspects have got to be there, but we know that development proceeds much later than the original organization. Even though there are preceding symptoms, they are not nearly as striking as ones that come on in late adolescence, and they are temporally focal: this is the time of onset. In human pathology, could the developmental aspects where neuregulin plays a role be not the global early setting up of the circuitry (which works well in pre-symptomatic individuals), but rather some plastic changes in subsystems that are especially changing during the second wave of the adolescent 'larval' state of humanity, where some of these affective systems are changing?

Harrison: Absolutely. It is entirely likely that it is the later function of these genes, not just their classic early ones, that is important. In schizophrenia, the cerebral cortex is essentially normal, certainly with no gross pathology. It has to be something much more subtle, perhaps molecular rather than 'structural'. The changes that characterize the onset of illness around adolescence would be candidates for that: the dopamine system is an obvious one.

Macklis: Is there a source of adolescent tissue to look at that phase of change?

Harrison: A limited amount. Our colleagues at NIH have been looking at a large series of human brains through development, mapping the ontogeny of the dopamine system, including dopamine receptors, tyrosine hydroxylase, and catechol-*O*-methyl transferase (COMT) (Weickert et al 2007, Tunbridge et al 2007).

Interestingly, different dopaminergic genes have a differing profile. Notably, the increase in COMT, a molecule important for dopamine metabolism in the prefrontal cortex, increases later than most other markers. So it may be that there are periods of development when your cortical dopamine signalling is set up a bit differently. Ongoing PET imaging studies of the dopamine system seem to show that people developing early symptoms of schizophrenia begin to show a hyperactive dopamine system at this time. Maybe it is the late adolescent maturation of the cortical dopamine system that is vulnerable, and helps trigger the onset of a psychotic state in those with the schizophrenia trait.

Tan: This is probably semantics: how can you call schizophrenia a neurodevelopmental disorder when you say the cortex is basically normal?

Harrison: I agree: what do we mean by neurodevelopment? I think psychiatrists (and some interested neuroscientists) have been guilty of using the term loosely, sometimes meaning nothing more than that there are differences in children who will get the illness, and that there is no evidence of degenerative pathology.

Tan: Why do a large proportion of women in postmenopausal years develop schizophrenia?

Harrison: It is not a large subgroup, but it is true that there is a second blip in women in their forties and fifties. This is almost entirely uninvestigated. Almost everyone studies the classic earlier onset illness. It could be very different in its origins.

O'Leary: I'd like to echo one of the themes of this meeting, which is that if you don't find it, it doesn't mean that it is not there. You say the cortex is relatively normal. I appreciate how difficult it is to analyse human cortex and find subtle abnormalities. We have been analysing ErbB4 knockout mice. Defects in ErbB4/neuregulin 1 signalling has been suggested to be related to some forms of schizophrenia. The Taiwan group has reported they do not find radial defects in cortex in the ErbB4 mutant, but we have analysed the ErbB4 heart rescue mutant made by Martin Gasmann and find there are radial defects, but they are subtle. We need to be patient and use the right markers to detect them. Then we can find radial defects that are scattered across the cortex in a random fashion. They include subpial heterotopias and some disorganization in the laminar distribution of cells, whereas the adjacent cortex appears completely normal. This is in a complete absence of ErbB4, so if the signalling is more subtly altered the defects will likely be less severe. Nonetheless they could give rise to dysfunctional circuits.

Crino: I'd agree. Certainly, autism is a neurodevelopmental disorder yet the hunt for reproducible or overt brain pathology in autism has been a long one. It is only recently that Manny Casanova's group published a similar finding: in autism, there are very subtle but highly reproducible alterations in the columnar organization of the cortex (Casanova et al 2002). We have to be careful not to insist that a neurodevelopmental disorder is necessarily a robust phenotype with devastated arealization or cytoarchitecture. It could be incredibly subtle, perhaps something as subtle as altered apical dendritic arborization. Unless these analyses are done rigorously, it would be easy to miss subtle alterations in laminar or dendritic architecture.

Tan: Subtle differences are also present in the general population as well, but don't result in disease. We need to explain this.

References

Anton ES, Marchionni MA, Lee K-F, Rakic P 1997 Role of GGF/neuregulin signaling in interactions between migrating neurons and radial glia in the developing cerebral cortex. Development 124:3501–3510

Casanova MF, Buxhoeveden DP, Switala AE, Roy E 2002 Minicolumnar pathology in autism. Neurology 58:428–432

Law AJ, Shannon Weickert C, Hyde TM, Kleinman JE, Harrison PJ 2004 Neuregulin-1 (NRG-1) mRNA and protein in the adult human brain. Neuroscience 127:125–136

Tunbridge E, Weickert C, Kleinman J et al 2007 Catechol-O-methyltransferase enzyme activity and protein expression in human prefrontal cortex across the postnatal lifespan. Cereb Cortex 17:1206–1212

Weickert CS, Webster MJ, Gondipalli P et al 2007 Postnatal alterations in dopaminergic markers in the human prefrontal cortex. Neuroscience 144:1109–1119

Focal brain malformations: a spectrum of disorders along the mTOR cascade

Peter B. Crino

PENN Epilepsy Center, Department of Neurology, University of Pennsylvania Medical Center, 3 West Gates Bldg, 3400 Spruce St., Philadelphia, PA 19104, USA

Abstract. Focal cortical dysplasia with balloon cells (FCDIIB), hemimegalencephaly (HMEG), and ganglioglioma (GG) are sporadic focal malformations of cortical development that are highly associated with epilepsy. Histologically, all three malformations are characterized by disordered cortical lamination and the presence of markedly enlarged cell types known as balloon cells in FCDIIB and HMEG and atypical ganglion cells (AGCs) in GG. These cells are similar to giant cells in the tuberous sclerosis complex (TSC). Recent work has shown that there is enhanced activation of the mTOR cascade in TSC, FCD, HMEG and GG, suggesting a common pathogenesis for these disorders. We propose that these malformation types reflect a spectrum of disorders along the mTOR cascade. The mTOR pathway is known to regulate cell growth and thus is an ideal candidate to study in malformations associated with aberrant cell size. We hypothesize that focal brain malformations form as a consequence of a somatic gene mutation occurring within a progenitor cell during brain development. Our work has implemented several strategies to investigate FCD, HMEG and GG. First, we use single nucleotide polymorphism (SNP) arrays and gene sequencing to identify mutations in candidate genes that would lead to activation of the mTOR cascade. Second, we are using gene and protein expression profile techniques to understand how mTOR activation affects the developing cortex.

2007 Cortical development: genes and genetic abnormalities. Wiley, Chichester (Novartis Foundation Symposium 288) p 260–275

Malformations of cortical development (MCD) are a common cause of epilepsy resistant to anti-epileptic drug polytherapy. A recent classification scheme has divided MCD into disorders of proliferation, migration, and organization on the basis of putative pathogenic mechanisms (Barkovich et al 2005). Disorders of cell proliferation including tubers in the tuberous sclerosis complex (TSC), focal cortical dysplasia with balloon cells (FCDIIB), hemimegalencephaly (HMEG) and ganglioglioma (GG) are of particular interest because these focal MCD affect restricted regions of the cerebral cortex, share histopathological features (Mischel et al 1995, Andermann 2000, Flores-Sarnat 2002), affect children and adults, and are highly associated with intractable epilepsy (Palmini et al 1995). TSC is an autosomal dominant disorder resulting from mutations in one or two genes (*TSC1* and *TSC2*; for review see Crino et al 2006) whereas FCDIIB, HMEG and GG are sporadic disorders that form by unknown mechanisms (for review, see Crino et al 2002a, Senol et al 2000). FCDIIB and GG are among the most common causes of paediatric epilepsy and HMEG is highly associated with severe intractable neonatal seizures and infantile spasms, a devastating epilepsy syndrome in infants, and often requires urgent neurosurgical intervention. Thus, TSC, FCDIIB, HMEG and GG are associated with significant healthcare impact for patients including cost, morbidity and even mortality (Raymond et al 1995).

A unifying histopathological feature of FCDIIB, HMEG, and GG is the presence of unique cells that exhibit dramatic cytomegaly known as balloon cells (BCs) in FCDIIB and HMEG, and atypical ganglion cells (AGCs) in GG (see below). Morphologically, BCs and AGCs are not similar to any known cell types in the normal human cortex and are not seen in any other disorders. Thus, we have proposed that the molecular events leading to cytomegaly in focal brain malformations are likely to be central to understanding how these lesions form. Indeed, an exciting recent discovery is that in addition to the morphological similarities between these cell types, there is solid evidence that enlarged cells in all three malformation types exhibit aberrant activation of the mTOR cascade, a cellular pathway known to govern regulation of cell size and which is abnormally and constitutively activated in TSC. We propose that like TSC, FCDIIB, HMEG and GG result from mutations in genes that modulate the mTOR cascade. However, in view of the sporadic nature of these lesions, we propose that that FCDIIB, HMEG and GG result from a somatic mutation occurring within a neuroglial progenitor cell during brain development (for review, see Crino et al 2002a). Several pivotal questions remain that, if answered, will aid in proving this hypothesis. First, what is the lineage of cells in these lesions? Second, how does the cell morphology help understand the pathogenesis of focal MCD? Third, what candidate genes are over expressed, underexpressed or contain deleterious mutations in focal MCD? Using the histopathological similarities between TSC and FCDIIB, HMEG and GG as a conceptual starting point, several studies over the past three years have revealed potential candidate cell pathways that may be activated in FCDIIB, HMEG and GG.

Histopathology in tubers, FCDIIB, GG and HMEG

Tubers, FCDIIB, and HMEG are characterized histologically by disorganized or absent cortical lamination, loss of radial neuronal orientation in relation to the pial surface, and the presence of dysmorphic cells that exhibit dramatic cytomegaly. For example, large, dysmorphic neurons (**cytomegalic** neurons; Andre et al 2004) extend axons and dendrites but have marked enlarged cell somas (110–1300 μm in size). These large neurons express neuronal marker proteins such as neurofilament and TuJIII. In contrast, BCs in FCDIIB (Fig. 1) and HMEG, and AGCs in GG are morphologically similar to giant cells (GCs) in TSC (Fig. 2). Morphologically, BCs, AGCs and GCs are distinct from any known cell types in the normal human cortex. BCs and AGCs are larger than dysmorphic, cytomegalic neurons and lack the neuronal morphology characteristic of dysmorphic, cytomegalic neurons. The cell soma of BCs/AGCs is quite large (often exceeding 120 µm) and exhibits a laterally displaced nucleus that may even be multi-nucleated (Mischel et al 1995, Crino et al 2002a). Unlike cytomegalic neurons, BCs/AGCs do not extend clearly defined axons or dendrites and BCs express marker proteins suggestive of a mixed astrocytic and neural lineage. For example, the expression of neural protein markers such as neurofilament and α-internexin (Duong et al 1994, Crino et al 1997,

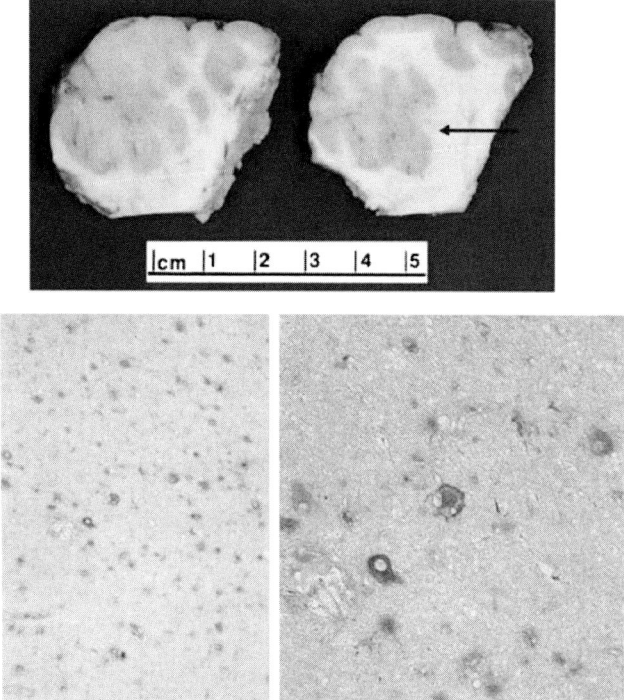

FIG. 1.   Top, resected FCDIIB specimen from a patient with intractable epilepsy. Note extension of thickened cortex into the subcortical white matter (arrow). Bottom, low (left) and high (right) magnification views of balloon cells in FCDIIB. Note large cell body and complete loss of normal laminar structure. Cells are immunolabelled with P-S6 antibodies, thus supporting the hypothesis that there is cell selective activation of the mTOR cascade in FCDIIB.

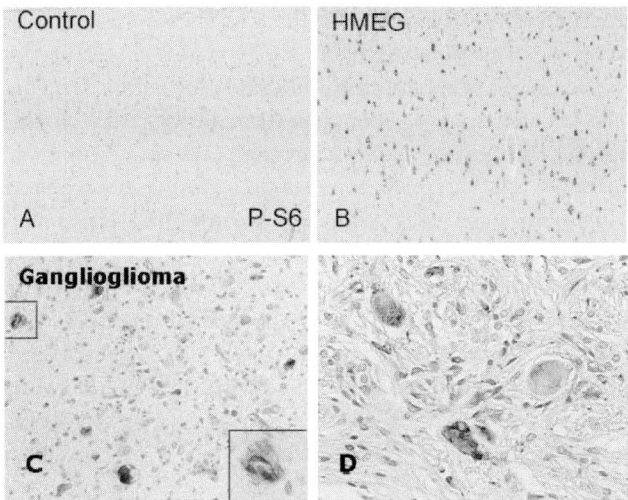

FIG. 2.   (A) Near absence of P-S6 expression in control cortex. (B) extensive expression of P-S6 in HMEG cortex. (C, and inset, D) P-S6 expression in AGCs in GG. Note selectivity of labelling in cytomegalic cells versus smaller surrounding cells.

Yamanouchi et al 1998, Taylor et al 2001) as well as neurotransmitter receptor subunits and neurotransmitter uptake sites (Ying et al 1998, Mikuni et al 1999, Najm et al 2000, Crino et al 2001, 2002b) suggest a neural phenotype for BCs/ AGCs whereas expression of glial markers such as S-100, vimentin, and glial fibrillary acidic proteins (De Rosa et al 1992, Mischel et al 1995, Taylor et al 2001) in BCs/AGCs suggest an astrocytic phenotype. In some cases, BCs and GCs express both neuronal and glial markers, suggesting a mixed glioneuronal lineage or that these cell types can be further delineated into glial- or neural-derived lineage subtypes. The highly aberrant morphologies and phenotype of BCs, AGCs and GCs suggest to many investigators that genesis of these cells may be critical to all other histological findings in FCDIIB, GG, tubers and HMEG.

TSC as a model disease for FCDIIB, HMEG and GG: activation of the mTOR cascade

There are no animal models for FCDIIB, HMEG and GG. However, in view of the morphological similarities between enlarged cells in TSC and FCDIIB, HMEG and GG, recent insights into the roles that the TSC genes play in governing cell size and growth have shed light on cytomegaly as a feature of FCDIIB, HMEG and GG. The *TSC1* gene encodes the protein hamartin and the *TSC2* gene encodes the protein tuberin. Hamartin and tuberin form a functional protein–protein complex that modulates several pivotal cellular pathways that govern cell size, cell proliferation and organ growth (van Slegtenhorst et al 1998, Gao & Pan 2001, Mak et al 2003). For example, the hamartin–tuberin heteromer is a pivotal regulatory complex in the mTOR cascade. Hamartin–tuberin functions downstream from insulin- and insulin-like growth factor receptors, activated PI3 kinase, and Akt to constitutively inhibit the phosphorylation of proteins that contribute to ribosomal assembly and protein translation including the mTORC (mammalian target of rapamycin complex), p70S6 kinase and ribosomal S6. Phosphorylation of mTORC may occur as a direct stepwise cascade or via yet to be identified intermediary steps. Negative modulation of this cascade by the hamartin–tuberin complex results in growth suppression and restricted cell size.

When tuberin is inactivated by Akt-mediated phosphorylation *in vitro*, hamartin and tuberin dissociate resulting in Rheb (**R**as **h**omolog **e**xpressed in **b**rain)-mediated phosphorylation of mTORC, p70S6 kinase and ribosomal S6 proteins (Fig. 3).

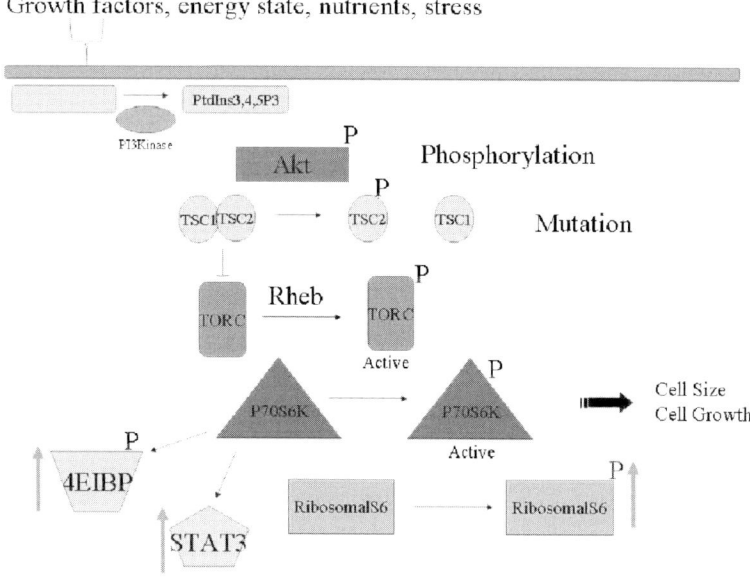

FIG. 3. Schematic of the mTORC cascade. Upstream activators include growth factors, nutri-ent conditions, energy state, and environmental stressors acting on the cell. In normal cells, TSC1 and TSC2 function as a heteromeric complex to inhibit or negatively modulate mTORC function. In the setting of TSC2 phosphorylation, TSC1 and TSC2 uncouple and there is dynamic activation of mTORC via Rheb. In TSC, in the setting of *TSC1* or *TSC2* mutations, the assembly of the complex is impaired, leading to constitutive or unopposed activation of mTORC. The downstream consequences include activation (phosphorylation) of several kinases or translational proteins (P-S6), ribosome, biogenesis, protein synthesis and cell growth.

Loss of mTOR inhibition leads to enhanced protein synthesis and progressive cellular enlargement. A similar effect occurs in the setting of loss of function mutations in either *TSC1* or *TSC2* that lead to failed heteromer complex formation. For example, loss of function mutation in the *Drosophila dTSC2* mutant *gigas* leads to enlarged ommatidia, head and wings (Potter et al 2001). Constitutive phosphorylation of mTORC, p70S6 kinase and ribosomal S6 is observed in neuroepithelial cells derived from *Tsc2* knockout mice (Onda et al 2002) and increased phospho-ribosomal S6 in particular has been identified in renal angiomyolipoma cells from TSC patients (El Hashemite et al 2003). Phosphorylation (activation) of mTORC, p70S6 kinase, and ribosomal S6 is observed in cells lacking tuberin or hamartin *in vitro* (Tee et al 2002, Inoki et al 2002). In human tuber specimens, there is robust expression of phospho-p70S6 kinase and phospho-ribosomal S6 proteins that is largely confined to cytomegalic GCs and neurons (Baybis et al 2004, Miyata et al 2004).

We and others have recently demonstrated that there is enhanced and aberrant phosphorylation of ribosomal S6 protein in BCs and AGCs (Baybis et al 2004, Miyata et al 2004, Ljungberg et al 2006; Fig. 1) suggesting that the mTOR cascade is activated in these malformations. Similar to giant cells in TSC, phospho-S6 expression was confined to cytomegalic neurons and BCs/AGCs. In addition, a recent study demonstrated that there is enhanced phosphorylation of Akt in FCDIIB (Schick et al 2006). Despite evidence that there is increased mTOR cascade activation in these malformation, loss of function mutations in TSC1 and TSC2 do not occur in FCDIIB (Becker et al 2001), ganglioglioma (Samadani et al 2007; Fig. 2), and HMEG (Ljungberg et al 2006; Fig. 2). Thus, mTOR cascade activation in FCDIIB, HMEG and GG likely occurs by an alternate mechanism.

Pretzel Syndrome (PS) is a recently described autosomal recessive disorder identified among old-order Mennonite families in Lancaster, PA that is clinically characterized by macrocephaly, stereotyped facial features, cognitive disability and intractable epilepsy. PS patients have multiple brain areas containing FCDIIB that at postmortem brain analysis reveal aberrant hexalaminar organization and, in particular, cytomegaly. PS results from a large deletion in the *LYK5* gene in affected individuals. The protein product of *LYK5*, STRAD, becomes a pivotal partner in the heterotrimeric LKB1-STRAD-MO25 complex upon binding to the kinase domain of LKB1. The LKB1-STRAD-MO25 complex acts as an upstream kinase of AMPK, and simultaneously activates the TSC2/TSC1 complex and inhibits mTOR. Loss of LKB1 function leads to mTOR activation. Indeed, phospho-S6 protein is robustly expressed in cytomegalic cells in PS brain tissue (Puffenberger et al 2007). These findings provide further support to our overarching hypothesis that loss of function mutations in genes that modulate the mTOR cascade (e.g. hamartin, tuberin and LYK5), lead to enhanced mTOR signalling, as evidenced by P-S6 expression and progressive enhancement of cell size.

The β-catenin signalling pathway

The TSC1–TSC2 heteromer has been shown to act as a constitutive inhibitor of the β-catenin pathway (Mak et al 2003), which is pivotal in regulation of cell growth. β-catenin is a transcriptional signalling modulator of the Wnt pathway via effects on T cell factor (TCF)-mediated gene transcription (for review, see Miller et al 1999). When activated, non-phosphorylated β-catenin is translocated to the nucleus where it functions as a transcriptional activator. Several downstream genes including cyclin D1 and c-myc have been identified that are transcriptionally activated by the β-catenin cascade. β-catenin is active when dephosphorylated but is targeted for degradation in the cytoplasm after phosphorylation by a GSK3/APC/axin1 complex at critical residues encoded in exon 3. Binding of Wnt to frizzled leads to phosphorylation of disheveled (Dvl1), which inhibits GSK activity. Total β-catenin protein levels were found to be elevated in TSC2-related renal tumours (Mak et al 2003). The TSC1–TSC2 complex inhibits Wnt1-stimulated Tcf/LEF luciferase reporter activity and hamartin and tuberin co-immunoprecipitated with GSK3 and axin, suggesting a direct effect of hamartin–tuberin on β-catenin (Mak et al 2003).

Enhanced expression of cyclin D1 mRNA has been reported in single microdissected GCs in tubers supporting the role of tuberin in β-catenin modulation (Crino et al 1996). Transgenic mice in which the β-catenin pathway was constitutively activated were shown to exhibit megalencephaly and abnormal cortical gyral patterning (Chenn & Walsh 2002). The *PTEN* gene negatively modulates both mTOR and β-catenin pathway activity and targeted knockout of the *Pten* gene in mice leads to megalencephaly and seizures (Backman et al 2001). *PTEN* mutations have been identified in Proteus syndrome associated with HMEG although *PTEN* mutations are not detected in sporadic HMEG (Aronica et al 2007). Interestingly, elevated levels of cytoplasmic β-catenin have been demonstrated immunohistochemically in BCs in FCDIIB (Cotter et al 1999). In HMEG, recent evidence demonstrates that there is enhanced expression of β-catenin and increased levels of cyclin D1 and c-myc mRNA, supporting increased transcriptional activity of β-catenin. A compelling finding was that the expression of cyclin D1 and phospho-S6 protein was co-localized in BCs in HMEG, suggesting that there may be parallel activation of these cascades. Thus, select cell cascades whose direct cellular function is to control cell size provide compelling candidate molecules to study that may lead to enhanced cell size in FCDIIB and HMEG.

Molecular mechanism of FCDIIB and HMEG formation

Clearly, cytomegaly is a pivotal histopathological feature of FCDIIB and HMEG and likely arises by persistent or aberrant activation of a cell cascade such as the mTOR and β-catenin, (MAPK)/p42/p44, JNK or Jak-Stat pathways. However, FCDIIB does not result from mutations in either *TSC1* or *TSC2* (Becker et al 2002) and so instead it seems plausible that altered regulation either up- or downstream of the mTOR or β-catenin pathways may be implicated in the formation of FCDIIB and HMEG and, specifically, generation of BCs during brain development.

How might FCDIIB and HMEG form during embryogenesis? We propose that FCDIIB and HMEG likely arise from an abnormality occurring during cell proliferative stages of cortical development (see Barkovich et al 2005). Several aetiologies including hypoxia–ischaemia, intrauterine infection, nutritional deficit, or environmental toxins such as cocaine or ethanol have been considered as inciting events responsible for FCDIIB and HMEG (Lombroso 2000, Andermann 2000). Recent studies have suggested that cytomegaly can reflect a reactive process in response to tissue injury (Marin-Padilla 2000). For example, 'mega-neurons' have been reported around the perimeter of porencephalic cysts. These cells are clearly neuronal in that they express neuronal marker proteins and are quite morphologically distinct from BCs. While reactive hypertrophy may reflect an injury process, the cell specific activation in TSC of a cascade clearly linked to enhanced cell size provides highly plausible evidence that activation of these cascades, or a related parallel cascade, contributes to cytomegaly in FCDIIB and HMEG.

Recent speculations have suggested the compelling hypothesis that FCDIIB and HMEG result from a somatic gene mutation occurring during cortical development (Fig. 4). No FCDIIB or HMEG pedigrees have been identified and monozygotic twins appear to be independently discordant for FCDIIB and HMEG (Senol et al 2000, Breillmann et al 2001). Thus, FCDIIB and HMEG may result from a mutational event occurring in a neural or astrocytic progenitor cell. Based on lineage analysis experiments in our lab (Hua & Crino 2003), we postulate that a somatic gene mutation occurs in a single progenitor cell at a critical stage in cortical development and that this cell continues to divide and yield progeny carrying a mutant gene. The effect of the gene mutation on these daughter cells would include cytoarchitectural abnormalities, cellular dysmorphism and laminar disorganization observed in FCDIIB and HMEG.

Somatic mosaicism provides a plausible explanation for the salient features of FCDIIB and HMEG including sporadic occurrence, histological heterogeneity, and why these lesions are confined to restricted brain regions or to one hemisphere. First, since a somatic mutation occurs long after meiotic recombination, these events are not germline mutations and thus are confined only to the progeny of cells derived from the original progenitor cell sustaining a mutation. From a practical perspective, this means that causative mutations in FCDIIB and HMEG are not likely to be identified in leukocyte DNA. In addition, these disorders are not heritable. Second, the cellular heterogeneity found in FCDIIB and HMEG

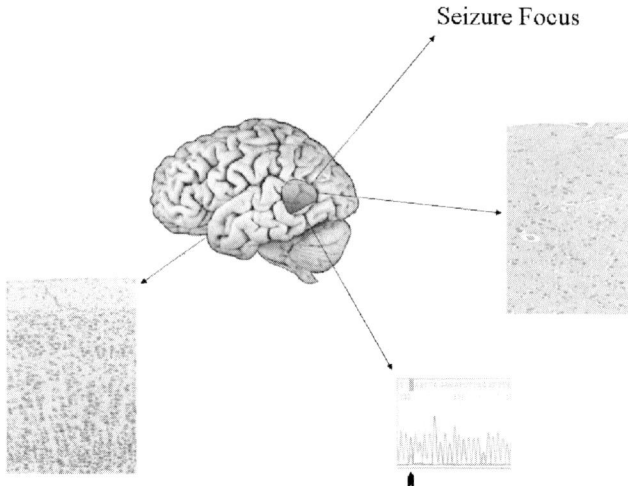

FIG. 4. Schematic depicting how a focal brain malformation (highlighted in dark grey), perhaps occurring in a progenitor cell during brain development, could result from a somatic gene mutation (lower sequencing tracing). The mutation leads to the structural alterations found in the focal lesion (right, cresyl violet stained section of FCDIIB), with seizures developing as a consequence of the disrupted network. Since the mutation affects only the lesion, the surrounding cortex exhibits normal structure (left, cresyl violet stained section).

suggests that distinct populations of cells may be derived by separate molecular mechanisms and thus, it is likely that FCDIIB and HMEG contain cells derived from the initial progenitor that sustained a gene mutation that are directly adjacent to other cells derived from distinct precursors cells that did not sustain a somatic gene mutation. Third, because the effects of a gene mutation are confined to a restricted population of cells, cytoarchitectural abnormalities remain within a focal brain region. The differences in the amount of cortex that is affected in these FCDIIB and HMEG can be accounted for by when a somatic mutation might occur. For example, an earlier occurring mutation could in theory render an effect on cell morphology and migration over a longer developmental time epoch and thus, lead to a more broad region of abnormality in the cortex as observed in HMEG. In contrast, a later occurring mutation might lead to a more focal or restricted region of structural abnormality as in FCDIIB.

Embryonic lineage of cells in cortical dysplasia

Defining the embryonic origin and pharmacological phenotype of BCs and DNs would provide important insights into how CDBC forms during brain development. A pivotal question regarding the developmental pathogenesis of CDBC as well as understanding the mechanisms leading to seizure onset is whether BCs and DNs derive from cells within the ventricular zone (VZ) or the medial ganglionic eminence (MGE) during cortical development and whether these cells are derived from progenitor cells destined to be excitatory or inhibitory neurons. Indeed, several studies have suggested that there may be reduced numbers of inhibitory interneurons in CDBC (Spreafico et al 1998, Garbelli et al 1999, Calcagnotto & Baraban 2005). The majority of cortical excitatory neurons derive from progenitors in the VZ whereas the majority of inhibitory cells derive from the MGE (see McConnell 1995, Anderson et al 1997). In human cortex, inhibitory interneurons are derived from both the MGE as well as the VZ. In a recent analysis, we analysed the origin and lineage of CDBC, using a panel of antibodies that define select cell types in the developing cortex. Collapsin response mediator protein (CRMP4) is a marker for newly generated neurons (Seki 2002). Dlx1 and Dlx2 are transcription factors expressed by inhibitory neurons that have their origin in the MGE. Otx1 is a transcription factor that is expressed by cells in layers V and VI of cortex that were generated in the VZ (Acampora et al 2001). The vesicular glutamate transporters (VGLUT1, VGLUT2) are expressed by glutamatergic neurons and facilitate synaptic glutamate re-uptake into neurons (Minelli et al 2003). The vesicular GABA transporter (VGAT) mediates reuptake of GABA into presynaptic vesicles and serves a marker for GABAergic neurons (McIntire et al 1997).

We find differential expression of these proteins in CDBC and propose that BCs and DNs arise from the ventricular regions rather than the medial ganglionic eminence (Lamparello et al, unpublished observations). In particular, the detection of OTX1 and absence of Dlx1 labelling, for example, suggests that BCs and DNs within CDBC derive from the VZ and not the MGE. Similarly, the expression of VGLUTs in BCs and DNs suggests that these cells may be excitatory or at least derive from cells that are within the VZ. The observation of CRMP4 expression in BCs raises the intriguing possibility that a subpopulation of cells in CDBC may be newly generated.

Summary

Focal brain malformations are the most common cause of intractable paediatric epilepsy yet their molecular basis remains unknown. Ongoing studies to define candidate gene loci play pivotal roles in understanding how these malformations form during brain development. Current evidence suggests that activation of the mTORC cascade is associated with altered cell growth and abnormalities of cell size that is characteristic of FCDIIB, HMEG and GG. These findings provide an exciting possibility for anti-epileptic drug development that may target the mTORC cascade. One such drug, rapamycin, which is known to directly inhibit mTORC is being implemented in clinical trials in TSC.

Acknowledgements

This work was supported by NS045877 (P.B.C). We thank our collaborator Eleonora Aronica MD PhD, Academic Medical Center, The Netherlands, for supply of resected tissue samples.

References

Acampora D, Gulisano M, Broccoli V, Simeone A 2001 Otx genes in brain morphogenesis. Prog Neurobiol 64:69–95

Andermann F 2000 Cortical dysplasias and epilepsy: a review of the architectonic, clinical, and seizure patterns. Adv Neurol 84:479–496

Anderson SA, Eisenstat DD, Shi L, Rubenstein JL 1997 Interneuron migration from basal forebrain to neocortex: dependence on Dlx genes. Science 278:474–6

Andre VM, Flores-Hernandez J, Cepeda C et al 2004 NMDA receptor alterations in neurons from pediatric cortical dysplasia tissue. Cereb Cortex 14:634–46

Aronica E, Boer K, Baybis M, Yu J, Crino P 2007 Co-expression of cyclin D1 and phosphorylated ribosomal S6 proteins in hemimegalencephaly. Acta Neuropathol (Berl) 114:287–293

Backman SA, Stambolic V, Suzuki A et al 2001 Deletion of Pten in mouse brain causes seizures, ataxia and defects in soma size resembling Lhermitte-Duclos disease. Nat Genet 29:396–403

Barkovich AJ, Kuzniecky RI, Jackson GD, Guerrini R, Dobyns WB 2005 A developmental and genetic classification for malformations of cortical development. Neurology. 65:1873–1887

Baybis M, Yu J, Lee A et al 2004 mTOR cascade activation distinguishes tubers from focal cortical dysplasia. Ann Neurol 56:478–487

Becker AJ, Lobach M, Klein H et al 2001 Mutational analysis of TSC1 and TSC2 genes in gangliogliomas. Neuropathol Appl Neurobiol 27:105–114

Becker AJ, Urbach H, Scheffler B et al 2002 Focal cortical dysplasia of Taylor's balloon cell type: mutational analysis of the TSC1 gene indicates a pathogenic relationship to tuberous sclerosis. Ann Neurol 52:29–37

Briellmann RS, Jackson GD, Torn-Broers Y, Berkovic SF 2001 Causes of epilepsies: insights from discordant monozygous twins. Ann Neurol 49:45–52

Calcagnotto ME, Baraban SC 2005 An examination of calcium current function on heterotopic neurons in hippocampal slices from rats exposed to methylazoxymethanol. Epilepsia 44:315–321

Chenn A, Walsh CA 2002 Regulation of cerebral cortical size by control of cell cycle exit in neural precursors. Science 297:365–369

Cotter D, Honavar M, Lovestone S et al 1999 Disturbance of Notch-1 and Wnt signalling proteins in neuroglial balloon cells and abnormal large neurons in focal cortical dysplasia in human cortex. Acta Neuropathol (Berl) 98:465–472

Crino PB, Dichter MA, Trojanowski JQ, Eberwine J 1996 Embryonic neuronal markers in tuberous sclerosis: single cell molecular pathology. Proc Nat Acad Sci USA 93:14152–14157

Crino PB, Trojanowski J, Eberwine J 1997 Internexin and nestin in focal cortical dysplasia and hemimegalencephaly as developmental markers. Acta Neuropathol (Berl) 93:619–627

Crino PB, Duhaime A-C, Baltuch G, White R 2001 Expression of glutamate and GABA A receptor subunit genes is distinct in dysplastic and heterotopic neurons. Neurology 57:904–913

Crino PB, Miyata H, Vinters HVV 2002a Neurodevelopmental disorders as a cause of seizures: neuropathologic, genetic, and mechanistic considerations. Brain Pathol 12:212–233

Crino PB, Jin H, Robinson M, Coulter D, Brooks-Kayal A 2002b Increased expression of the neuronal glutamate transporter (EAAT3/EAAC1) in hippocampal and neocortical epilepsy. Epilepsia 43:211–218

Crino PB, Nathanson KL, Petri-Henske E 2006 Medical progress: tuberous sclerosis complex. New Engl J Med 355:1345–1356

De Rosa MJ, Secor DL, Barsom M, Fisher RS, Vinters HV 1992 Neuropathologic findings in surgically treated hemimegalencephaly: immunohistochemical, morphometric, and ultrastructural study. Acta Neuropathol (Berl) 84:250–260

Duong T, De Rosa MJ, Poukens V, Vinters HV, Fisher RS 1994 Neuronal cytoskeletal abnormalities in human cerebral cortical dysplasia. Acta Neuropathol (Berl) 87:493–503

El-Hashemite N, Zhang H, Henske EP, Kwiatkowski DJ 2003 Mutation in TSC2 and activation of mammalian target of rapamycin signalling pathway in renal angiomyolipoma. Lancet 361:1348–1349

Flores-Sarnat L 2002 Hemimegalencephaly: part 1. Genetic, clinical, and imaging aspects. J Child Neurol 17:373–384

Gao X, Pan D 2001 TSC1 and TSC2 tumor suppressors antagonize insulin signaling in cell growth. Genes Dev 15:1383–1392

Garbelli R, Munari C, De Biasi S et al 1999 Taylor's cortical dysplasia: a confocal and ultrastructural immunohistochemical study. Brain Pathol 9:445–461

Hua Y, Crino PB 2003 Single cell lineage analysis in human focal cortical dysplasia. Cereb Cortex 13:693–699

Inoki K, Li Y, Zhu T, Wu, J, Guan KL 2002 TSC2 is phosphorylated and inhibited by Akt and suppresses mTOR signalling. Nat Cell Biol 4:648–657

Ljungberg MC, Bhattacharjee MB, Lu Y et al 2006 Activation of mammalian target of rapamycin in cytomegalic neurons of human cortical dysplasia. Ann Neurol 60: 420–429

Lombroso CT 2000 Can early postnatal closed head injury induce cortical dysplasia. Epilepsia 41:245–253

Mak BC, Takemaru K, Kenerson HL, Moon RT, Yeung RS 2003 The tuberin-hamartin complex negatively regulates β–catenin signaling activity J Biol Chem 278:5947–5951

Marin-Padilla M 2000 Perinatal brain damage, cortical reorganization (acquired cortical dysplasias), and epilepsy. Adv Neurol 84:153–172

McIntire SL, Reimer RJ, Schuske K, Edwards RH, Jorgensen EM 1997 Identification and characterization of the vesicular GABA transporter. Nature 389:870–876

McConnell SK 1995 Strategies for the generation of neuronal diversity in the developing central nervous system. J Neurosci 15:6987–6998

Minelli A, Edwards RH, Manzoni T, Conti F 2003 Postnatal development of the glutamate vesicular transporter VGLUT1 in rat cerebral cortex. Brain Res Dev Brain Res 140:309–314

Miller JR, Hocking AM, Brown, JD, Moon RT 1999 Mechanism and function of signal transduction by the Wnt-beta-catenin and Wnt/Ca2+ pathways. Oncogene 18:7860–7872

Mikuni N, Babb TL, Ying Z et al 1999 NMDA-receptors 1 and 2A/B coassembly increased in human epileptic focal cortical dysplasia. Epilepsia 40:1683–1687

Mischel PS, Nguyen LP, Vinters HV 1995 Cerebral cortical dysplasia associated with pediatric epilepsy: review of neuropathologic features and proposal for a grading system. J Neuropathol Exp Neurol 54:137–153

Miyata H, Chiang AC, Vinters HV 2004 Insulin signaling pathways in cortical dysplasia and TSC-tubers: tissue microarray analysis. Ann Neurol 56:510–519

Najm IM, Ying Z, Babb T et al 2000 Epileptogenicity correlated with increased N-methyl-D-aspartate receptor subunit NR2A/B in human focal cortical dysplasia. Epilepsia 41:971–976

Onda H, Crino PB, Zhang H et al 2002 Tsc2 null murine neuronal epithelial cells are a model for human tuber giant cells, and show activation of an mTOR pathway. Mol Cell Neurosci 21:561–574

Palmini A, Gambardella A, Andermann F et al 1995 Intrinsic epileptogenicity of human dysplastic cortex as suggested by corticography and surgical results. Ann Neurol 37:476–487

Potter CJ, Huang H, Xu T 2001 *Drosophila Tsc1* functions with *Tsc2* to antagonize insulin signaling in regulating cell growth, cell proliferation, and organ size. Cell 105:357–368

Puffenberger E, Strauss KA, Ramsey KE et al 2007 Syndromic cortical dysplasia caused by a homozygous 7 kilobase deletion in LYK5. Brain 130:1929–1941

Raymond AA, Fish DR, Sisodiya SM, Alsanjari N, Stevens JM, Shorvon SD 1995 Abnormalities of gyration, heterotopias, tuberous sclerosis, focal cortical dysplasia, microdysgenesis, dysembryoplastic neuroepithelial tumour and dysgenesis of the archicortex in epilepsy. Clinical, EEG and neuroimaging features in 100 adult patients. Brain 118:629–660

Samadani U, Judkins AR, Akpalu A, Aronica E, Crino PB 2007 Differential cellular gene expression in ganglioglioma. Epilepsia 48:646–653

Schick V, Majores M, Engels G et al 2006 Activation of Akt independent of PTEN and CTMP tumor-suppressor gene mutations in epilepsy-associated Taylor-type focal cortical dysplasias. Acta Neuropathol (Berl) 112:715–725

Seki T 2002 Expression patterns of immature neuronal markers PSA-NCAM, CRMP-4 and NeuroD in the hippocampus of young adult and aged rodents. J Neurosci Res 70:327–334

Senol U, Karaali K, Aktekin B, Yilmaz S, Sindel T 2000 Dizygotic twins with schizencephaly and focal cortical dysplasia. Am J Neuroradiol 21:1520–1521

Spreafico R, Battaglia G, Arcelli P et al 1998 Cortical dysplasia: an immunocytochemical study of three patients. Neurology 50:27–36

Taylor JP, Sater R, Baltuch G, Crino PB 2001 Expression of intermediate filament genes is enhanced in focal cortical dysplasia. Acta Neuropathol (Berl) 102:141–148

Tee AR, Fingar DC, Manning BD, Kwiatkowski DJ, Cantley LC, Blenis J 2002 Tuberous sclerosis complex-1 and -2 gene products function together to inhibit mammalian target of rapamycin (mTOR)-mediated downstream signaling. Proc Natl Acad Sci USA 99:13571–13576

van Slegtenhorst M, Nellist M, Nagelkerken B et al 1998 Interaction between hamartin and tuberin, the TSC1 and TSC2 gene products. Hum Mol Genet 7:1053–1057

Yamanouchi H, Jay V, Otsubo H, Kaga M, Becker LE, Takashima S 1998 Early forms of microtubule-associated protein are strongly expressed in cortical dysplasia. Acta Neuropathol (Berl) 95:466–470

Ying Z, Babb TL, Comair YG, Bingaman W, Bushey M, Touhalisky K 1998 Induced expression of NMDAR2 proteins and differential expression of NMDAR1 splice variants in dysplastic neurons of human epileptic neocortex. J Neuropathol Exp Neurol 57:47–62

DISCUSSION

Kriegstein: In discussing hemimegalencephaly, you described it as a second hit phenomenon, where there's a general mutation and a later somatic hit. But since the entire hemisphere is affected, wouldn't there be other examples where that second hit might come a little bit later on, or affect a smaller region, rather than a whole hemisphere?

Crino: Hemimegalencephaly is a problem. Unlike focal cortical dysplasia where we have a restricted area, in hemimegalencephaly the whole hemisphere is affected. We wonder about non cell-autonomous effects and secreted factors such as growth or trophic factors. What are the changes that happen in surrounding cortex if only a limited number of progenitors are affected by a gene mutation, for example? This is definitely a challenge. The other possibility is that hemimegalencephaly doesn't necessarily need to result from a single gene defect. For example, there is plenty of evidence in sporadic human cancers that there is an initiating molecular event followed by an accumulation of mutational events over time. For example, microsatellite instability assays looking at focal cortical dysplasia have not found any evidence that the genome is more or less unstable. Thus, it doesn't seem that hemimegalencephaly or cortical dysplasia are like glioma where all kinds of mutations can be picked up as the cells get more and more bizarre. In terms of a developmental epoch, if the hypothesis is true, does this mean that there is an ultimate progenitor for each hemisphere? This would have to be a pretty early mutational event. I suspect that hemimegalencephaly may be more complicated than a single cell mutation. In addition, I don't want to imply that any one of these malformations necessarily results from the same gene mutation. All we have is two phenotypic markers: epilepsy to identify the tissue, and histopathology. The phenotypic marker is cytomegaly, but there are varying degrees of this. There are some cases of cortical dysplasia where 5% of the cells are balloon cells, and others in which 50% are balloon cells. These histopathologies may not be a spectrum and instead could be completely different disorders. It is entirely possible that it is a late second hit event; it could also be a late third or fourth hit event. Our evidence for a heterozygous deletion affecting normal and aberrant hemisphere suggests that there must be some second hit. Also, second hits don't always have to be mutational. There are gene silencing effects, for example, which could come into play.

Davies: Some of the instability may be copy number deficits. Even your deletion could just be a polymorphism. You may get gene dosage effects with some of these other genes.

Crino: Standard karyotyping assays do not reveal anything in the typical sporadic hemimegalencephaly patients. There are other syndromic associations, for example with Proteus syndrome or even tuberous sclerosis.

Richards: Just to follow up on the potential differences between the hemispheres, does the hemimegalencephaly always occur in the same hemisphere?

Crino: No, it is random. It could be either side, and it doesn't have to affect the whole hemisphere. There is some issue of whether hemimegalencephaly blurs the line with hemilissencephaly or so-called lobar cortical dysplasia in which a malformation of the cortex spans just the frontal lobe or a portion of a hemi-sphere, but the rest of the brain is intact.

Goffinet: The asymmetry is probably not a problem of early precursors. It seems that there are aberrant cells invading normal tissue, and this generates hemi-megalencephaly.

Crino: That is possible. It doesn't have to be that the hemisphere is born out of the progenitor. It may be that corticogenesis is ongoing and there is essentially an invasion of this progenitor.

Goffinet: It is not too far from an invasive glioma.

Crino: Exactly right. This gets to the issue of tissue heterogeneity. My view is that the lesion is an admixture of cells that contain a molecular hit, and adjacent cells that probably die or may be normal.

Chelly: Your study on focal dysplasia focused on the mTOR pathway. Did you investigate the other genes or components involved in neuronal migration, such as *LIS1*?

Crino: We have looked at *LIS1* and doublecortin because we thought they were interesting candidate genes. However, the pathology of focal cortical dysplasia or hemimegalencephaly is distinct from lissencephaly or subcortical band heterotopia. For example, cytomegaly is not a feature of lissencephaly or subcortical band heterotopia. I believe that the presence of cytomegalic cells is an important histological finding, especially if we look at tuberous sclerosis complex as a model: it is the same cascade, a very similar phenotype and the cells are also gigantic. Thus, I would have been surprised if we had found mutations in these genes.

Rubenstein: A large fraction of autistic children develop large brains transiently between 12 months and 3 years of age. What is your opinion on this? Also, the frequency of autism in tuberous sclerosis (TS) is 10 times higher than in non-TS patients. The heterozygous state of TS without the second mutation effect does have a phenotype and could be contributing to increased brain size.

Crino: In this regard, TS complex (TSC) is unlike focal cortical dysplasia, hemi-megalencephaly and ganglioglioma. TSC is an autism syndrome whereas the others are defined by epilepsy as their clinical phenotype. Investigators have suggested that *TSC1* and *TSC2* may be independent autism gene susceptibility loci. However, I am not aware of evidence that there is an increased incidence of *TSC1* or *TSC2* gene mutation in autism without TSC. In TSC, it is a different kind of big brain. Histologically I have looked at plenty of autism brains from post-mortem studies, and there is nothing in the autism brain that looks like what you see in hemimega-lencephaly and cortical dysplasia in terms of balloon cells, and in autism the lamination is generally preserved.

Rakic: Talking about autism and developmental disorders, you mentioned the association with the Amish population in Pennsylvania. They are not using drugs and ultrasound and other medical diagnostic and therapeutic interventions, so do they have a similar rise in autism as is present in California?

Crino: I don't claim to be an expert in the epidemiology of neurological diseases in the Amish. I know that some diseases, such as cystic fibrosis, have never been reported in the Amish. There are certain diseases that, by virtue of their allelic frequency and this restricted gene pool, have been excluded. I am not aware that the incidence of autism is any different than it is in the general population, and nor is the incidence of epilepsy.

Rakic: Does this explain the increase in autism? 25% of autistic children have childhood epilepsy.

Crino: It is true that there's an overlap between autism and epilepsy. In fact, if we take migraine, headache, schizophrenia, bipolar disorder and autism, there is an overlapping Venn diagram with epilepsy as well. Is this a primary defect or just some broad change in the nervous system with epilepsy and related disorders as phenotypic disease 'read outs'? There's another possibility: some of epilepsy may have nothing to do with gene mutation, but rather SNP polymorphism in a population. Just the right combination may make you high susceptibility low threshold or vice versa.

Molnár: What is the size of the Amish population?

Crino: It's approximately 47 000 in Lancaster, but there are also groups in upstate New York and Ohio.

Final Discussion

Parnavelas: I would like to ask all participants to say one or two things about future directions in their work and what they consider important issues in the field of cortical development at present.

Harrison: As a clinician, the key developmental question is the extent to which primates and humans are different from mice. Are the mice a good enough model for human psychiatric disorders? We need more primate research.

Keays: I have been trying to work out whether I want to spend the next 20 years of my life working in this field. In doing so I find myself asking; are there seminal discoveries that remain in this field? I don't know. Part of me feels that the big advances have been through the application of genetic technology. This has happened: perhaps the best years have passed. Are there still big discoveries in cortical development to be found?

Macklis: Just questions like, 'How does it get built?' and 'How does it work?'

Kennedy: I think that if we do the sort of analysis that Jeff Macklis has reported on layer V at the level of the microcircuitry in the supragranular layers then we will be tackling what the cortex actually does. Then we'll gain a lot of insight.

Keays: Do you think the focus should be more on individual cells?

Kennedy: Yes, I think this will happen. The analysis that Jeff Macklis has shown on layer V, where we take this up in the supragranular layers, is going to be absolutely fascinating in terms of the algorithm which is being implemented in these layers.

Wilson: My origins are as a molecular biologist. One of the things that has impressed me has been that a lot of the studies are on transcription factors and mutations of transcription factors. I look forward to trying to understand the downstream players that determine the phenotype of the cells. Which cells talk to which cells, and how is this communication made? My own interest is whether it is neurons humming to each other or not. What is determining the real phenotype of the cell in terms of the cascades of genes and patterns of genes that are expressed under the supervision of these transcription factors?

Tan: I am impressed that over the last few years we have made huge progress in basic questions about lineage. Twelve years ago we had huge arguments about cell lineage and cell migration; it is all solved now, at least in rodents. I feel that it is time to do more primate studies to go forward, despite the difficulty and expense.

Hevner: In this meeting we have heard some evidence of how transcription factors regulate the formation of the axon pathways, but there is still a paucity of data on what the molecules involved are. Ephrins, neuropilins and semaphorins are all expressed in the cortex: maybe it will be combinations of these that regulate pathways. Even for the major pathways, such as thalamocortical, we lack clear understanding of steps in the formation of the pathways or the molecules that are regulating them at each step: the fingers on the hand that does the handshake.

O'Leary: I'd like to echo this general sentiment. We are just scratching the surface. Thinking back to the meeting held here in 1999, we had no evidence for any gene that controlled any aspect of areal patterning. Now we have a few candidates, but there is debate about whether they have been validated as true candidates and what their real roles are. These are just the beginnings: we have little idea about the co-operativity, the interactions, the other candidates and the downstream candidates. This is in just one field of cortical development. I also have a comment about the mouse as a model. Personally, I'd love to work on human tissue and try to extend some of our ideas and findings into humans, but this is difficult and costly to do, and requires the right collaborations. I still think the mouse is a valuable model. One can't really study gene and molecular function in humans like one can in the mouse. One can't design experiments to validate hypotheses in humans, such as rescue experiments.

Stoykova: With the help of molecular and mouse genetic approaches novel interesting features of cortical neurogenesis have been discovered in the past few years. The specific molecular patterning of neurons in distinct layers and functional areas and the limited neurogenic capacity of a few zones of the adult brain are among them. Most of the known molecular determinants of corticogenesis turned out to be evolutionary conserved transcription factors that exert their roles during embryogenesis. I would expect that in the next years we will gain a deeper knowledge about molecular players that act in dependence of such factors or control their function during development in establishment of the subtype neuronal identity of the mammalian cortex. I hope, furthermore, that we will get better ideas about whether and how developmental regulators with neurogenic capacity could possibly modulate the extremely limited neurogenic ability of the adult brain.

Chelly: We have heard a lot about transcription factors in brain development. This is an area where huge progress has been made over the last few years. But from my short experience in human genetic disorders, and the focus on developmental disorders and mental retardation, one of the major aspects that we will need to work out in order to understand common disorders like autism and schizophrenia will be synaptic activity and its regulation. Huge progress has been made in early development, but now we need to better understand the synapses and connectivity, and how neurons talk to each other.

Friocourt: I know a lot of progress has still to be made, but I am really impressed how quickly our knowledge is progressing and also, how every aspect of brain development (brain patterning, neurogenesis, neuronal migration or connectivity) seems more and more to fit together. We are now beginning to understand some of the basic mechanisms underlying autism, mental retardation, epilepsy, dyslexia or schizophrenia, which is very exciting.

Goffinet: There are different scales to this problem. On the small scale, I want to know what Reelin does to neurons. Then I would like to know whether this idea that planar cell polarity as a concept will help us understand more about patterning of the brain. On the global scale, what fascinates me is what makes us human: how genetics and the epigenetic landscape impact on the development of specific human behaviour and its variations.

Kriegstein: From a global perspective, this is a meeting on genes and brain development, and the most exciting thing for me is the way human disease is being unravelled from two directions: from animal models that yield insights into the mechanism of disease, and from human diseases where identification of genes informs the animal model. This is a wonderful synergy. On a smaller scale, the mechanism(s) regulating symmetrical and asymmetrical cell division in the brain is a tractable problem that has yet to be solved, and has implications for stem cell biology as well as for development.

Walsh: A lot of people are debating whether the field of developmental biology in general is one where the principal actors have been pretty well worked out. In simpler models it becomes harder to find key insights because the tools are so good, and so many good experiments have already been done. In terms of where we are going, there's no threat that the field will stop any time soon. In human genetics, of the 25 000 genes in the genome, we have perhaps 2000–3000 genetic disorders. That represents 10% of the genome that has been functionally annotated, as such. There's still a lot of grist for the mill. More importantly, we really don't know how genes map onto behaviour. This is what got us into this room: what defines behaviour? I don't know whether to think about genes as defining the cortex, or genes that define some vague initial conditions of the brain.

Guillemot: Like other people working on the genetic mechanisms of neural development, I am struck and a bit frustrated that although we have learned a lot about what transcription factors do, we still don't know how they do it. I am confident that in the next five years we'll learn a lot more about the genetic programmes controlled by transcription factors during neural development. I think we will understand much better the key steps that control the generation of neurons by stem cells.

Yamamoto: For the past 10–20 years molecular and genetic approaches have uncovered various aspects of development. In particular, in the cortex, our understanding of the relationships between transcription factors and morphogenesis have been revealed greatly with knockout animals. I can't imagine how many transgenic mice are present in the world! However, our knowledge about neural circuit formation is still small, although fundamental mechanisms including attractive and repulsive factors were revealed. I think that this issue will be important again in future.

Crino: I was struck by a comment Dennis O'Leary made regarding rescuing a phenotype in animal research. In human research you can't rescue the phenotype. But I'd submit that in clinical medicine rescuing the phenotype might be called 'therapy'. One could make a strong case for the fact that we have amassed a mountain of data trying to understand how what we see in the mouse may allow us to understand phenotypes seen in the patients. Yet, there are few examples of how we can turn these data around the other way and make a statement about how we can indeed reverse the phenotype or cure disease. An important mission of doing mouse genomic work to induce and understand mutants is to try to translate this to the human condition. There is a substantial portion of the population that has a significant neurological disorder with varying degrees of neurodevastation for which we offer essentially no cures or therapy. I'd like to see a therapeutic intervention based on the mountain of science.

O'Leary: By 'rescue the phenotype' I meant to 'validate the correlation', not to correct the mouse's behaviour.

Richards: We need an approach in which we look at mouse and also consider how this translates into human fetal brain development. It isn't difficult to get this tissue even in the USA. This is a resource that could be better utilized. In terms of the corpus callosum field, it is a minefield with respect to figuring out the huge spectrum of disorders that cause agenesis of the corpus callosum. We are trying to understand what developmental mechanism might be disrupted in this huge variation of disorders. We are trying to understand the underlying genes, so we can provide pediatric neurologists with some kind of screening mechanism for their patients. We now have wonderful ultrasound imaging techniques in which we can identify these disorders very early in gestation, but we need to provide patients with accurate prognostic data in these cases.

Rakic: I was surprised but also pleased to see how many people recognize as a primary goal the understanding of the human brain. I entered the field because I was interested in human brain development. We started to study mice because we can do many interesting experiments in them, and then as a bridge from mice to humans I started research in monkeys. I recognized that we needed to learn the differences, and believe that it's these differences that hold the secret of what make us human. This is what makes me want to continue working in this field, because of the possibility of understanding evolution from these differences. I don't think it will be philosophy, literature or religion that will give us answers about where we come from and what we are and where we are going, but science.

Davies: I enjoyed this meeting and have learned a lot. The answer, from my geneticists viewpoint, lies in evolution and comparative genomics. The reason I'm here is not because of the 25000 genes that are hardly different between human and mouse, but the rest of the genome which is similar between human and mouse but is evolving quickly. Over the next few years, we intend to take some non-coding RNAs that are highly conserved but evolving quickly and look at what these non-coding regions are doing. They are expressed in an exquisitely specific way in the developing brain. We need to define their function and find out what they tell us about disease and behaviour.

Price: I agree that for transcription factors we need to know about their targets and how they work. This is a huge challenge. Working towards this we need to improve our use of transgenic technologies. The knockout is like an aeroplane crash: something went wrong and you end up with a mess. Using this to analyse gene function can be misleading. I feel strongly that we need to start understanding gene function in a more quantitative way. We need to think more about levels of expression and whether these matter. Splice variants are numerous and genes act combinatorially. These are ways of gaining immense complexity from a relatively small set of molecules.

Mallamaci: If I compare the situation today with the time I started working in brain development, about eight years ago, I'm impressed not only by the increased sophistication of the techniques, but also by the much more intricate interconnection among people interested in different subfields of brain development. Collaboration is much wider now than in the past. This will be one of the major driving forces in the acceleration in our field. For the future, the reconstruction of the algorithms of the mechanisms leading to morphogenesis and function of brain circuits is an important issue. As a molecular biologist, I'm extremely interested in the mechanisms regulating neuronal plasticity from the point of view of laminar and areal specification. There is a lot of work to do in terms of analysing how regulation of the chromatin state is important in these phenomena, and how non-coding genes are involved.

Molnár: I also have the impression that the field has mellowed and we are much more humble. We had great fights here at the first meeting on issues that we couldn't answer with the techniques available at the time. Many of these are now settled. Also we are 10 years older! For example the time-lapse movies of Arnold Kriegstein can answer questions related to clonal relationships and lineage that we debated for hours. The big problem we have is that the brain develops in a way that relies on constant interactions with the environment in a highly complex manner, and it has evolved to constantly evolve further. Some of these complex developmental mechanisms proved to be a bit more difficult to understand than we thought 10 years ago. I think we still have a relatively narrow view on how the brain is developing. But, now we have a real opportunity to collaborate with bioinformaticians, look at comparative genomics, and interact with evolutionary developmental biologists and clinicians to look at various forms of pathologies and eventually get a much broader perspective.

Rubenstein: A lot of the diseases of cognition are side effects of our evolution from animals that don't have computational skills as advanced as humans. We need to consider the neuroanatomical inventions that have occurred during the evolution of humans. It is likely that parts of the prefrontal cortex and its connections are among the newer nuts and bolts of those evolutionary steps. I am interested in understanding the neural systems that the prefrontal cortex is integrated with, and that are probably disrupted in autism and schizophrenia.

Macklis: I only have two things to add. We are at the level of needing a massively more sophisticated cell-type specific dissection of microcircuitry. We are at the level of talking about 6 layers and 49 areas. I agree that we need to know how the transcription factors function. One thing they will do is give us a little handle to go deeper down to functional classifications, more like haematopoiesis, but much deeper. We will then know what kind of cells do what. Second, we have disparate bits of information. I went to the National Portrait Gallery a few days ago and there was a complex sketch that was made for a big mural. Once they had the general outlines, they put these tiny fiduciary marks that helped the transfer to the wall, so they could then spend the next 14 months painting the picture. I think we are at the stage of the sketch: we have the general outline, but we don't have any of it filled in.

Murakami: This is my first experience of one of these meetings, and I have been deeply impressed. The cortex is very complicated and important, but it is not everything. What we learn from the cortex might be generalized to other places, but it might not be. People working on cortex should also keep an eye on what is happening in other parts of the brain.

Parnavelas: On a small scale, since the focus of my lab is interneuron migration, I would like us to understand, as much as we can, the signalling mechanisms that guide these cells to their locations in the cortex. We have hardly begun this long journey. On a larger scale, through the work of everybody in this field, it would be nice to one day understand what makes us human. Thank you all for your participation.

Contributor Index

Non-participating co-authors are indicated by asterisk. Entries in bold indicate papers; other entries refer to discussion contributions

Subject Index